ELECTRICAL POWER

Motors, Controls, Generators, Transformers

by

Joe Kaiser, B.S.E.E.
K and M Electronics, Inc.

South Holland, Illinois
THE GOODHEART-WILLCOX COMPANY, INC.
Publishers

Library of Congress Catalog Card Number 90-46776
International Standard Book Number 0-87006-834-2

4567890-91-98765

Library of Congress Cataloging in Publication Data

Kaiser, Joe.
  Electrical power: motors, controls, generators, transformers / by Joe Kaiser.

    p.   cm.
  Includes index.
  ISBN 0-87006-834-2
  1. Electric machinery. I. Title
TK2000.K33   1991
621.31'042--dc20   90-46776
                                              CIP

INTRODUCTION

ELECTRICAL POWER provides an understanding of the principles of operation of motors, generators, transformers and motor controls—the fundamentals and practical applications.

The text is organized in three parts: The first five chapters describe the rules governing the behavior of electricity and magnetism. This portion can serve as a review for those who already have a thorough understanding of electricity and electrical components.

Chapters 6 through 12 deal with the machines and devices that generate, transform, and use electrical power. The final three chapters are devoted to the control of motors. Both electromechanical and solid state electronic control devices and systems are covered.

ELECTRICAL POWER is written for secondary and postsecondary students preparing for careers in the electrical trades; men and women in apprenticeship programs; students in electrical or engineering technology programs; anyone being trained in automation and computer control of electrically powered machinery; mechanical, civil, and other engineering students who must also learn the rudiments of electrical power.

ELECTRICAL POWER is written in a manner that interests and motivates students, with emphasis being placed on the practical rather than the theoretical. Mathematical equations are used principally to illustrate physical concepts. Higher math has been avoided.

The field of electrical power is an exciting and challenging one. Vast opportunities are open to men and women with knowledge, understanding, and skill in this area.

JOE KAISER

CONTENTS

4

CONTENTS

SAFETY

Safety is more than a topic in a book; it should be a way of life. We naturally try to act in a safe manner no matter what we are doing. If we want to be safe when walking, riding a bike or driving a car, we follow the "rules of the road" and expect everyone else to do the same. If we become just as aware of the possible dangers associated with electrical power and rotating machinery, we will be more than willing to follow common sense rules of safety.

There are five types of hazards associated with the operation of electrical power equipment:
1. Electrical shock.
2. Electrical burns.
3. Fire and explosion.
4. Heat buildup.
5. Mechanical hazards.

ELECTRICAL SHOCK

There is no way to tell if an electrical terminal is "live" by looking at it. You should always test it yourself using proper test equipment. Further, make absolutely sure that it cannot be energized by someone else while you are working on it.

Electricity affects the body in two ways. It contracts muscles and it overrides impulses from the brain. Never test a wire with your fingers. Your fingers may tighten around the wire and you will not be able to let go.

Shocks that are merely uncomfortable when only one hand is involved may be fatal if the electricity passes through your body. Electricity that flows through the body from hand to hand or hand to feet can tighten the chest muscles. The victim finds it impossible to breathe. Further, the heart may stop if the impulses control-ing it become scrambled. Safe procedure involves using insulated tools, working with one hand only and standing on an insulated surface.

Shocks tend to be more severe when the skin is wet. Your skin may have 1000 times more resistance to the flow of electricity when dry. Not only should you work with dry hands, you should also remove rings, watches, bracelets and any other metal objects that could come in contact with energized wires or terminals.

Another source of electrical shock is improperly grounded equipment. The metal housings of some hand tools, for example, should be grounded through a three-prong plug. If instead, it is plugged into a two-wire receptacle, the housing could be electrically "live." The tool user then provides the electrical path to ground. Make sure the tools you use are in good repair and, if required, are properly grounded.

Never intentionally cause anyone else to receive a shock. They may be especially susceptible to electricity. There is never any excuse for horseplay when working around electricity.

ELECTRICAL BURNS

Normally, air does not conduct electricity. However, when electrically "hot" wires are not properly insulated, the air can break down and form a conducting path between them or to ground. The electricity forms a white-hot arc, like a miniature lightning bolt. Inspect power cords frequently for wear or fraying. Replace them if necessary.

Electrical arcs injure people in two ways:
1. The heat causes a conventional burn.
2. The electricity can destroy skin tissue.

An electrical burn requires immediate and special treatment.

Coils and capacitors in a circuit cause another type of hazard. They can store electrical energy and release it after power has been turned off. It is best to test such circuits before working on them, even if you are sure they are "dead."

FIRE AND EXPLOSION

Equipment used in flammable or explosive surroundings has explosion-proof housings. Its purpose is to prevent any sparking or arcing from taking place in the atmosphere. Equipment brought into the area however, may not be so housed. Hand drills, for example, often arc and thus should not be used in an explosive atmosphere.

Electrical arcing of power cords have been known to start fires. Never throw water on an electrical fire. First, shut off the power, if possible. Then, use only a "class C" fire extinguisher, a type approved for electrical fires.

HEAT BUILDUP

The flow of electricity may produce heat. Normally, you will not notice it except in heat-producing devices. An electric heater, for example, is supposed to get hot. So is an incandescent lamp and an electric motor. Most motors are rated for a temperature rise of 40°C (72°F) above room temperature. In a 25°C (78°F) room the housing of a motor can easily be 66°C (150°F)—too hot to rest your hand on it.

Heat can build up in wires, too. A lightweight extension cord gets hot when used for heavy-duty service. Avoid using extensions at all, if possible. If you must use them, be sure they can carry the current without overheating. Do not string them overhead, across aisles or under mats where heat can build up.

MECHANICAL HAZARDS

Electricity is often used to run machinery. Spinning shafts and moving parts are always sources of danger. Be sure that necessary guards are in place. Wear safety glasses, if required. Do not allow loose clothing or long hair to become entangled in the equipment. Make sure that the machine you are working on cannot be turned on without your knowledge.

SUMMARY

Many organizations are dedicated to making your shop a safe place to work and learn. They have established rules and guidelines for your protection. Among these organizations are:

National Safety Council
National Fire Protection Association
Underwriters Laboratories
American Vocational Association
National Electrical Manufacturers
  Association
Occupational Safety and Health
  Administration

They cannot do it alone, however. They need your help. There is an often-used safety slogan which says, "Accidents don't just happen; they are caused." This does not necessarily mean "deliberately caused." Accidents are often caused by carelessness, lack of knowledge or failure to follow safety rules. By being aware of the possible hazards and avoiding them, you can use electrical power safely.

Electrical Power

Numerous substations, such as this 765 kilovolt transformer installation, dot the landscape to help distribute and control the electricity which power factories, cities, and homes. Maintaining this huge distribution network, as well as the electrical machines and devices which use electrical power, requires thousands of skilled workers. These highly trained people must understand the nature of electricity and electrical circuitry.
(Power Systems Div., McGraw-Edison Co.)

1

ELECTRICITY AND ELECTRICAL CIRCUITS

After studying this chapter, you will be able to:

☐ Define the terms voltage, current, resistance and power.

☐ List the factors affecting the resistance of a wire.

☐ Calculate current, voltage, resistance and power, given two of these factors.

☐ Determine the equivalent resistance of resistances in series and parallel.

When you turn on a light switch, the room gets brighter. An electric heater makes you warm. A fan or air conditioner cools you. We have become so used to electrical devices like these that we seldom think about the electricity that makes them work. The power company, however, thinks about it 24 hours a day. Their equipment must be on the job every minute, producing and delivering electricity. Fig. 1-1 shows generation, transmission, and distribution of electricity by a utility.

Unlike other products, electricity cannot be stored. The utility must generate it only when it is needed. When you turn on a light, the elec-

tricity it uses was produced the instant before. STORAGE BATTERIES, like those installed in cars, do not store electricity. They merely convert chemical energy to electrical energy and vise versa.

NATURE OF ELECTRICITY

Electrical energy is somewhat like KINETIC ENERGY which is the energy of motion. See Fig. 1-2.

Fig. 1-2. A ball moving through the air has energy of motion known as kinetic energy. Electrical energy is somewhat like that. It moves unseen through a wire conductor to do work. (Progress Lighting, subsidiary of Walter Kidde & Co., Inc.)

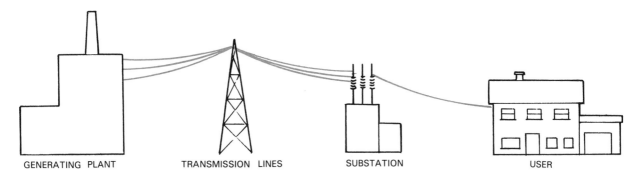

GENERATING PLANT   TRANSMISSION LINES   SUBSTATION   USER

Fig. 1-1. Electricity cannot be stored. It is generated as it is used.

This may seem strange at first. After all, you can see a ball moving, but you cannot see electricity moving through a wire. Still, electricity *is* the result of motion. The reason we cannot see anything move in a wire is that the particles, called ELECTRONS, are so tiny. In fact, they cannot even be seen with the most powerful microscope.

ELECTRON THEORY

Benjamin Franklin, who was a scientist as well as a statesman, made many electrical discoveries. He thought there were two kinds of ELECTRICAL FLUID, as he called electricity. The names Franklin suggested, POSITIVE and NEGATIVE, are still in use today. We speak of positive electrical charges and negative electrical charges. The behavior of electricity was still not understood, however, until the early 1900s. It was then that scientists came up with the ELECTRON THEORY.

According to the electron theory, all matter is made up of ATOMS. The atoms, in turn, have a center, around which ELECTRONS move in orbits (circular paths). As shown in Fig. 1-3, it is

a little like our solar system where planets orbit around the sun. In the center are PROTONS, which contain positive electrical charges. The center, called the nucleus, also contains NEUTRONS, which have no charge. The orbiting electrons contain negative electrical charges. Atoms normally contain the same number of electrons as protons.

CONDUCTORS AND INSULATORS

In the atoms of some materials (glass, for example), electrons are held tightly in their orbits. They cannot easily move from one atom to another. Such materials make poor electrical conductors. They are called INSULATORS.

In good conductors, like copper, electrons can easily move from one atom to another. When there is an ELECTRICAL PRESSURE, millions of electrons move through a conductor. Electrical pressure is really just a buildup of electrons.

THE FLOW OF ELECTRICITY

Electrical pressure is built up in what is called a POWER SOURCE. The power source uses chemical energy (or some other form of energy) to force electrons from one side (terminal) to the other. Then, electricity can flow from the terminal that has extra electrons to the terminal that has fewer electrons.

This flow takes place outside the source if

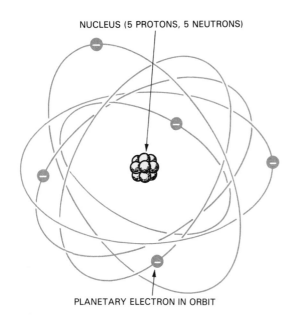

NUCLEUS (5 PROTONS, 5 NEUTRONS)

PLANETARY ELECTRON IN ORBIT

Fig. 1-3. Electrons orbit nucleus of atom like planets of solar system orbiting the sun.

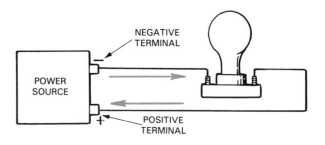

NEGATIVE TERMINAL

POWER SOURCE

POSITIVE TERMINAL

Fig. 1-4. Electricity flows from negative terminal of power source to positive terminal if there is a conductor between them. The path is called a circuit.

there is an electrical path (conductor) between them. This is illustrated in Fig. 1-4. Negative electrical charges rush through the conductor, trying to balance out the positive charges. Electrons, themselves, move along the conductor rather slowly. Negative charges, however, move extremely fast — approximately 300 000 000 metres per second (186,411 miles per second).

If it is hard for you to visualize the difference between electron flow and charge flow, Fig. 1-5 may help. It shows several billiard balls in line. Assume that a cue ball strikes the first ball in line. The last ball in the line shoots out almost immediately. The energy has been transmitted through each and every ball, front to back. Yet, the rest of the balls themselves hardly move at all. In the same way, negative charges speed through the conductor, while the electrons move slowly.

ELECTRICAL TERMS

In a way, the term ELECTRICITY is like the term "wind." Wind is defined as "air in motion." It is a happening. For wind to "happen," there must be a pressure to cause the air to flow. In the same way, electricity is a happening. It is "negative charges in motion." For electricity to happen, there must be an electrical pressure to cause the charges to flow. Electrical pressure is called VOLTAGE. Charge flow is called CURRENT.

VOLTAGE

To help you visualize how electrical charges move along a wire look at Fig. 1-6. Imagine that the electrons are as big as marbles. For a power source, assume we have a paddle wheel that takes marbles from one side and forces them to the other side. The paddle wheel is to the marbles what voltage is to electricity — a force or pressure.

It is important to realize that an accumulation of electrons is *not* the same as the flow of

electricity. The displacement of electrons inside a power source produces electrical pressure, or *voltage*.

Voltage is measured in *volts*, named for Alessandro Volta, inventor of the electrical storage battery. The symbol for voltage is either E (for electromotive force) or V (for voltage). When we talk about the voltage of a power source, we are really referring to the amount of energy necessary to move electrons from their normal positions toward one of the terminals.

This energy may come from:
1. Friction (static electricity).
2. Pressure (piezoelectric).
3. Heat (thermocouple).

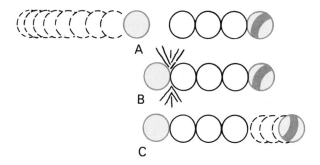

Fig. 1-5. Electrical energy moves through a wire much like energy is transmitted along a row of billiard balls. A—All except cue ball are at rest. B—Cue ball strikes first ball at rest. Cue ball's energy of motion is passed along line of balls. C—Last ball in line receives energy and is moved.

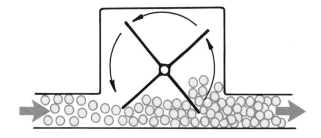

Fig. 1-6. Picture a paddlewheel turned by some force and moving marbles through a pipe. Wheel causes buildup of pressure much like an electrical power source creates electrical pressure also called voltage.

4. Chemical (battery).
5. Light (photoelectric).
6. Or electromagnetism (generator).

Although all of these ways of producing electricity are important, this book will cover only the electromagnetic method. At this point, however, we need not worry about the kind of energy used. Assume simply that energy has been given to the electrons and that a voltage exists. The terminal where the electrons accumulate is electrically NEGATIVE. The other terminal, where the electrons have left, is electrically POSITIVE.

DIRECTION OF CURRENT FLOW

Before the electron was discovered, people thought it was the positive charge that moved. The conventional way of describing current direction was plus to minus, or positive to negative. It is still easier to think of current that way. A positive number, for example, is higher than a negative number. It is logical for flow to be from the higher (plus) terminal to the lower (minus) one.

Energy is generated in the power source and used by the load. It does not really matter which way we describe current direction. The electrical energy, itself, flows from the power source to the load. This book, therefore, will use *conventional current direction*. That is, current will always be shown moving from the plus terminal to the minus terminal. This is common practice in industry. Where electron movement, not current, is meant, it will be described as such.

CURRENT INTENSITY

Let us continue to think of electrons in a circuit as marbles in a pipe. Fig. 1-7 shows the terminals connected by a tube completely filled with marbles. Think of the marbles as being a part of the tube. If the tube were cut, the marbles would not spill out the end. The tube represents an electrical circuit.

When the tube is connected to the power source, one of three things will happen:
1. There will be no movement because of an interruption in the tube.
2. There will be movement in an unobstructed tube.
3. A barrier or restriction in the tube can slow down the movement.

As shown in Fig. 1-8, it is possible that there will be no marble flow at all. You see, each electron in a conductor must be able to push the one in front of it. If there is a gap, like an open switch, anywhere in the path, no electrons will move. The voltage stays the same, but there is no flow of electricity.

Suppose the tube is large and smooth, as shown in Fig. 1-9. The pressure makes many marbles flow. In electrical circuits, instead of talking about the number of electrons flowing,

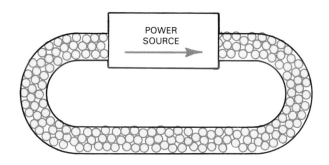

Fig. 1-7. Power source gives marbles energy. This energy is used up as marbles flow through pipe.

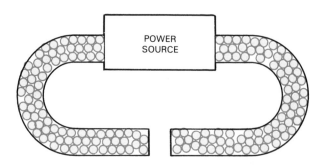

Fig. 1-8. Gap in tube prevents marble flow. In same way, gap in electrical circuit prevents flow of electricity.

we talk about the intensity of charge flow. Intensity of charge flow is what is meant by the term, CURRENT.

The symbol for current is I (for intensity). We measure current in amperes, named for Andre Ampere, a French scientist. The symbol for amperes is A. It is important to remember: *current is directly proportional to voltage.*

RESISTANCE

A third possibility is shown in Fig. 1-10. The tube may not be completely open its entire length. If there is a narrow section, not as many marbles would get through. The narrow section RESISTS the flow of marbles. The result is a less intense flow. Electrical circuits, while they

provide a path for current, also resist it. Usually, the resistance of a circuit comes from an electrical device. Heaters, lights and motors all have resistance. Also, the wires that carry current have a small amount of resistance. The symbol for resistance is R. Resistance is measured in OHMS. Often, the Greek letter omega (Ω) is used in place of the word "ohms."

The resistance of a device or of a wire depends on the following factors:

1. Type of material. Conductors like copper and aluminum, which have a large number of mobile electrons, have low resistance. Insulators like wood and glass, whose electrons are held firmly to their centers, have very high resistance.
2. Cross-sectional area. Fig. 1-11 shows the effect size has on resistance. The larger (thicker) the wire, the less the resistance. Wire sizes are given in standard AWG (American Wire Gage) numbers. You might expect the cross-sectional area to be given in square centimetres. However, a more convenient unit, "circular mils," is used. *The circular mil area of a wire is its diameter in mils squared.* A mil is 0.001 in. (0.00254 cm). One circular mil equals 0.00000507 square centimetres (cm²).

$$[A = \pi (\frac{d}{2})^2 = 3.14 \times (0.00127)^2$$
$$= 0.00000507 \text{ cm}^2]$$

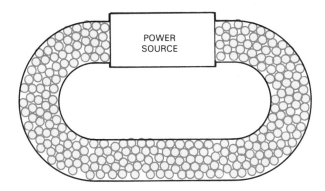

Fig. 1-9. Large, smooth tube permits high intensity of marble flow.

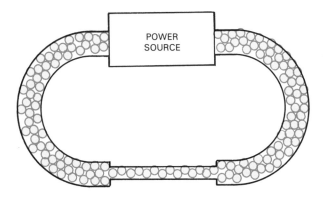

Fig. 1-10. Resistance in electrical circuit partially blocks electron movement the way this narrow section blocks marble movement.

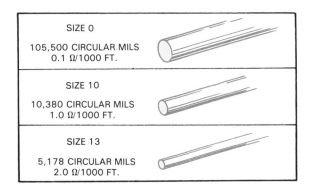

| SIZE 0 |
| 105,500 CIRCULAR MILS
0.1 Ω/1000 FT. |
| SIZE 10 |
| 10,380 CIRCULAR MILS
1.0 Ω/1000 FT. |
| SIZE 13 |
| 5,178 CIRCULAR MILS
2.0 Ω/1000 FT. |

Fig. 1-11. Larger diameter wires have less resistance.

3. Length. Fig. 1-12 gives the resistance of various sizes of copper wire in ohms per 1000 ft. A long wire has more resistance.

4. Temperature. The resistance of most conductors increases as the material warms up. Such materials are said to have a *positive temperature coefficient.*

AWG NO.	CIRCULAR MILS	Ω/1000 ft @20 °C(68 °F)
0000	221 600	0.05
000	167 800	0.06
00	133 100	0.08
0	105 500	0.10
1	83 690	0.12
2	66 370	0.16
3	52 640	0.20
4	41 740	0.25
5	33 100	0.31
6	26 250	0.40
7	20 820	0.50
8	16 510	0.63
9	13 090	0.80
10	10 380	1.00
11	8 234	1.26
12	6 530	1.60
13	5 178	2.00
14	4 107	2.53
15	3 257	3.19
16	2 583	4.02
17	2 048	5.06
18	1 624	6.39
19	1 228	8.05
20	1 022	10.15
21	810.1	12.80
22	642.4	16.14
23	509.5	20.35
24	404.0	25.67
25	320.4	32.36
26	254.1	40.82
27	201.5	51.47
28	159.8	64.89
29	126.7	81.83
30	100.5	103.19
31	79.7	130.12
32	63.21	164.07
33	50.13	206.91
34	39.75	260.89
35	31.52	328.95
36	25.00	414.77
37	19.83	523.01
38	15.72	659.63
39	12.47	831.95
40	9.888	1048.88

Fig. 1-12. Resistance of various wire sizes can be predicted. Chart shows resistance of different sizes of copper wire at constant temperature.

An ohm of resistance permits one ampere of current when the voltage is one volt. With a given voltage, a large resistance results in a low current. A low value resistance results in a high current. Therefore, we can say: *Current is inversely proportional to resistance.*

Besides the resistance of wires and of electrical devices, there are components built specifically to produce resistance. They are called RESISTORS. The ones shown in Fig. 1-13 have a fixed value of resistance. Others are made so that you can vary their resistance. Variable resistors, like the one shown in Fig. 1-14, are called RHEOSTATS. When resistors and rheostats are used, their purpose is to either produce a desired current or a desired voltage by adding resistance to a circuit.

OHM'S LAW

George S. Ohm, after whom the unit of resistance is named, put the relationship between voltage, current and resistance in the form of an equation called Ohm's Law:

$$I = \frac{E}{R}$$

Where I is the current in amperes
E is the voltage in volts
R is the resistance in ohms

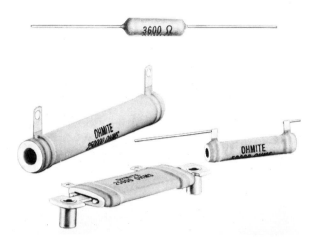

Fig. 1-13. These types of resistors have fixed value of resistance. (Ohmite Mg. Co.)

Example 1

The power source in Fig. 1-15 is supplying 100 volts to the 10 ohm resistor. Find the current.

$$I = \frac{E}{R} = \frac{100}{10} = 10 \text{ amperes (A)}$$

Example 2

The voltage is increased to 200 volts (V). Find the current.

$$I = \frac{E}{R} = \frac{200}{10} = 20 \text{ A}$$

Ohm's Law can also be used to find the resistance if you know the voltage and the current. The equation is:

$$R = \frac{E}{I} \qquad \begin{array}{l} \text{R is the resistance in ohms} \\ \text{E is the voltage in volts} \\ \text{I is the current in amperes} \end{array}$$

Example 3

In the circuit of Fig. 1-16, the power source is supplying 50 V and the current is 10 A. Find the resistance (ohms).

$$R = \frac{E}{I} = \frac{50}{10} = 5 \text{ ohms } (\Omega)$$

Very often, this is the only way of finding out exactly how much resistance there is in a circuit. You can guess from the material, size and length. But the real measure of resistance is how much current results when a specific voltage is applied to the circuit.

You can also compute the voltage when you know both the current and resistance. The equation is:

$$E = I \times R \qquad \begin{array}{l} \text{E is the voltage in volts} \\ \text{I is the current in amperes} \\ \text{R is the resistance in ohms} \end{array}$$

We use this equation in two ways. Voltage, remember, means energy. The voltage of a power source represents the energy *picked up* by

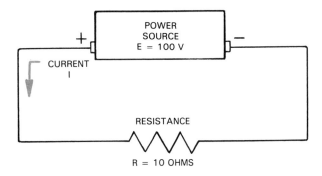

Fig. 1-15. Current is directly proportional to voltage and inversely proportional to resistance.

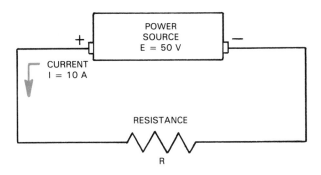

Fig. 1-16. Resistance and power can be calculated from voltage and current.

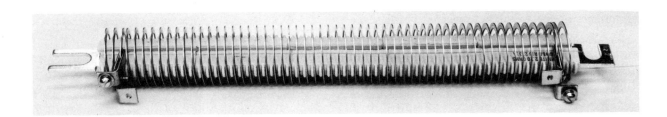

Fig. 1-14. Variable resistors are made so their resistance to flow of electricity can be changed (varied).

the electrons as they are forced to the negative terminal. But, how about the energy *used* by the electrons as they move through the circuit? This is called voltage, too. The energy used by the electrons (called VOLTAGE DROP in the external circuit) always exactly equals the energy they picked up (source voltage). Therefore, if we know the resistance between two points in a circuit, and the current, we can compute the voltage drop.

Example 4

The resistance between points A and B in Fig. 1-17 is 15 Ω. Current is 10 A. Find the voltage drop.

$$E = I \times R = 10 \times 15 = 150 \text{ V}$$

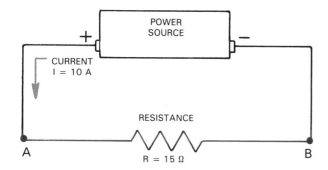

Fig. 1-17. Voltage and power can be calculated from resistance and current.

ELECTRICAL POWER

When we talk about mechanical power, we are describing the intensity of using energy. For example, it takes the same *energy* to lift ten 10-kilogram (kg) weights one at a time as it does to lift one 100 kg weight. But it takes 10 times the *power* to lift the 100 kg weight. The energy *times* the intensity equals power. It is the same with electrical power. The voltage drop across a resistance is a measure of the energy used. Current is a measure of the intensity of that usage. Electrical power (P) is measured in watts (W), for James Watt, an American inventor. *One watt is the power of a current of one ampere when the voltage drop is one volt.*

To compute electrical power, multiply voltage times current. This is known as Watt's Law. The equation is:

$$P = E \times I \qquad \begin{array}{l} P \text{ is the power (in watts)} \\ E \text{ is the voltage (in volts)} \\ I \text{ is the current (in amperes)} \end{array}$$

Example 6

In the circuit of Fig. 1-16, the voltage drop across the resistor is 50 V. The current is 10 A. Find the power used by the resistor.

$$P = E \times I = 50 \times 10 = 500 \text{ watts (W)}$$

There will be times when you know only the current and resistance and you need to know the power used. You could first find the voltage, but that is not necessary. You can simply substitute in the power equation as follows:

$$P = E \times I \text{ but } E = I \times R.$$

Therefore, $P = I \times R \times I$ or $P = I^2 \times R$.

Example 7

In the circuit of Fig. 1-17, the current is 10 A. Resistance is 15 Ω. Find the power used.

$$P = I^2 \times R = 10^2 \times 15 = 100 \times 15$$
$$= 1500 \text{ watts (W)}$$

TWO OR MORE RESISTANCES

So far we have been looking at circuits with only one resistance. A flashlight circuit is like this. There is:

1. A power source (usually two dry cells).
2. A switch to open and close the electrical path.
3. A resistance (the light bulb).

Most circuits, however, are more complicated. When there are two or more resistances in a circuit, they can be connected in *series* or in *parallel*.

SERIES CIRCUITS

Let us go back to the tube full of marbles, as shown in Fig. 1-18. Having two resistances in series is like having two narrow places in the tube. One narrow section reduces the marble flow in the entire tube. Two narrow sections reduce the flow still further. In an electrical circuit, a specific voltage causes a specific current to flow. The value of that current depends on the EQUIVALENT RESISTANCE of the circuit. If you know the current and voltage you can compute equivalent resistance by Ohm's Law:

$$R = \frac{E}{I}$$

You see, any circuit, no matter how complicated, is electrically the same as a single equivalent resistance connected to a power source. To find the equivalent resistance of a series circuit, simply add the values of the individual resistances. The equation for two resistances is:

$$R_{EQUIV} = R_1 + R_2$$

Example 8

A circuit consists of two resistances in series as shown in Fig. 1-19. R_1 has a resistance of 20 Ω and R_2, a resistance of 40 Ω. The applied voltage is 60 V. Find the equivalent resistance, the current, the voltage drop across each resistance, and the total power used by the circuit.

a. Compute the equivalent resistance.
$R_{EQUIV} = R_1 + R_2 = 20 + 40 = 60$ Ω
b. Compute the current.
$$I = E/R_{EQUIV} = \frac{60}{60} = 1 \text{ A}$$
c. Compute the voltage drop across the 20 Ω resistance.
$E_1 = I \times R_1 = 1 \times 20 = 20$ V
d. Compute the voltage drop across the 40 Ω resistance.
$E_2 = I \times R_2 = 1 \times 40 = 40$ V
e. Compute the total power.
$P = I^2 \times R_{EQUIV} = 1 \times 60 = 60$ W

Notice that the sum of the individual voltage drops ($E_1 + E_2$) is equal to the applied voltage (E). Actually, any number of resistances may be connected in series. The equivalent resistance will always equal the sum of the individual resistances. Likewise, the sum of the individual voltage drops will always equal the applied voltage.

PARALLEL CIRCUITS

Connecting a second resistance in parallel is like having a second tube joined to the first. This is illustrated in Fig. 1-20. Some of the marbles go through one tube section and some through the other. As a result, more marbles flow from the source. In electrical circuits, too, adding a resistance in parallel always increases current.

Just exactly how the current divides depends

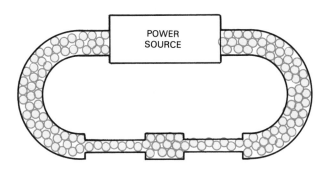

Fig. 1-18. Marble flow is less intense with two narrow places than with one.

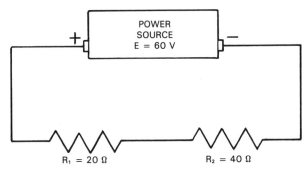

Fig. 1-19. In series circuit, equivalent resistance is sum of individual resistances.

on the resistance of each path. Current, remember, is inversely proportional to resistance. The larger the resistance of a path, the less the current in that path. The Ohm's Law equation, $I = \dfrac{E}{R}$, applies to each resistance path.

Refer to Fig. 1-21. The total circuit current flows from point A at the power source to point X where it splits. Part of the current flows through R_1. The rest of it flows through R_2.

Suppose you want to compute the total current. One way would be to compute the current in R_1 and the current in R_2 and add the two together.

How about the voltage in a parallel circuit? In Fig. 1-21, points A, B, C, and X are at the same electrical potential. This assumes the points are

close together so that any voltage drop in the wires is too small to measure. Likewise, points F, Y, E, and D are at the same electrical potential. The voltage drop across R_1, then, must be equal to the voltage drop across R_2. Of course, this drop is equal to the applied voltage, E.

Example 9

The voltage applied to the circuit in Fig. 1-21 is 60 V. R_1 is 30 Ω; R_2 is 30 Ω. Find the current through R_1, the current through R_2, and the total current.

a. Compute I_1, the current through R_1.
$$I_1 = \frac{E}{R_1} = \frac{60}{30} = 2 \text{ A}$$

b. Compute I_2, the current through R_2.
$$I_2 = \frac{E}{R_2} = \frac{60}{30} = 2 \text{ A}$$

c. Compute I, the total current.
$$I = I_1 + I_2 = 2 + 2 = 4 \text{ A}$$

Notice that 4 A of current flows into the junction, X. There it splits. Two amperes flow through R_1 and the other two amperes flow through R_2. This shows us a very important feature of parallel circuits: *The current leaving a junction always equals the current entering that junction.*

EQUIVALENT RESISTANCE OF A PARALLEL CIRCUIT

The other way to find total current is to compute first the equivalent resistance. Total current, then, equals the applied voltage divided by the equivalent resistance.

Computing the equivalent resistance of a parallel circuit can be easy or hard, depending on the value of the resistances. The easiest is when all parallel legs have the same resistance. Then you simply divide the resistance of one leg by the number of legs.

Example 10

Using the same circuit and values as in Example 9, find the equivalent resistance,

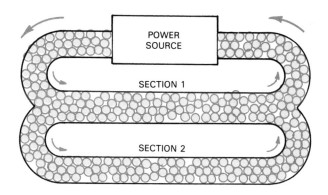

Fig. 1-20. Adding second section or pathway causes greater intensity of marble flow from source.

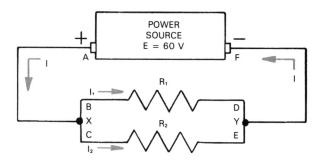

Fig. 1-21. Total resistance of resistors in parallel is less than the smallest resistor.

total current and total power.

a. Compute the equivalent resistance.

$$R_{EQUIV} = \frac{30}{2} = 15 \ \Omega$$

b. Compute the total current.

$$I = \frac{E}{R_{EQUIV}} = \frac{60}{15} = 4 \ A$$

c. Compute the total power.

$$P = I^2 \times R_{EQUIV} = 4^2 \times 15$$
$$= 16 \times 15 = 240 \ W$$

It is a little harder to compute the equivalent resistance of a parallel circuit when the two legs have different resistance. We use what is called the *product over sum method*. First, you multiply R_1 by R_2. This gives you the *product*. Then you add R_1 and R_2. This gives you the *sum*. Now divide the product by the sum.

$$R_{EQUIV} = \frac{R_1 \times R_2}{R_1 + R_2}$$

Example 11

The voltage applied to the circuit of Fig. 1-21 is 60 V. R_1 is 20 Ω; R_2 is 30 Ω. Find the equivalent resistance, the total current and the total power.

a. Compute the equivalent resistance.

$$R_{EQUIV} = \frac{R_1 \times R_2}{R_1 + R_2} = \frac{20 \times 30}{20 + 30}$$
$$= \frac{600}{50} = 12 \ \Omega$$

b. Compute the total current.

$$I = \frac{E}{R_{EQUIV}} = \frac{60}{12} = 5 \ A$$

c. Compute the total power.

$$P = I^2 \times R = 5^2 \times 12 = 25 \times 12$$
$$= 300 \ W$$

The most difficult parallel circuit to deal with has more than two legs and none are equal. The way to solve it is known as the *reciprocals method*. Actually, this method works on all kinds of parallel circuits. However, it is avoided when one of the other methods can be used.

The *reciprocal* of a number is the answer you get when you divide that number into 1. For

example, the reciprocal of 0.5 is 2 ($\frac{1}{0.5} = 2$).

The reciprocals methods can be stated as: *The reciprocal of the equivalent resistance is equal to the sum of the reciprocals of the individual resistances.*

Mathematically, the equation is:

$$\frac{1}{R_{EQUIV}} = \frac{1}{R_1} + \frac{1}{R_2} + \frac{1}{R_3} + \ldots$$

Example 12

The applied voltage to the circuit in Fig. 1-22 is 10 V. R_1 is 20 Ω; R_2 is 5 Ω; R_3 is 4 Ω. Find the equivalent resistance, total current, total power and the power used by each resistance.

a. Compute the equivalent resistance.

$$\frac{1}{R_{EQUIV}} = \frac{1}{R_1} + \frac{1}{R_2} + \frac{1}{R_3}$$
$$= \frac{1}{20} + \frac{1}{5} + \frac{1}{4}$$

$$\frac{1}{R_{EQUIV}} = 0.05 + 0.2 + 0.25 = 0.5$$

$$R_{EQUIV} = \frac{1}{0.5} = 2 \ \Omega$$

b. Compute the total current.

$$I_T = \frac{E}{R_{EQUIV}} = \frac{10}{2} = 5 \ A$$

c. Compute the current in R_1.

$$I_1 = \frac{E}{R_1} = \frac{10}{20} = 0.5 \ A$$

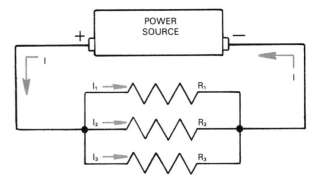

Fig. 1-22. As resistors are added in parallel, total resistance becomes less while current increases.

d. Compute the current in R_2.
$$I_2 = \frac{E}{R_2} = \frac{10}{5} = 2 \text{ A}$$

e. Compute the current in R_3.
$$I_3 = \frac{E}{R_3} = \frac{10}{4} = 2.5 \text{ A}$$

(Note: $I_1 + I_2 + I_3 = I_T$;
$0.5 + 2.0 + 2.5 = 5$ A.)

f. Compute the total power.
$$P = I^2 \times R_{EQUIV} = 5^2 \times 2 = 25 \times 2$$
$$= 50 \text{ W}$$

g. Compute the power used by R_1.
$$P_1 = I_1{}^2 \times R_1 = 0.5^2 \times 20$$
$$= 0.25 \times 20 = 5 \text{ W}$$

h. Compute the power used by R_2.
$$P_2 = I_2{}^2 \times R_2 = 2^2 \times 5 = 4 \times 5$$
$$= 20 \text{ W}$$

i. Compute the power used by R_3.
$$P_3 = I_3{}^2 \times R_3 = 2.5^2 \times 4$$
$$= 6.25 \times 4 = 25 \text{ W}$$

(Note: $P_1 + P_2 + P_3 = P_T$;
$5 + 20 + 25 = 50$ W)

There is a very important point about the equivalent resistance of a parallel circuit that you may have already noticed: *The equivalent resistance of a parallel circuit is always less than the smallest resistance leg.*

SERIES-PARALLEL CIRCUITS

Many electrical circuits are partly series and partly parallel. They are not hard to solve if taken a step at a time. First, find the equivalent resistance for each of the parallel parts. What remains is a simple series circuit.

Example 13

The circuit in Fig. 1-23 has the following values.

$$E = 60 \text{ V}; R_1 = 20 \text{ }\Omega$$
$$R_2 = 30 \text{ }\Omega; R_3 = 8 \text{ }\Omega$$

Find the equivalent resistance of the circuit, total current, the voltage drop across each resistance, the total power, and the power used by each resistance.

a. Compute the equivalent resistance of the parallel part.
$$R_{EQUIV}(P) = \frac{R_1 \times R_2}{R_1 + R_2} = \frac{20 \times 30}{20 + 30}$$
$$= \frac{600}{50} = 12 \text{ }\Omega$$

b. Redraw the circuit as shown in Fig. 1-24.

c. Compute the equivalent resistance of the total circuit.
$$R_{EQUIV} = R_{EQUIV}(P) + R_3 = 12 + 8$$
$$= 20 \text{ }\Omega$$

d. Compute the total current.
$$I = \frac{E}{R} = \frac{60}{20} = 3 \text{ A}$$

e. Compute the voltage drop across R_1 and R_2. This can be done two ways:
(i) $E_1 = E_2 = E - E_3 = 60 - 24$
$\quad\quad = 36 \text{ V}$
(ii) $E_1 = E_2 = I \times R_{EQUIV}(P) = 3 \times 12$
$\quad\quad = 36 \text{ V}$

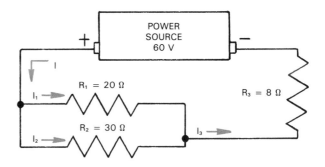

Fig. 1-23. Series-parallel circuit is solved by finding the equivalent resistance of parallel portion first.

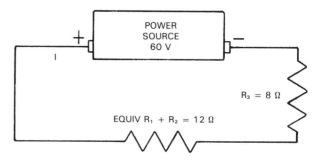

Fig. 1-24. This circuit is equal to the one in Fig. 1-23.

f. Compute the voltage drop across R_3.
$E_3 = I \times R_3 = 3 \times 8 = 24$ V

g. Compute the current in each resistance.
$I_1 = \dfrac{E_1}{R_1} = \dfrac{36}{20} = 1.8$ A

$I_2 = \dfrac{E_2}{R_2} = \dfrac{36}{30} = 1.2$ A

$I_3 = \dfrac{E_3}{R_3} = \dfrac{24}{8} = 3$ A

also $I = I_1 + I_2 = 1.8 + 1.2 = 3$ A

h. Compute the total power. This can be done three ways:

(i) $P = I^2 \times R_{EQUIV} = 3^2 \times 20$
$= 9 \times 20 = 180$ W

(ii) $P = \dfrac{E^2}{R_{EQUIV}} = \dfrac{60^2}{20} = \dfrac{3600}{20}$
$= 180$ W

(iii) $P = E \times I = 60 \times 3 = 180$ W

i. Compute the power used by each resistance.
$P_1 = I^2 \times R_1 = 1.8^2 \times 20 = 3.24 \times 20$
$= 64.8$ W

$P_2 = I_2^2 \times R_2 = 1.2^2 \times 30$
$= 1.44 \times 30 = 43.2$ W

$P_3 = I_3^2 \times R_3 = 3^2 \times 8 = 9 \times 8$
$= 72.0$ W

(Note: $P = P_1 + P_2 + P_3$
$= 64.8 + 43.2 + 72.0 = 180$ W)

TWO OR MORE POWER SOURCES

Some circuits will have more than one power source. Then you will have two things to consider. One is the *polarity* of the power source. The other is whether they are connected in series or in parallel.

Consider a flashlight, for example. Each of the dry cells is a power source. You must install them with the positive terminal of one touching the negative terminal of the other. This is known as a *series-aiding* connection. A circuit diagram is shown in Fig. 1-25. With a series-aiding connection, the total voltage is equal to the sum of the two. For example, the total voltage of the two flashlight batteries shown in Fig. 1-25 is (1.5 + 1.5) 3 volts.

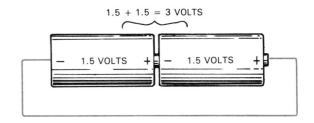

Fig. 1-25. Voltages of power sources are added together when they are connected series-aiding.

Now, suppose you had a three-cell flashlight. If you put one cell in backward, it would be connected *series-opposing*. The total source voltage in that case equals the sum of the voltages connected series-aiding *minus* the one connected series-opposing.

$E = E_1 + E_2 - E_3$

$E = 1.5 + 1.5 - 1.5 = 3 - 1.5 = 1.5$ V

Power sources can also be connected in parallel. When they are, their positive terminals must be connected to the same point. Also, their voltages ought to be equal. If two unequal power sources are connected parallel, current drains from the higher to the lower one. Having two or more power sources in parallel does not make the total voltage any larger. What it does is to increase the amount of current that can be supplied. Current from each of the sources combines to produce total circuit current.

Assume you have four 1.5 V dry cells connected in parallel. The total voltage would still be 1.5 V. However, if each dry cell could supply one ampere, the total current available would be four amperes. This is illustrated by Fig. 1-26.

SUMMARY OF ELECTRICITY AND ELECTRICAL CIRCUITS

Electricity can be thought of as a flow of negative electrical charges.

Electrical pressure exists between two points in an electrical circuit when one point has more

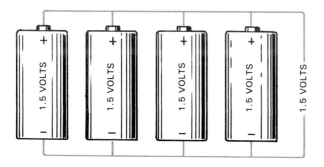

Fig. 1-26. All power sources must have the same voltage when they are connected in parallel.

electrons than the other point.

Electrical pressure is measured in volts and is represented by the symbols E and V.

Electrical current requires a closed path between two points that have a voltage drop across them.

Current is the intensity of charge flow. It is measured in amperes, and is represented by the symbol I.

Resistance is the opposition to current. It is measured in ohms, and is represented by the symbol R.

Resistance is the ratio between voltage and current, $R = \dfrac{E}{I}$.

In a series circuit, the equivalent resistance is equal to the sum of the individual resistances.

In a series circuit, the sum of the voltage drops around the circuit is equal to the applied voltage.

In a parallel circuit the equivalent resistance is computed by the equation

$$\frac{1}{R_{EQUIV}} = \frac{1}{R_1} + \frac{1}{R_2} + \frac{1}{R_3} \cdots$$

In a parallel circuit, the total current is equal to the sum of the current in each leg.

The current entering a junction in a parallel circuit equals the current leaving the junction.

The power, P, supplied by a current, I, to a resistance, R, is computed by the equation $P = I^2 \times R$.

The power, P, resulting from a voltage, E, across a resistance, R, is computed by the equation $P = \dfrac{E^2}{R}$.

The power, P, supplied by a current, I, resulting from a voltage, E, is computed by the equation $P = E \times I$.

TEST YOUR KNOWLEDGE

1. Circle correct answers: (Conductors, insulators) allow electricity to flow through them, while (conductors, insulators) block the flow of electricity.
2. Complete the following three Ohm's Law equations:
   I =
   R =
   E =
3. Complete the following Watt's Law equations:
   (a) In terms of current and resistance: P =
   (b) In terms of current and voltage: P =
   (c) In terms of voltage and resistance: P =
4. If you added a lamp in series with two lamps that are connected in series would the current become larger or smaller? What would that do to the brightness of the lamps?
5. Is the equivalent resistance of a number of resistors in series smaller or larger than the largest resistance?
6. Is the equivalent resistance of a number of resistors in parallel smaller or larger than the smallest resistance?

PROBLEMS

1. Given: A voltage of 100 V is applied to a 5 Ω resistance. Find the current.
2. Given: A 12 V lamp draws 0.5 A. Find the resistance of the lamp.
3. Given: Three resistors, having resistance values of 10 Ω, 15 Ω, and 25 Ω, are connected in series. Find the equivalent resistance.
4. Given: Two resistors, having resistance values of 35 Ω and 65 Ω are connected in series. A voltage of 120 V is applied across the two of them. Find the equivalent resistance, current, and the voltage drop across each reactor.
5. Given: Two resistors, each having a resis-

tance value of 30 Ω, are connected in parallel. Find the equivalent resistance.

6. Given: Three resistors, having resistance values of 40 Ω, 20 Ω, and 200 Ω, are connected in parallel. A voltage of 125 V is applied across the parallel circuit. Find the equivalent resistance, total current, and the total power used.

2 | MAGNETISM

After studying this chapter you will be able to:

☐ State the fundamental rules of magnetic attraction and repulsion.

☐ Describe the relationship between the direction of current in a conductor and the direction of the magnetic field surrounding it.

☐ Describe the relationship between the direction of current in a coil and its magnetic field.

☐ Define the terms magnetomotive force, flux, reluctance, permeability, and flux density.

☐ Explain two ways in which voltage can be induced into a conductor.

If magnets were not so common, you might think magnetism was something right out of a science fiction movie. Magnets can stick together without glue. They can lift many times their own weight. Magnets can even make objects move without touching them.

The force field that surrounds a magnet can pass through anything. Yet, you cannot see, hear, or feel it. Nor can you taste or smell magnetic fields. But we know that they do exist—and we know how to use them.

In generators, we use magnetic fields to *change motion into electricity*. In motors, we use magnetic fields to *change electricity into motion*. And, in transformers, we use magnetic fields to step voltages up and down.

MAGNETIC POLES

However, the first use for magnetism was in guiding ships. Fig. 2-1 shows a piece of magne-

tite or lodestone. The Chinese were using similar pieces around 5000 years ago.

If you hang a piece of lodestone on a string, it will always line up in a north-south direction. Since the same end always points north, early navigators called that end of the lodestone the NORTH POLE. Then, in the late 1800s, the tungsten steel compass was invented. Since then, the only use for lodestone is in making iron. You see, magnetite is really iron ore that has been struck by lightning.

A compass needle, like that shown in Fig. 2-2, is simply a small magnet. It is free to rotate

Fig. 2-1. Piece of magnetite or magnetized iron ore. Ancient sailors called them lodestones and used them as primitive compasses.

Fig. 2-2. Compass needle is piece of magnetized steel.

and it will line up in a north-south direction. That is because the earth itself is a huge ball-shaped magnet. Imagine that it has a giant bar magnet in the center, as illustrated by Fig. 2-3. There is a magnetic field around the planet and there are magnetic poles. Notice, however, that the earth's *magnetic south pole* is located near the spot we call the North Pole (the geographic north pole). Likewise, the *magnetic north pole* is down near the geographic South Pole.

The mix-up in naming the earth's magnetic poles resulted from a wrong guess. At first, people naturally thought that the end of a compass needle that pointed in the northerly direction should be called NORTH. Later it was learned that the north end of any magnet is attracted to the south end of another. Therefore, the magnetic pole to which the north end of a compass needle points has to be called SOUTH.

MAGNETIC LINES OF FORCE

When people talk about magnets, we usually think of either the bar, shown in Fig. 2-4, view A, or the U-shaped magnet in Fig. 2-4, view B. To get an idea of what a magnetic field looks like, we use small bits of iron, called iron filings. If you were to lay a piece of glass on top of the magnets and sprinkle iron filings on the glass, you would see the patterns shown in Fig. 2-5. We say magnetic fields are made up of lines of force. These lines do not flow like electric

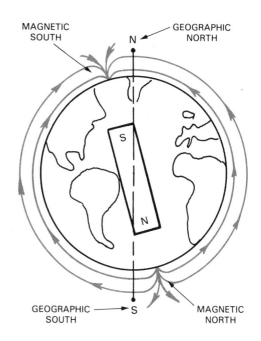

Fig. 2-3. The Earth is a giant permanent magnet. We can imagine its center as a giant bar magnet.

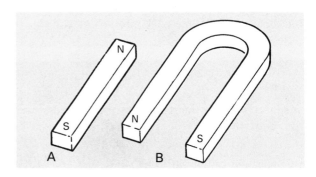

Fig. 2-4. Common permanent magnet shapes. A—Bar magnet. B—U-shaped magnet.

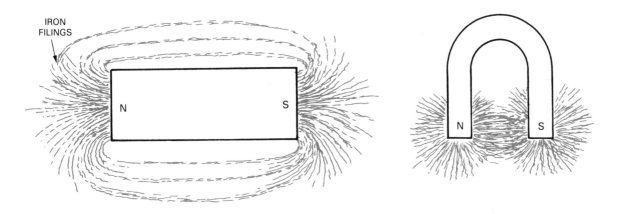

Fig. 2-5. Patterns formed by iron filings show shape of magnetic field.

current but they do point in a definite direction. They point outward from the north pole and inward toward the south pole.

MAGNETIC ATTRACTION

Since generators, motors, and transformers all need two magnetic fields, we ought to see what happens when two fields get together. Look at Fig. 2-6. The two bar magnets have been set with the north pole of one near the south pole of the other. The pattern made by the iron filings shows us that some of the lines of force coming out of the north pole of one magnet are going into the south pole of the other. The two fields have combined into one big field. This field tries to pull the two magnets together.

One of the basic rules of magnetism is: *Unlike poles attract*. The north pole of any magnet exerts a force of attraction on the south pole of any other magnet.

MAGNETIC REPULSION

Fig. 2-7 shows the result of turning one of the magnets around. The pattern for this north-north setup is quite a bit different. The two fields seem to be fighting each other. Notice that the magnetic lines of force do not cross. In the space directly between the two magnets, the lines cancel each other out because they are headed in different directions. But off to the

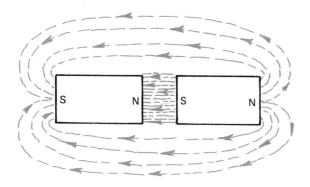

Fig. 2-6. Unlike poles exert a force of attraction on each other.

side, they find themselves going in the same direction and they combine. The result is a force that tries to keep them apart. This is the second rule of magnetism: *Like poles repel*.

The north pole of any magnet exerts a force of repulsion on the north pole of any other magnet. Likewise, the south pole repels any other south pole that happens to get close.

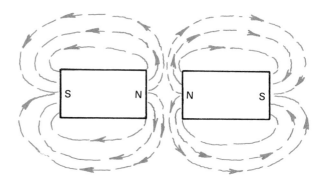

Fig. 2-7. Like poles exert a force of repulsion against (push away from) each other.

ORIGIN OF MAGNETIC FIELDS

Where do magnetic fields come from? Why is it that only certain materials can be magnetized? As in electricity, the basic cause is the electron.

Electrons, as we know, orbit the center of their atoms like the Earth orbits the sun. Electrons are also spinning on their axis, like the Earth does. It is this spinning that makes each one of them a tiny magnet. See Fig. 2-8.

The direction of the field around each electron depends on whether it is spinning clockwise or counterclockwise. The materials that can not be magnetized, like wood, have atoms with the same number of electrons spinning in each direction. This cancels out all magnetic fields.

MAGNETIC DOMAINS

On the other hand, atoms of the so-called MAGNETIC MATERIALS, like iron, have more electrons spinning in one direction than in the other. Thus, each atom ends up with a field. This is illustrated in Fig. 2-9. When a lot of these magnetic atoms get together (about 1,000,000,000,000,000 or 10^{15} of them) with their fields oriented in the same direction, you have what is called a DOMAIN.

Fig. 2-10, view A shows an unmagnetized piece of iron. The domains are helter-skelter, causing all of the fields to cancel each other out. But when you stroke the iron with a magnet, the domains line up in the same direction, producing a field around the iron. This is shown in Fig. 2-10, view B.

PERMANENT AND TEMPORARY MAGNETS

Some people think a magnet is like a battery — that it will run down sooner or later. You might think so too, if you magnetized a piece of iron. It does not stay magnetized. Iron quickly gives up most of its magnetism. But it does not run down. The domains just go back to their helter-skelter positions.

This does not happen with a piece of hard steel or other permanent magnet material. Its domains get "locked-in" and can stay that way forever.

It takes some form of energy to line up the magnetic domains. This is true of both perma-

nent and temporary magnets. Once the domains are lined up, however, it does not take any additional energy to keep them that way.

To get energy out of generators, motors, or transformers, you have to put energy into them. Even though they all use magnetism, their power output comes from the external power you put in, not from the magnetic field.

In addition to iron and steel, magnetic materials include cobalt, nickel, metallic alloys

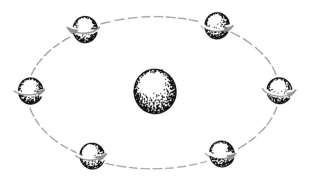

Fig. 2-9. In atoms making up material like iron there are six electrons in the same orbit. Five are spinning clockwise. One is spinning counterclockwise. The result is a magnetic field around the atom.

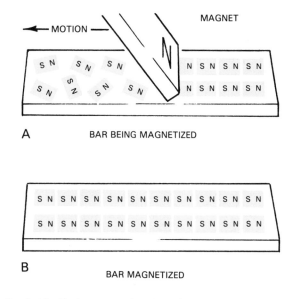

A   BAR BEING MAGNETIZED

B   BAR MAGNETIZED

Fig. 2-10. Each atom or domain in ferrous material is like a tiny magnet. Magnetization lines up the domains as shown in view B.

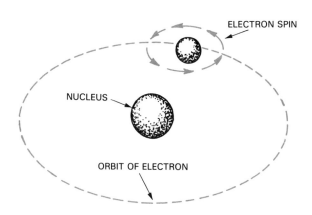

Fig. 2-8. Each electron spins on its own axis creating tiny magnetic field.

Electrical Power

such as Alnico and ceramic materials known as ferrites. There is a long list of nonmagnetic materials, including air, glass, and wood. Metals like copper and aluminum make very good electrical conductors. However, they are such weak magnets that they can be called non-magnetic, too.

ELECTROMAGNETISM

Magnetic fields are produced in two ways.
1. From the domains of magnetic materials.
2. By electricity.

Any electrical conductor can be used to create magnetism. Current passing through the conductor sets up a magnetic field.

In 1820, Hans Christian Oersted performed an experiment that you can easily do. First, hook up a wire with a switch to a battery. See Fig. 2-11. With the switch open, lay a compass on top of the wire. Be sure the wire is lined up north-south, so the compass needle is exactly parallel with it. When you close the switch you will see the needle swing, as shown in view A, Fig. 2-12. Next, reverse the battery terminals as in view B, Fig. 2-12 and then close the switch again. The needle will swing the other way. Oersted had discovered that current produces a magnetic field around the wire like an invisible sleeve. This field is caused by the movement of the electrons through the wire. With many elec-

trons moving in the same direction, all their little fields combine. The result is a field strong enough to move the compass needle.

DIRECTION OF MAGNETIC LINES OF FORCE

Fig. 2-13 shows the magnetic field around a current-carrying wire. Notice that the lines of force are in closed loops. There are no points we can call poles.

Actually, all magnetic fields are made up of lines in closed loops. The reason bar magnets have poles is that their lines are partly in air and partly in the magnet itself. A pole is simply the dividing line between the magnet and air.

What would happen if we put two U-shaped

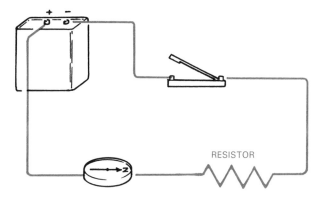

Fig. 2-11. Oersted's experiment. He put down a wire circuit as shown. Then he laid a compass on top of the wire carefully aligning the compass needle with the direction of the wire.

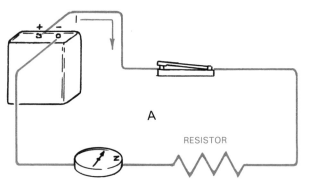 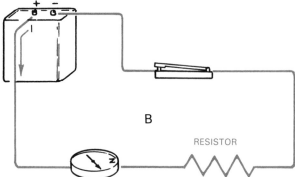

Fig. 2-12. Oersted's experiment with power applied to circuit. A—When switch closes compass needle moves indicating presence of magnetic field. B—With current flow reversed, magnetic field reverses polarity.

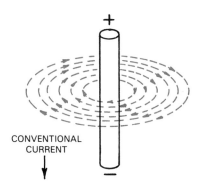

Fig. 2-13. Magnetic field is produced around every point of a wire when current is flowing through it.

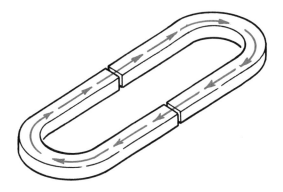

Fig. 2-14. With two U-shaped magnets placed end to end, magnetic field is entirely inside of the steel.

magnets together, as in Fig. 2-14? Now the field is all within the steel. The direction of the lines is the same, even though we have eliminated the poles.

What we need is a new rule for the direction of magnetic lines of force. This rule must hold good for a field inside a magnet, in air, or partly in each.

The direction of magnetic lines of force is the direction that would be taken by an isolated north pole as it is made to travel around a closed field.

With a simple bar magnet, the magnetic lines of force move from north to south in air and south to north inside. Around a current-carrying wire, they move counterclockwise when conventional current is flowing toward you and clockwise when the current is flowing away from you.

RIGHT HAND RULE FOR CONDUCTORS

There is an easier way to remember the direction of the field around a wire. It is called the "right hand rule for conductors." As Fig. 2-15 shows, wrap your right hand around a bar with your palm facing you. With your thumb pointing in the direction of the current, your fingers will be curved in the direction of the magnetic lines of force.

Fig. 2-15. Right hand rule for conductors. Thumb points in direction of current flow. Curled fingers curve in direction of magnetic lines of force.

The field around a straight wire conductor is not very strong. We can, however, increase its strength by coiling the wire. The weak fields from every point on the wire all combine into a single strong field. Then, if you put an iron core in the center, the coil's field will magnetize the iron, producing a very strong field. This is an ELECTROMAGNET. Its field is often called an ELECTROMAGNETIC FIELD, although it is the same in every way as the field of a permanent magnet. The field of an electromagnet, however, disappears when current stops flowing.

RIGHT HAND RULE FOR COILS

The field of an electromagnet is partly outside and partly in the center of the coil. This produces definite magnetic poles.

The direction of current around the coil determines which end is which pole. The rule for this is known as the "right hand rule for coils." Again, wrap your right hand around the conductor, with your palm facing toward you. This time your fingers will be curved in the direction that conventional current is flowing around the wire. Your thumb is pointing in the direction of the magnetic lines inside the coil. That is, toward the north pole of the coil. This is illustrated in Fig. 2-16.

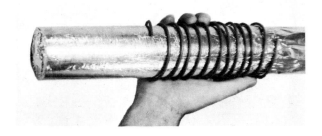

Fig. 2-16. Right hand rule for coils. Fingers curve in direction of current travel in coil. Thumb points in direction of magnetic lines inside coil.

CHARACTERISTICS OF MAGNETISM

There are magnetic fields at work in all motors, generators, and transformers. These fields have certain characteristics that we will be talking about throughout this book. At this point, it would help to understand the more important of these characteristics.

There are several systems of measurement used for magnetic quantities. The one used in this book is the SI metric system. (SI stands for "Le Systeme International d Unites" or International System of Units. It is the system used by most of the countries of the world.) For a complete discussion of the magnetic units presently in use, refer to the section on Technical Information at the end of the text.

MAGNETOMOTIVE FORCE

Magnetomotive force (mmf) is the "push" behind the magnetic field. It is to magnetism what electromotive force, or voltage, is to electricity.

Magnetomotive force is directly proportional to two things:

1. The number of turns (N) on the coil.
2. The amount of current passing through the coil.

The equation for mmf is:

$$\text{mmf (ampere-turns)} = \text{N (turns)} \times \text{I (amperes)}$$

Example 1

The coil shown in Fig. 2-17 has 10 turns. Two amperes of current is flowing. Find the mmf.

$$10 \text{ turns} \times 2 \text{ A} = 20 \text{ ampere-turns}$$

Two magnetic fields in the same space always combine into a single field. Sometimes we talk as if there were two separate fields, even though there is really only one. It is the mmf's that either add or subtract. The resultant mmf, then, creates the single field. In an electrical circuit, this would be like connecting two batteries in series. There is only one current flow, resulting from the sum (or difference) of the two voltages.

FIELD INTENSITY

The intensity of the field of an electromagnet depends on two things:

1. The mmf. The greater the magnetomotive force, the more intense the field.
2. The length of the coil form. The shorter the coil's length, the more intense the field.

The symbol for field intensity is the letter H. (Do not confuse field intensity with the unit of induction, both of which use the symbol, "H." They have nothing to do with one another.)

Field intensity is expressed in ampere-turns per metre. The equation is:

$$\text{H (ampere-turns/m)} = \frac{\text{mmf (ampere-turns)}}{\text{length (metres)}}$$

Fig. 2-17. Mockup represents a 10-turn coil. Such coils have an iron core and are a basic element of motors, generators and transformers.

Example 2

The coil shown in Fig. 2-17 has an mmf of 20 ampere-turns. Its length is 0.1 metres. Find the field intensity.

$$H = \frac{mmf}{length}$$

$$H = \frac{20}{0.1} = 200 \text{ ampere-turns/metre}$$

Field intensity is also called MAGNETIZING FORCE. If you think of the coil's iron core as being magnetized temporarily (while current is flowing through the coil), you can see that the ampere-turns/metre is the force that does the magnetizing.

FIELD STRENGTH

A magnetic field is made up of imaginary lines of force. Its strength is gauged by the number of lines it contains. Field strength is also referred to as FLUX. Flux is measured in WEBERS. One weber equals 100,000,000 (10^8) lines. The symbol for flux is the same as the Greek letter phi (ϕ).

RELUCTANCE

Magnetic flux forms a closed loop called a magnetic circuit. When the flux path is in air, the magnetizing force cannot create any more lines. However, in magnetic materials, like an iron core, the magnetizing force causes the domains in the material to line up, creating additional flux.

In electrical circuits, the ratio between applied voltage and the resulting current is called resistance ($R = \frac{E}{I}$). It is like that in magnetic circuits, too. The ratio between the magnetomotive force and the resulting flux is called RELUCTANCE ($\mathcal{R}$).

The equation is:

$$\mathcal{R} \left(\frac{\text{ampere-turns}}{\text{weber}}\right) = \frac{\text{mmf (ampere-turns)}}{\phi \text{ (webers)}}$$

If you want to, you can think of reluctance as a characteristic of material, like resistance. Reluctance, then, is the opposition to flux. Even though additional flux in the core results from its magnetization, you can pretend it is due to a low reluctance.

PERMEABILITY

The FIELD INTENSITY (H) tells us how much magnetizing force is available to produce the flux in the core. The amount of flux that actually gets produced, however, depends on how easily the domains of the core can be lined up. This is what is called PERMEABILITY. The symbol for permeability is the Greek letter mu (μ). The value of the permeability of air (μ_o) is 1.26×10^{-6}. All other materials are related to air. For example, if an iron core has a RELATIVE PERMEABILITY (μ_r) of 500, the density of the field in it is 500 times that of air, with the same magnetizing force applied.

$$\mu = \mu_r \times \mu_o = 500 \times 1.26 \times 10^{-6} = 0.00063$$

FLUX DENSITY

So far we have seen that the mmf of a coil produces a magnetizing force (H), which, in turn, creates a flux (ϕ). The density of the resulting field is proportional (increases or decreases at same rate) to both the magnetizing force and the permeability (μ) of the core.

The symbol for flux density is the Greek letter β. The unit of measure is the TESLA. (In electric machine calculations found later in this text, the Greek letter, ϕ, is used for "field strength." This term means the same thing as flux density. It also is measured in teslas.) One tesla is equal to one weber per square metre. Giving the flux density is the same thing as saying how many lines pass through the cross section of the material. The equation is:

$$\beta \text{ (in teslas)} = \mu \times H \text{ (ampere-turns} \times \text{metre)}$$

Or, if you know the flux (ϕ):

$$\beta \text{ (in teslas)} = \frac{\phi \text{ (webers)}}{\text{area (square metres)}}$$

Example 3

The coil in Fig. 2-17 produces a magnetizing force of 200 ampere-turns per metre. The iron core has a relative permeability (μ_r) of 800, and has a cross-sectional area of 0.004 square metres (m²). Find the flux density and the total flux in the core.

a. Compute the flux density (β):
$$\beta = \mu_r \times \mu_o \times H$$
$$= 800 \times 1.26 \times 10^{-6} \times 200$$
$$= 0.2 \text{ teslas}$$
b. Compute the flux (ϕ):
Since $\beta = \frac{\phi}{A}$, then $\phi = \beta \times A$
$$= 0.2 \times 0.004 = 0.0008 \text{ webers}$$

(Note: 0.0008 webers [wb] is equivalent to 80,000 lines.)

CHARACTERISTICS OF MAGNETIC MATERIALS

Generators, motors, and transformers have one thing in common. They all contain coils that are wound on cores. These cores are usually made of thin sheets of steel (called laminations) which are fastened together. During operation, the cores become magnetized. The magnetic characteristics of the core, then, are very important to the understanding of electrical power devices.

SATURATION

When you put current through the coil in Fig. 2-17, a certain flux density results. If you increase the current, you would expect the flux to become more dense. The flux density, β, *does* increase in proportion to the current, but only up to a point. This point we call SATURATION. After all, there is a limited number of domains in the iron core. When you reach the saturation point, you have so many of the domains lined up that additional current has little or no effect on the flux density.

Since permeability is a measure of the flux density produced by the magnetizing force, it follows that the permeability of a material beyond saturation is different from its permeability before saturation. When you read that a magnetic material has a certain permeability, that means before saturation. Fig. 2-18 shows a magnetization curve. It demonstrates how the flux density increases with field intensity. It is also called a saturation curve, since it shows when saturation occurs.

RESIDUAL MAGNETISM AND RETENTIVITY

After an iron core has been temporarily magnetized, what happens when you take away the magnetizing force? That depends on the material's RETENTIVITY. *This is the ability of a material to retain magnetism when the current flow drops to zero.* "Hard" steel, from which permanent magnets are made, retains almost all of it. The domains are "locked-in." "Soft" iron, on the other hand, retains only a little of its magnetism. It is referred to as RESIDUAL MAGNETISM.

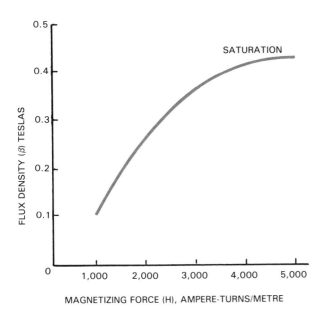

Fig. 2-18. This magnetization curve is also known as a saturation and as a "B-H" curve.

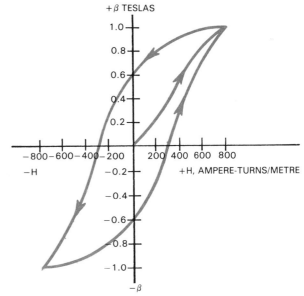

Fig. 2-19. This hysteresis curve is a B-H curve covering full cycle of alternating current.

HYSTERESIS

When you turn off the current flowing through a cell, its magnetizing force falls to zero. However, there is still residual magnetism present in its core. Then, if you were to apply voltage with the opposite polarity, the magnetizing force has the opposite magnetic polarity. It tends to oppose the residual magnetism. First the magnetism *does* fall to zero. Then it reverses its direction.

Since the flux in the iron is always lagging the current in the coil, we say the iron has HYSTERESIS. Hysteresis comes from a Greek word meaning "to lag." The curve shown in Fig. 2-19 is known as a hysteresis curve.

It takes electrical energy to reverse the domains in the coil's core. Therefore, hysteresis can cause power losses in generators, motors, and transformers.

INDUCTION-TRANSFORMER EMF

We have seen that magnetism is produced by electricity. What about the electricity that magnetism produces? The generation of electricity by magnetism is called INDUCTION. Voltage that is induced by magnetism is called ELECTROMOTIVE FORCE, abbreviated emf.

It was Michael Faraday who made two very important discoveries about emf. In the first one, he wound a wire on a coil form and connected it to a GALVANOMETER. *A galvanometer is a meter that can measure small currents.* Next, he wrapped insulation over the first winding. Then he wound a second coil on the same form. This coil he connected to a direct current source through a switch. This setup is shown in Fig. 2-20.

When he closed the switch, sending current through the second coil, he saw the galvanometer needle move. When he opened the switch, the needle moved the other way.

There was no electrical connection between the two coils. Where did the current in the insulated coil come from? Why did it flow for such a short time? Finally, why did the current flow one way when he closed the switch and the opposite way when he opened the switch? These

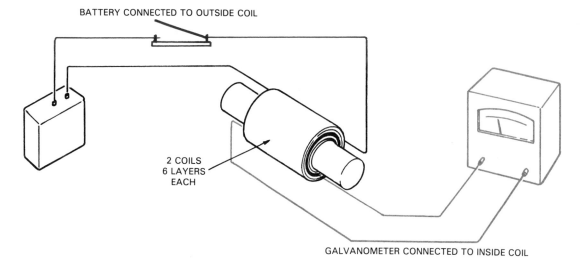

BATTERY CONNECTED TO OUTSIDE COIL

2 COILS
6 LAYERS
EACH

GALVANOMETER CONNECTED TO INSIDE COIL

Fig. 2-20. Faraday set up this experiment to demonstrate electromotive force of a transformer.

questions were some of the things Michael Faraday had to figure out.

If you were to do the same experiment, you would get the same result. Refer to Fig. 2-21. Here is what happens:

1. Current begins to flow the instant you close the switch. It starts at a value of zero amperes and grows to its full value.
2. While this current is building up, the magnetic flux lines are expanding outward.
3. As they expand, they link up with the wires of the insulated coil.
4. These changing flux linkages exert a force on the electrons of the insulated wire, making them move toward one end.

    With extra electrons at one end and a shortage at the other, we have an INDUCED VOLTAGE. Current flows because there is a closed electrical path (through the galvanometer) between the two ends. However, the current flows in the insulated coil only while the magnetic flux is changing.
5. As soon as the current in the first coil reaches a steady value, the magnetic flux stops expanding. No further voltage is induced in the insulated coil. If, at this point the switch is opened, the current will start to drop.

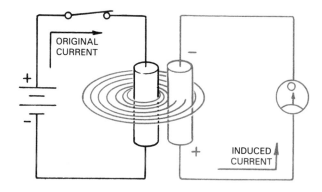

ORIGINAL CURRENT

INDUCED CURRENT

Fig. 2-21. Setting up your own transformer experiment.

6. Next, the magnetic field starts to collapse.
7. Again the flux lines are cut by the wires of the insulated coil. This time they exert a force on the electrons in the opposite direction. The polarity of the induced voltage is reversed.
8. This reverses the current flow and the galvanometer needle moves the other way. Voltage induced by expanding and collapsing magnetic fields is called TRANSFORMER EMF.

INDUCTION-MOTION EMF

In another experiment, Faraday mounted a wire on a board and connected one end of the

Magnetism

wire to a galvanometer. He then took a second wire and connected it to a battery. This setup is shown in Fig. 2-22. When he moved the current-carrying wire toward the first wire, the galvanometer needle moved. Induced electricity was flowing in the first wire. Then, when he moved the second wire away, the galvanometer indicated current flowing in the opposite direction. Current flowed, however, only while the wire was moving.

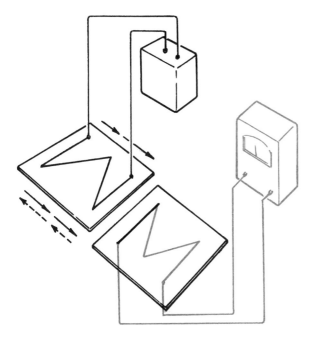

Fig. 2-22. Faraday's demonstration of motion emf. Passing one wire over or under the other will cause current flow.

You can easily duplicate Faraday's experiment. Here is what happens:

1. The current-carrying wire has a steady magnetic field around it. When you move the wire, you also move its magnetic field. As you move this field toward the second wire, its flux lines get cut by the second wire.
2. A force is exerted on the electrons in the wire. They begin to move, producing an emf, or voltage. Current is able to flow because the galvanometer provides a closed path.
3. When you move the current-carrying wire away from the second wire, the force on the

electrons is reversed. Induced voltage builds up with the opposite polarity. This is what makes current flow in the opposite direction.

You will get the same result if you move the first wire toward and away from the current-carrying wire. Voltage is induced in a wire as long as there is relative motion between the wire and a magnetic field.

Faraday discovered that the amount of voltage induced depends on two things:

1. The strength of the magnetic field.
2. How fast the magnetic lines are cut by the wire. Voltage induced this way is called MOTION EMF.

SUMMARY OF MAGNETISM

Magnetic fields are made up of lines forming closed loops.

A group of atoms with their fields oriented in the same direction make a magnetic domain.

A piece of magnetic material is magnetized by lining up the magnetic domains in the same direction.

The two points at which magnetic lines of force enter and leave the material are called respectively the south and north poles.

Like poles repel and unlike poles attract.

The direction of a magnetic field is from south pole to north pole inside the material and from north pole to south pole outside the material.

Electricity flowing through a wire sets up a magnetic field like a sleeve around a wire.

Magnetomotive force (mmf) equals the number of turns on a coil times the current. The units of mmf are ampere-turns.

When the thumb of your right hand is pointed in the direction of conventional current flow in the wire, the fingers are curved in the direction of the magnetic field.

A coil with current flowing through it is an electromagnet, whose north and south poles are the ends of the coil.

37

If you curve the fingers of your right hand in the direction of conventional current around a coil, the extended thumb points towards the north pole.

TEST YOUR KNOWLEDGE

1. Give the two laws of magnetic attraction and repulsion.
2. Residual magnetism is (select the correct answer):
   a. The magnetism produced by current in a wire.
   b. The magnetism remaining after the magnetizing force has been removed.
3. When current through a coil increases, its magnetizing force (H) (increases, decreases, remains the same).
4. In order to have "motion" emf induced in a conductor, there must be relative _____ between the conductor and a magnetic field.
5. In order to have "transformer" emf induced in a conductor, the conductor must be in the presence of a magnetic field that is either _____ or _____.
6. Only magnetic materials have magnetic domains. When a number of domains are lined up with their poles in the same direction, the material is said to be _____.

PROBLEMS

1. Given: A coil of 50 turns has a current of 2 A. Find the magnetomotive force (mmf).
2. Given: The coil of *Problem 1* is 5 cm (0.05 metres) long. Find the field intensity (H).
3. Given: A coil with an area of 0.0025 square metres (m^2) has a flux density of 0.2 teslas. Find the flux.
4. Given: The reluctance of the core of the coil of *Problem 3* is 1.5 × 10^6 (1,500,000) ampere-turns per weber. Find the magnetomotive force.
5. Given: A current of 5 A through the coil of *Problem 3* produces 0.2 T. Find the number of turns on the coil.
6. Given: The coil of *Problem 3* is 0.4 m long. Find the field intensity.

3   ALTERNATING CURRENT

After studying this chapter, you will be able to:

☐ Explain alternating current by using a sine wave.
☐ Define the following values of alternating current: RMS (effective), instantaneous, average, and peak.
☐ Calculate the ac ripple of a fluctuating dc voltage.
☐ Determine the sum of two 90 degree out-of-phase voltages using a phasor diagram and trigonometry.

Current flows between two points if:

1. There is a conducting path between them.
2. Each of the points has a different electrical potential.

Current always flows from the point with the higher potential to the point with the lower potential. The higher potential we call positive, or plus. The negative potential we call negative, or minus.

MEANING OF DIRECT CURRENT

In the circuits discussed so far, one terminal of the power source was plus at all times. The other was always minus. This is illustrated in Fig. 3-1. The current moved through the wire in one direction only. *Current that flows in one direction only is called direct current.* It is abbreviated dc. Voltage across terminals that never change polarity is called dc voltage.

The first electrical power source was the VOLTAIC CELL. The voltaic cell works like an automobile battery or dry cell. It converts chemical energy to dc electrical energy.

Later, when electrical generators were developed, they naturally generated direct current. Electric motors, heaters, lights, and other devices used dc.

There is still a great deal of direct current used today. All battery-operated tools, toys, and radios operate on direct current. Even large factories may use dc motors to drive machinery. If they do, they usually produce the electricity right in the plant. The reason they do so is to avoid transmitting dc power over long distances. The resistance of the wires would cause too much voltage drop. The power loss could be reduced by keeping the current low ($P = I^2R$). However, changing the voltage and current levels of dc is very difficult.

Normally, direct current must be used at the generated voltage. If you want to run a 240 V motor, for example, the generator has to put out 240 V. It is not easy to operate a 120 V lamp from the same source. Of course, you could put

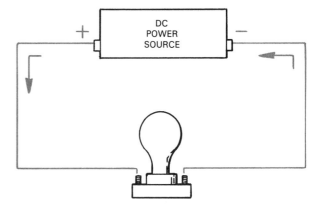

Fig. 3-1. Direct current voltages may change value but polarity never changes.

a resistance in series to drop the extra 120 volts. You would not want to, though. The resistance would waste as much power as the lamp uses. This is illustrated in Fig. 3-2.

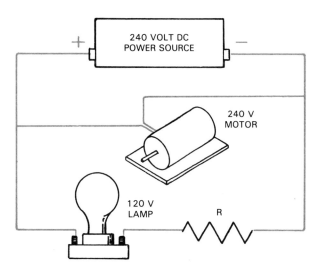

Fig. 3-2. Direct current voltages cannot be easily transformed to different values.

THE MEANING OF ALTERNATING CURRENT

In the early 1900s, a Yugoslav-born American inventor, Nichola Tesla, came up with the idea of ALTERNATING CURRENT. Alternating current, abbreviated ac, flows first in one direction through the wires, then in the other. See Fig. 3-3, view A. First, terminal 1 is more positive than terminal 2. Current flows out from 1 and back into 2. Then, as shown in view B, Fig. 3-3, terminal 2 becomes more positive than 1. This causes current to flow out from 2 and back into 1.

Offhand, it does not seem to make much sense to have the current flowing back and forth. Many people, including the famous inventor, Thomas Edison, were actually against alternating current. They did not understand it and they saw no reason why they should. Today, of course, all of the electric utilities generate alternating current. It runs our houses, schools, businesses, and manufacturing plants.

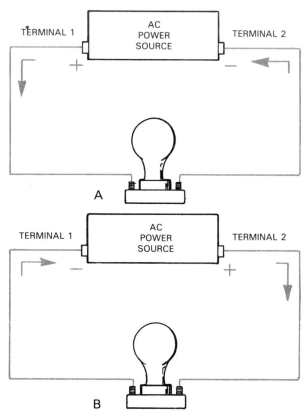

Fig. 3-3. Alternating voltage is constantly changing polarity. A—Current flowing counterclockwise. B—Current flowing clockwise. Polarity changes many times during a second.

Why did alternating current take over, when so many people were against it? The main reason is that ac can be TRANSFORMED to different voltage and current levels. Then, too, alternating current does not waste a lot of power when it is sent over long distances.

AC WAVE FORM

The current from an alternating source starts out at zero and rises to what is called its PEAK value. Then it goes down to zero again. At that instant, the current starts flowing in the opposite direction. It increases in value in this direction until it reaches the same peak value it had when flowing the other way. Then it decreases to zero again. What was just described is one complete CYCLE of alternating current. See Fig. 3-4.

Notice that the current does not suddenly jump to its peak and then drop. It follows a very definite pattern when plotted on paper. The pattern is called a SINE WAVE. The current is said to have a SINUSOIDAL wave form. (It is so called because it is directly related to the sine of an angle.) We call the part above the zero line the POSITIVE half of the cycle. The part below the zero line is the NEGATIVE half of the cycle.

For alternating current calculations, a cycle is divided into 360 units, called electrical degrees. This is similar to the way a circle is divided into 360 angular degrees. Portions of a cycle are measured in "electrical angles." This allows us to draw geometric angles to represent ac and to use trigonometry to solve ac problems.

An electrical degree is a measure of time. To get an idea of the relationship between electrical angles and time, consider the 60-hertz ac in your home and school. It goes through a complete cycle every 1/60th of a second (0.0167 sec). The base line of Fig. 3-5 shows the actual time in seconds, while the degrees are marked on the sine wave.

Seldom, however, will you be interested in how many seconds it takes for an ac cycle. Of greater interest is the frequency.

AC FREQUENCY

Three hundred sixty electrical degrees represent one full cycle of alternating current. What we need to know is the number of cycles completed in a time period of one second. This is called FREQUENCY (its symbol is f).

Frequency or cycles per second is expressed in hertz (Hz), named for Heinrich R. Hertz, a German physicist. *An alternating current going through one complete cycle in one second, has a frequency of one hertz.*

Standard alternating current in the United

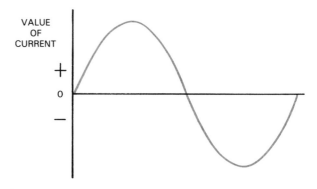

Fig. 3-4. During one cycle, alternating current rises, falls, changes direction, rises and falls again.

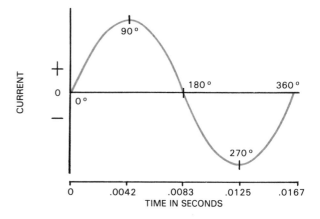

Fig. 3-5. The value of alternating current changes with time.

States is produced and used at 60 Hz. The current completes 60 cycles in one second.

By way of comparison, the frequencies of radio broadcast signals, called radio frequencies, are from 30,000 to 30,000,000 Hz (30 to 30,000 kilohertz). Fig. 3-6 shows a comparison of a 60 Hz wave form and an 180 Hz wave form on the same time base.

EFFECTIVE (RMS) VALUE OF AC

Alternating current changes in value from one instant to another. Yet, we can and do talk about specific values of voltage and current, such as a line voltage of 120 V or a load current of 7 A. What we are talking about is the *effective value* of ac. A voltage of 120 V ac has the

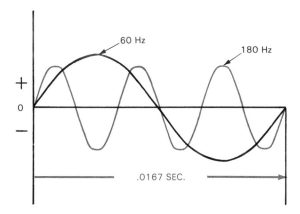

Fig. 3-6. A 180 Hz current completes three cycles in same time that it takes a 60 Hz current to complete one cycle.

Average ac value (AVG) means the instantaneous values averaged over half a cycle. The relationships between peak, average ac (AVG), and RMS values are shown in Fig. 3-7 and as follows:

MULTIPLY	TIMES	TO GET
Peak value	0.707	RMS value
Peak value	0.637	AVG value
RMS value	1.414	Peak value
AVG value	1.57	Peak value
AVG value	1.11	RMS value
RMS value	0.9	AVG value

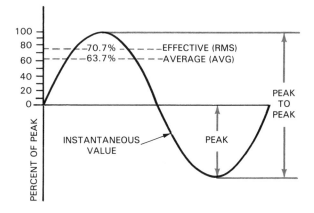

Fig. 3-7. The value of an ac voltage or current is the RMS value unless one of the other values is specified.

same *effect* as 120 V dc. It is a little easier to understand with current. The current flowing through a lamp causes as much light in one direction as it does in the other. The 7 amperes of ac delivers as much power as 7 amperes dc. The effective value of voltage or current is also called the RMS value. RMS stands for ROOT MEAN SQUARED. It is computed on the basis of electrical power. Power, remember, is proportional to the current squared ($P = I^2R$), or the voltage squared ($P = \dfrac{E^2}{R}$). The RMS value is computed by:

1. Squaring all of the instantaneous values.
2. Averaging them.
3. Extracting the square root.

As it turns out, the RMS value of voltage or current is about 70 percent (.707, to be exact) of the peak value.

INSTANTANEOUS, AVERAGE, AND PEAK VALUES

Normally, we will be working with the RMS value. Instantaneous value, peak value, peak-to-peak value, average dc and average ac value are seldom used in power work.

Average dc means the average of the instantaneous values over a full cycle. The average dc value of a full cycle of a sinusoidal ac wave form is always zero.

If you are talking about an instantaneous, average, or peak value, always specify which is meant. If the value is given by itself, without specifying, everyone understands it to mean the RMS, or effective value.

Example 1

An ac voltage with a sinusoidal wave form has an RMS value of 120 V. Find the peak value and average value.

a. Compute the peak value.

$E_{peak} = E_{RMS} \times 1.414 = 120 \times 1.414$
$= 170\ V$

b. Compute the average value.

$E_{AVG} = E_{RMS} \times 0.9 = 120 \times 0.9$
$= 108\ V$

$E_{AVG} = E_{peak} \times 0.637 = 170 \times 0.637$
$= 108\ V$

AC RIPPLE

The output of a rectifier or a dc generator may not be a steady voltage. We know that dc voltage does not change polarity. However, it can change in value. It may follow a sine wave pattern like that shown in Fig. 3-8.

To study this type wave form, we have to break it down into two parts, a steady part and a fluctuating part. The steady part is simply the average dc value. As shown in Fig. 3-9, the average value can be represented by a horizontal line that splits the fluctuations in half. To compute the average, add the highest value to the lowest value; then divide by two.

The fluctuating part is called AC RIPPLE, although it is not alternating current in the strict sense of the term. It is merely higher and lower than the average dc value. For most dc supply

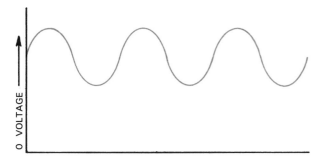

Fig. 3-8. This is a dc voltage because it does not change direction. It is called pulsating dc.

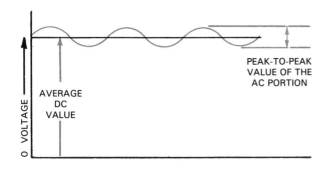

Fig. 3-9. Ripple voltage is the RMS value of the ac portion of a pulsating dc voltage.

voltages, the less the ripple content the better. Ripple is computed by dividing the RMS value of the ripple voltage by the average value. The resulting quotient is then multiplied by 100 to express it as a percentage. The equation is:

$$\% \text{ Ripple} = \frac{\text{RMS of ripple voltage}}{\text{average (dc) value}} \times 100$$

Example 2

The dc wave form shown in Fig. 3-9 varies between a high of 110 V and a low of 90 V. Find the average (dc) value, the RMS value of the ripple voltage and the percent ripple.

 a. Compute the average value (dc part).
  $$E_{AVG} = \frac{E_{high} + E_{low}}{2} = \frac{110 + 90}{2}$$
  $$= 100 \text{ V}$$
 b. Compute peak value of ripple voltage.
  $$E_{high} - E_{AVG} = 110 - 100 = 10 \text{ V}$$
 c. Compute RMS value of ripple voltage.
  $$E_{RMS} = E_{peak} \times 0.707 = 10 \times 0.707$$
  $$= 7.07 \text{ V}$$
 d. Compute percent ripple.
  $$\% \text{ Ripple} = \frac{7.07}{100} \times 100 = 7.07\%$$

AC PHASE RELATIONSHIP

Combining two dc voltage sources is easy. Each terminal is always either positive or negative. You can connect them series-aiding or series-opposing. The total voltage is determined by simple addition or subtraction. Combining ac voltages is not quite as simple, however. The value of an ac voltage changes from one instant to the next. If you need to add two ac voltages together, you have to know their PHASE RELATIONSHIP.

THE MEANING OF PHASE

Phase is one of the most important terms in ac theory. It always refers to the relationship between two wave forms having the same frequency. They can be two voltages, two currents, or a current and a voltage. The two quantities are said to be either IN PHASE or a certain number of (electrical) degrees OUT OF PHASE.

First we will discuss the meaning of the term, "in phase." Assume you have a double power source. Its three terminals are labeled A, N (for neutral), and B, as shown in Fig. 3-10. A voltmeter across terminals A and N reads 120 V. This is the RMS value of the wave form shown in Fig. 3-11. A voltmeter between N and B reads 60 V. This is the RMS value of the wave form shown in Fig. 3-12.

Both wave forms are shown in Fig. 3-13 plotted on the same time (electrical degrees) scale. Both start out at zero at the same instant. Both reach their positive and negative peaks at the same instant. In fact, they are at the same part of their cycles at every instant of time. Such wave forms are said to be in phase.

The total voltage of two in-phase voltages is simply their sum. Fig. 3-14 shows the combined wave form. At every instant, the total equals the sum of the instantaneous values. Note the result in a sine wave whose RMS value is (120 + 60) 180 V.

PHASOR DIAGRAMS

Fortunately, we do not have to draw out sine waves every time we want to add ac quantities. We can use PHASOR DIAGRAMS. Phasor diagrams are drawn to show phase relationships. The electrical degrees of time become simple angles. There are 360 electrical degrees in a cycle just as there are 360 angular degrees in a circle. Phasor diagrams are a powerful tool for solving ac circuits.

The lines which make up phasor diagrams are called VECTORS. *A vector is a line having a specific length and specific direction.* The length

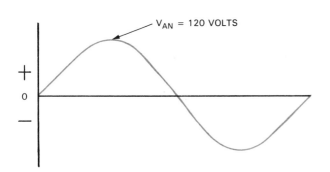

Fig. 3-10. Voltage V_{AN} is in phase with voltage V_{NB}.

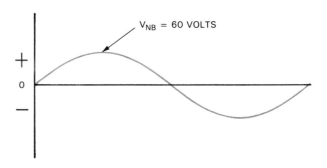

Fig. 3-12. Wave form of voltage V_{NB} is drawn on the same time base as wave form V_{AN} in Fig. 3-11.

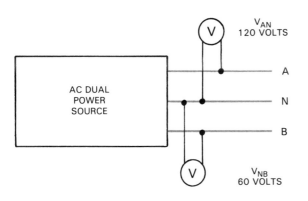

Fig. 3-11. This is the wave form of voltage V_{AN}.

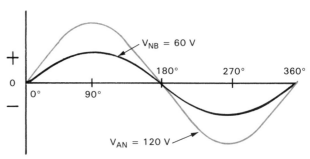

Fig. 3-13. Voltages V_{AN} and V_{NB} are at the same point in their respective cycles at every instant.

of the line represents the RMS value of voltage or current. The direction of the line depends on PHASE ANGLE.

In constructing a phasor diagram, you will start with a reference line drawn horizontally at the "3 o'clock" position. This is the line from which the angles are taken for other quantities. In phasor diagrams, the angles are measured in a counterclockwise direction.

SINGLE-PHASE VOLTAGE

Fig. 3-15 is an example. A vector 120 units long is drawn along the horizontal line. This represents the voltage from A to N (V_{AN}). The arrow on the right end shows that the vector is at 0 degrees. To add the voltage from N to B (V_{NB}) we draw a second vector. Since the voltage is in phase, its vector is also drawn horizontally. This is a phase angle of 0 degrees with voltage V_{AN}. Note that the V_{NB} vector starts at the end of V_{AN}. Its length must be measured in the same units as vector V_{AN}. It is therefore 60 units long.

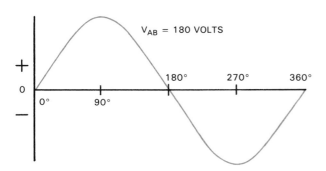

Fig. 3-14. The sum of two in-phase voltages is their algebraic sum at every instant of time.

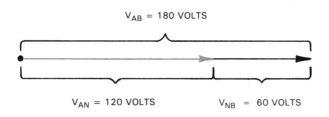

Fig. 3-15. The RMS value of the total of two in-phase voltages equals the algebraic sum of the RMS values of each.

The solution of a phasor diagram is a line drawn from the beginning of the first vector to the end of the second. In this case it is a line 180 units long, representing 180 volts. Further, this voltage, V_{AB}, is in phase with both V_{AN} and V_{NB}.

Most homes are supplied single phase voltage over three wires. It is similar to the example given. The difference is that each line-to-neutral voltage is 120 V. The lighting and convenience outlets are 120 V. The 240 V service is available for electric ranges, clothes dryers, and air conditioners.

TWO-PHASE VOLTAGE

Adding two in-phase voltages is really too simple for phasor diagrams. They are more useful when the voltages are out of phase. An example is TWO-PHASE voltage, illustrated by Fig. 3-16. The wave form labeled "Phase B" lags "Phase A" by 90 degrees. At one time, two-phase power was the only kind of ac generated.

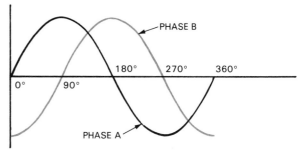

Fig. 3-16. Phase B voltage lags phase A voltage by 90 degrees.

A two-phase, three-wire system is shown in Fig. 3-17. Phase A voltage, measured between lines A and N (V_{AN}) is 120 V. Phase B voltage, measured between line N and B (V_{NB}) is also 120 V. With a phasor diagram we can determine the voltage between lines A and B (V_{AB}).

Fig. 3-18, view A, shows the beginning of this phasor diagram. Vector V_{AN} is drawn along the horizontal reference line. It is 120 units long and is at zero degrees. The vector V_{NB} is also 120 units long. It is drawn straight up. As shown in view B, Fig. 3-18, this 90 degree (right angle) represents 90 electrical degrees. To add these two vectors, we draw a line from the start of V_{AN} to the end of V_{NB}. The length of this line is voltage V_{AB} in the same units used for V_{AN} and V_{NB}. Vector V_{AB} is 170 units long, representing 170 volts. As Fig. 3-18, view C, shows, the completed phasor is a right triangle.

SOLVING RIGHT TRIANGLES

The rules that govern right triangles were discovered about 500 BC by a Greek mathematician named Pythagoras. Notice in Fig. 3-18, view C, that the side opposite the right angle is longer than the other two sides. It is called the hypotenuse. The main rule of right triangles (called the Pythagorean Theorem) is as follows:

The square of the hypotenuse equals the sum of the squares of the other two sides.

If you know the length of the two short sides, the length of the long side can be determined as follows:
1. Square the length of each of the two short sides.
2. Add the two squares together.
3. Extract the square root of the sum. This is the length of hypotenuse.

A simple example will show how this works. Then we shall look at the special case where the two short sides have the same length, like the two-phase voltage.

Example 3

In the right triangle shown in Fig. 3-19: Side a = 3 cm. Side b = 4 cm. Find the hypotenuse, Side c. The equation is $c^2 = a^2 + b^2$ or $c = \sqrt{a^2 + b^2}$

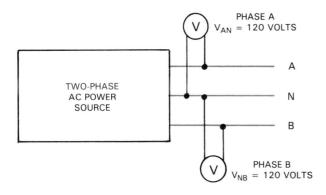

Fig. 3-17. Diagram shows two-phase, three-wire system.

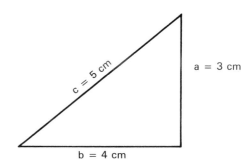

Fig. 3-19. In any right triangle, $c^2 = a^2 + b^2$.

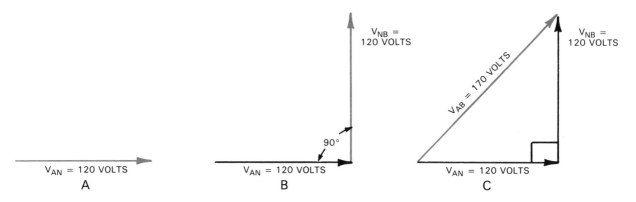

Fig. 3-18. In phasor diagrams, length of each line represents a value; its direction represents a phase angle.

a. Square side a.

   $3^2 = 9$

b. Square side b.

   $4^2 = 16$

c. Add the square of the sides.

   $9 + 16 = 25$

d. Extract the square root of the sum

   $c = \sqrt{25} = 5$ cm.

Example 4

Refer to the right triangle shown in Fig. 3-20. The two short sides both equal 1 unit. Find the hypotenuse.

$$c = \sqrt{a^2 + b^2} = \sqrt{1^2 + 1^2} = \sqrt{1 + 1}$$
$$= \sqrt{2} \text{ or } 1.414 \text{ units}$$

This means that whenever two sides of a right triangle are equal, the hypotenuse can be computed by multiplying one of the sides by 1.414.

Now, let us go back to the two-phase voltage. We can compute the line-to-line voltage, V_{AB}, by multiplying 120 times 1.414. This works out to be 170 V. This is the same answer we got by measuring the hypotenuse.

USING TRIGONOMETRY

Phasor diagrams can also be worked out using trigonometry. The name "trigonometry" scares many people. But it really is not as hard as it sounds. It is simply another way of talking

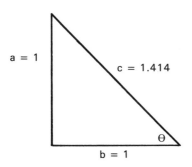

Fig. 3-20. When a = b, the hypotenuse, c, equals 1.414 times the length of a or b.

about the relationships of the sides and angles of right triangles. You need to know a little "trig" to understand alternating current.

ANGLES OF A RIGHT TRIANGLE

All triangles have one thing in common. The three angles of every triangle always add up to 180 degrees. The special point about a right triangle is that it has one 90 degree angle. From this, we know that the other two must add up to 90 degrees. Notice the angle marked Θ (theta) on the right triangle shown in Fig. 3-21. This is the "PHASE ANGLE" on ac phasor diagrams.

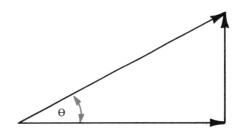

Fig. 3-21. The smallest Θ can be is 0 degrees; the largest it can be is 90 degrees.

The smallest it can get is zero degrees. That is what it was on Fig. 3-15. The two voltages were in phase. Therefore, the phase angle was 0. When the two sides are equal, as in Fig. 3-20, Θ (phase angle) is 45 degrees. The largest it can get is 90 degrees. We will therefore be dealing with angles between 0 and 90 degrees.

SIDES OF A RIGHT TRIANGLE

For every angle, Θ, there is a specific ratio between each pair of sides. The sides themselves can be long or short. The ratios, however, are determined by the angle. See Fig. 3-22.

The sides of a right triangle have specific names. The long side, marked c, as noted before, is the HYPOTENUSE. The vertical side marked a, is known as the SIDE OPPOSITE. It is opposite the angle, Θ. The horizontal side,

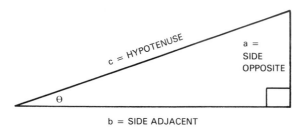

Fig. 3-22. The relationships of a right triangle are similar to the relationships in alternating current circuits.

marked b, forms angle Θ with the hypotenuse. It is called the SIDE ADJACENT. Ratios between sides of right triangles are known as functions. There are three functions used in solving ac problems. They are:
1. The SINE function (abbreviated sin).
2. The COSINE function (abbreviated cos).
3. The TANGENT function (abbreviated tan).

THE SINE FUNCTION

For every angle that Θ can be, there is a definite ratio between sides a and c.

The ratio of the side opposite to the hypotenuse is called the sine function of the angle Θ (sin Θ). The equation is:

$$\sin \Theta = \frac{\text{side opposite (a)}}{\text{hypotenuse (c)}}$$

The smallest number a sine function can be is zero. This occurs when Θ = 0 degrees. This happens when both factors are in phase. The length of the side opposite is zero. There is no vertical side.

The largest number a sine function can be is one. This occurs when Θ = 90 degrees.

When Θ is 90 degrees, the length of the side adjacent is zero. The side opposite has the same length as the hypotenuse. Therefore, when Θ = 90 degrees, the ratio of the side opposite to the hypotenuse (sin Θ) is 1.

Between 0 degrees and 90 degrees, the sine function is a decimal between 0 and 1. All of these ratios have been worked out and are given in the Table of Trigonometric Functions in the Technical Information section at the back. We find, for example, that sin 30 degrees = 0.5. This means that in a 30-60-90 right triangle, side "a" is half the length of side "c".

One of the ways the sine function is used, is to compute the hypotenuse when the side opposite and the angle are known.

Example 5
A current of 120 V is lagging another 120 V current by 90 degrees. Θ = 45 degrees. Find the phasor sum of the two voltages.
a. Construct the phasor diagram. This is shown in Fig. 3-23.
b. Look up sin 45 degrees in the Table of Functions.
$$\sin 45° = .707$$
c. Compute the phasor sum (V_T).
$$\sin 45 = \frac{\text{side opposite}}{\text{hypotenuse}}$$
$$.707 = \frac{120}{V_T} \qquad V_T = \frac{120}{.707}$$
$$V_T = 170 \text{ V}$$

THE COSINE FUNCTION

The ratio of the side adjacent to the hypotenuse is called the COSINE FUNCTION of the angle Θ. The equation is:

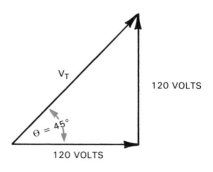

Fig. 3-23. V_T = 120 divided by sin Θ.

$$\cos \Theta = \frac{\text{side adjacent (b)}}{\text{hypotenuse (c)}}$$

The cosine function equals 1 when Θ is 0 degrees. When Θ is 90 degrees, $\cos \Theta = 0$. All of the values of $\cos \Theta$ between 0 degrees and 90 degrees are given in the Table of Functions in the back of the book.

THE TANGENT FUNCTION

The ratio of the side opposite to the side adjacent is called the TANGENT FUNCTION of the angle Θ. The equation is:

$$\tan \Theta = \frac{\text{side opposite (a)}}{\text{side adjacent (b)}}$$

The tangent function can often be used to solve phasor diagrams. In *Example 3,* we computed the length of the hypotenuse by squaring the sides, adding them, then extracting the square root. Now let us take the same example and solve it using trigonometric functions.

Example 6

Refer to the right triangle shown in Fig. 3-24. Side "a" = 3 cm. Side "b" = 4 cm. Find side "c."
a. Compute $\tan \Theta$. $\tan \Theta = \frac{3}{4} = .75$

b. Find Θ in the Table of Functions.
$\Theta = 36.9°$
c. Find $\sin \Theta$ in the Table of Functions.
$\sin \Theta = .6$

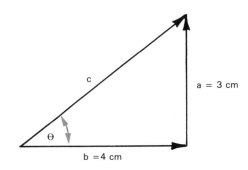

Fig. 3-24. The tangent of Θ is $\frac{a}{b}$.

d. Compute side c.

$$\sin \Theta = \frac{a}{c}$$

$$c = \frac{a}{\sin \Theta}$$

$$c = \frac{3}{.6} = 5 \text{ cm.}$$

SUMMARY

Direct current does not change its direction of flow.

Alternating current periodically changes its direction according to a regular pattern.

The pattern that alternating current follows is called a sine wave.

One complete sine wave pattern is called a cycle and is divided into 360 electrical degrees.

The number of cycles completed in one second is called frequency and is measured in hertz (Hz).

The value of alternating current most often used is its effective (RMS) value.

When either ac currents or ac voltages are combined, the phase angle between them must be taken into account.

Phasor diagrams are composed of vectors whose length represents a quantity and whose direction represents electrical degrees.

TEST YOUR KNOWLEDGE

1. Direct current may change in (value, direction) but does not change (value, direction).
2. Current which flows through a wire first in one direction and then in another direction is called:
   a. Abbreviated ac.
   b. Alternating current.
   c. Direct current.
   d. A sine wave.
3. The standard frequency for ac voltage in the United States is _____.
4. Arrange the following characteristics of the same sinusoidal wave form according to value (highest first):
   a. RMS (effective) value.

b. Average dc value
c. Peak-to-peak value.
d. Average ac value.
e. Peak value.

5. Explain what is meant by "ripple" when describing a pulsating dc voltage.

6. A diagram drawn to show _____ _____ is called a phasor diagram.

7. In a phasor diagram, a vector has both length and direction. What does its length represent? Its direction?

8. Define in-phase wave forms.

9. In constructing a phasor diagram, the line drawn horizontally at "3 o'clock" is known as the _____ line.

10. Define the following trigonometric functions (ratios):
a. Sine (sin).
b. Cosine (cos).
c. Tangent (tan).

PROBLEMS

1. Given: An alternating current completes two cycles in a time of one second. What is the frequency in hertz?

2. Given: An ac voltage has an RMS value of 120 V. Find its peak value.

3. Given: A pulsating dc voltage varies between 125 V and 115 V. Find its average dc value and the percent ripple.

4. Given: A 100 V voltage is 60 degrees out of phase with another 100 V voltage. Find the phasor sum.

5. Given: Two voltage are 180 degrees out of phase. One is 100 V and the other is 50 V. Find the phasor sum.

6. Given: A right triangle has a vertical side (side opposite) of 10 cm and a base (side adjacent) of 15 cm. Find the hypotenuse and the sine, cosine, and tangent of the angle, Θ.

4 INDUCTANCE

After studying this chapter, you will be able to:

☐ List four ways of producing changing flux linkages.
☐ Calculate the equivalent inductance of inductances in series and parallel.
☐ State the equation for computing inductive reactance.
☐ Describe the effect that inductance has on current with respect to voltage.
☐ Define impedance, apparent power, active power, reactive power, and power factor.

When Michael Faraday performed his electromagnetic experiments, he used direct current. What he learned, however, turned out to be very important to alternating current circuits. He discovered the principle of induction. *Induction means using magnetism to produce electricity.*

ELECTROMAGNETIC PRINCIPLES

Any time magnetic flux links a wire and that flux changes, a voltage is induced into the wire. This holds true no matter where the flux comes from. It can be the field of a permanent magnet. Or, the flux can be produced by an electromagnet. It can even come from current in the wire itself.

There are four ways in which flux linkages change. All four result in induction.
1. The magnetic field moves while the wire does not.
2. The wire is made to travel through a steady magnetic field.
3. Both move but at different speeds.
4. The ups and downs of alternating current will cause a magnetic field to expand and collapse.

As shown by Fig. 4-1, the flux lines start in the center of a wire and expand outward. As they expand, they are cut by that wire. It does not matter that the field is produced by the wire. The wire is magnetically linked to a changing flux. A small voltage is induced into it. A straight wire would have to be long, like transmission lines, before the voltage could make any difference.

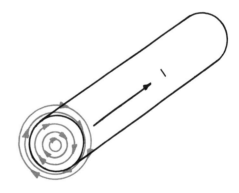

Fig. 4-1. Alternating current flowing through a wire causes a magnetic field (self-induction). Flux lines expand outward from center of wire. They are cut by the wire itself.

THE MEANING OF INDUCTANCE

Induced voltage is a big factor when coils are used in alternating circuits. See Fig. 4-2. Magnetic fields, remember, combine when they get a chance. As we learned in our study of electromagnetism, a coil has a stronger magnetic field than a straight wire. If it has an iron core, its field is even stronger.

When this field changes, it gets cut by all of the turns of the coil. Now we have a relatively large voltage being induced into the coil. In other words, the coil induces a voltage into itself. The ability to do this is called *self-inductance*, or simply INDUCTANCE.

LENZ'S LAW

It may sound strange that a coil can induce a voltage into itself. It seems like getting something for nothing. But that is not the case. If you think about it, you see that the polarity of a self-induced voltage would have to be opposite to the applied voltage. At every instant, during the ac cycle, the induced voltage opposes applied voltage. In fact, it is often called a BACK VOLTAGE.

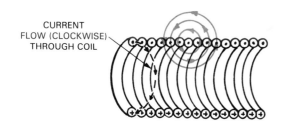

CURRENT FLOW (CLOCKWISE) THROUGH COIL

Fig. 4-2. Cutaway of coil through which current is flowing. Flux from every point in each turn of wire links all other turns.

This principle is summed up by Lenz's Law. In its simplest form, this law says: *inductance always tends to oppose whatever causes it.*

Induced (or back) voltage, then, is acting like a separate voltage source. See Fig. 4-3. You can think of it as being connected series-opposing with the source voltage. Unfortunately, you cannot measure this back voltage. You know that it is there only because of the reduced current flow.

UNITS OF INDUCTANCE

Faraday discovered that inductance depends on two things:
1. The strength of the magnetic field. Strength determines how many flux lines are being cut by the induced voltage.
2. How fast this cutting is done. In other words, the rate at which the flux is changing.

The unit of inductance is the henry (symbol: H). It is named for Joseph Henry, an American scientist.

A coil is said to have an inductance of so many henrys. *Each henry of inductance results in an induced voltage of one volt when the current is changing at a rate of one ampere per second.*

That sounds complicated. But, fortunately, we never have to compute a coil's inductance.

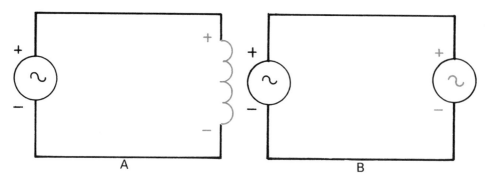

Fig. 4-3. Effect of induced voltage. A—Instantaneous polarity of induced back voltage is the same as applied voltage. B—An inductance acts like a voltage source connected series-opposing.

That is determined by the way a coil is made. It depends on:
1. The number of turns of wire.
2. Spacing of the turns.
3. The size of the core.
4. Permeability of the core. This is the property of the core which enables it to change the magnetic flux in a magnetic field.

INDUCTANCES IN SERIES AND PARALLEL

To find the total inductance of two or more coils in series, simply add them together. It is just like the total resistance of several resistors in series. For example, two 8-henry (H) coils in series have a total inductance of 16 henrys. See Fig. 4-4.

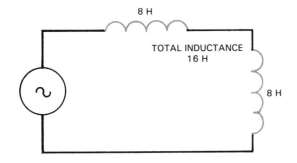

Fig. 4-4. Total inductance of coils in series is the sum of individual inductances.

Likewise, inductance in parallel is computed the same as resistance in parallel. See Fig. 4-5. If they are identical coils, you can divide the inductance of one by the number of coils. For example, the equivalent inductance of four 8-henry coils in parallel is 2 henrys ($\frac{8}{4} = 2$ H).

If there are only two coils, you can use the *product over sum method*. For example, the equivalent inductance of an 8-henry coil in parallel with a 2-henry coil is 1.6 henrys. ($\frac{8 \times 2}{8 + 2}$ = $\frac{16}{10}$ = 1.6 H). Three or more unlike inductances must be solved with the *reciprocals* meth-

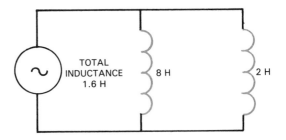

Fig. 4-5. Total inductance of coils in parallel is computed the same way as resistances in parallel.

od. Fortunately this does not have to be done very often.

INDUCTIVE REACTANCE

So far, we have seen that an alternating current passing through a coil creates a back voltage. It is like having two opposing voltage sources in the same circuit. Even so, a circuit can have only one current. *This current is proportional to the applied voltage minus the back voltage.*

However, we cannot measure the back voltage. So, instead of worrying about the value of back voltage, we concentrate on current. We pretend that a coil has a quality somewhat like resistance. It is known as INDUCTIVE REACTANCE.

Inductive reactance (X_L) is measured in ohms, like resistance. If you know the voltage across an inductance (E_L) and the current through it (I_L), you can compute inductive reactance from the equation $X_L = \frac{E_L}{I_L}$.

You can compute a coil's inductive reactance if you know the inductance. *Inductive reactance is proportional to both inductance and the frequency of the applied voltage.* The equation is:

$X_L = 6.28 \times f \times L$
X_L = inductive reactance in ohms (Ω)
f = frequency in hertz (Hz)
L = inductance in henrys (H)

53

Example 1

An 8 H (henry) coil is connected to a 120 V, 60 Hz supply. See Fig. 4-6. Find the inductive reactance and current.

a. Compute the inductive reactance, X_L.
$$X_L = 6.28 \times f \times L = 6.28 \times 60 \times 8$$
$$= 3014 \ \Omega$$

b. Compute the current.
$$I_L = \frac{E_L}{X_L} = \frac{120}{3014} = 0.04 \text{ A}$$

Example 2

The current through a 30 millihenry (mH) coil is 2 A. The frequency is 5 Hz. See Fig. 4-7. Find the inductive reactance and the voltage drop across the coil.

a. Convert millihenrys to henrys.
$$30 \times .001 = 0.030 \text{ H}$$

b. Compute the inductive reactance.
$$X_L = 6.28 \times f \times L = 6.28 \times 5 \times 0.03$$
$$= 0.942 \text{ or } 1 \ \Omega$$

Fig. 4-6. Inductive reactance can be computed from inductance and frequency. Current can be computed from inductive reactance and voltage.

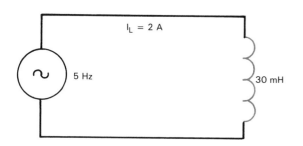

Fig. 4-7. Voltage drop across a coil can be computed from inductive reactance and current.

c. Compute the voltage drop.
$$E_L = I_L \times X_L = 2 \times 1 = 2 \text{ V}$$

PHASE SHIFT CAUSED BY INDUCTANCE

Inductance is like inertia. Inertia is the tendency of a moving object to keep moving. It is also the tendency of a motionless object to remain still.

This can be explained in terms of marbles moving through a tube. See Fig. 4-8. A coil acts like an oversized marble that is hard to start. Once it starts, however, it is hard to stop.

Assume that a "marble power source" applies a pressure to this "circuit." At first, the marbles will not move. As shown in Fig. 4-9, the pressure begins to build behind the big marble. If this pressure stays constant for a while, the big marble will start to move.

Once it starts, it moves along with the little marbles. Then, if the pressure is removed, the inertia of the big marble keeps them all moving for a while.

Now look at what happens if pressure is applied first in one direction, then the other. Just

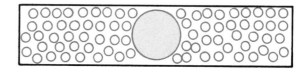

Fig. 4-8. If current flow were like marble flow, inductance of a coil would be like a large marble whose inertia opposes any change in flow.

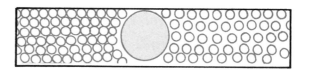

Fig. 4-9. Pressure builds to a peak before marbles start flowing along the tube.

when the big marble starts rolling, the pressure behind it starts decreasing. At the instant shown in Fig. 4-10, pressure has built up in the opposite direction. At that instant, however, the motion of the marbles has stopped.

Fig. 4-11 shows two curves. One shows how pressure changes with time. The other shows how motion changes with time. Notice that motion is out of step with pressure. In ac circuits, inductance causes current to get out of step with voltage.

ELECTROMAGNETIC INERTIA

Induction takes place only while current is changing. Further, induced voltage opposes applied voltage. Therefore, a coil opposes any change in current. That is why we can compare it to inertia. Inertia opposes any change in motion.

DC VOLTAGE APPLIED TO A COIL

In dc circuits, coils are sometimes used to "smooth out" pulsations. These coils, called CHOKES, are a part of filter circuits. Fig. 4-12 shows a pulsating dc wave form before and after filtering.

Coils affect dc circuits, too, even if the direct current is steady. Recall that the current starts at zero and builds to its final steady value. While it is growing, the flux lines are expanding. For a brief time, induced voltage holds the current back. Fig. 4-13 is known as a TIME CONSTANT CURVE. The shape of this curve is about the same for all coils. The difference is the amount of time involved.

One time constant is the number of seconds required for the current to reach 63 percent of its steady value. A coil's time constant can be computed by dividing its resistance by its inductance $(TC = \dfrac{R}{L})$.

TC is in seconds
R is in ohms
L is in henrys

When the dc current has reached its steady value, there is energy stored in its field. No additional energy is needed to keep it there. The coil appears to have no effect, other than the dc resistance of its wire. However, you may run into trouble when you open the circuit. The field collapses very quickly. The coil becomes a high voltage power source. This is known as an INDUCTIVE KICK. It can cause serious injury or

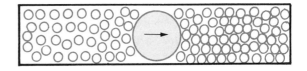

Fig. 4-10. While marble is still moving in one direction, pressure builds in opposite direction.

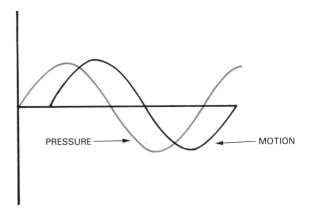

Fig. 4-11. Motion lags behind pressure because of inertia.

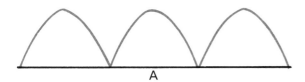

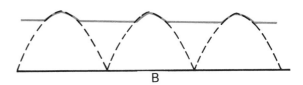

Fig. 4-12. Direct current wave form. A—Before filtering. B—After filtering.

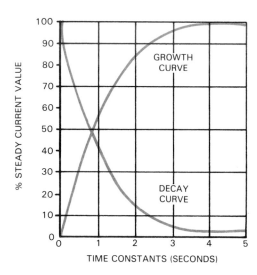

Fig. 4-13. Coil current grows or decays along these curves when direct current is applied.

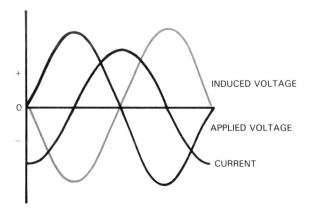

Fig. 4-14. Inductance makes current lag voltage by 90 degrees. Induced voltage is 180 degrees out of phase with applied voltage.

equipment damage. Even small coils cause an arc across the switch contacts.

AC VOLTAGE APPLIED TO A COIL

In ac circuits, inductance causes the current to lag 90 electrical degrees behind the voltage. The expanding and collapsing magnetic field produces an induced back voltage. This voltage, remember, is proportional to the *rate-of-change* of current. When current is zero, it is changing at its most rapid rate. Therefore, induced voltage is maximum when current is zero. When current is at its peak, its *rate-of-change* is zero. Thus, induced voltage is zero at that instant. Fig. 4-14 shows the phase relationship between the applied voltage, current and induced voltage.

RESISTANCE AND INDUCTIVE REACTANCE IN AC CIRCUITS

Many electrical devices are all resistive. Electric heaters and lights, for example, have no inductive reactance. For resistive ac circuits, you can use Ohm's Law and Watt's Law equations as you do in dc circuits. Household voltage is 120 volts. Therefore, a 60 watt lamp draws half an ampere ($I = \dfrac{P}{E}$). It has a resistance of 240 ohms

($R = \dfrac{E}{I}$). A 10 ampere heater has a resistance of 12 ohms and draws 1200 watts of current.

Electric motors, however, have inductive reactance as well as resistance. Solving reactive circuits really is not any harder. It is just different. Inductance causes current to lag 90 degrees behind voltage. Therefore, we have to use phasor diagrams.

PHASE RELATIONSHIPS

Phasor diagrams are used to solve problems in voltage, current, power and IMPEDANCE. *Impedance is the total opposition to current flow.* It is measured in ohms and is expressed with the symbol Z. Impedance is given as the ratio between ac voltage applied and the resulting current ($Z = \dfrac{E}{I}$). It is also the phasor sum of resistance and reactance in series. The phasor sum, remember, is the hypotenuse of a right triangle.

RESISTOR AND COIL IN SERIES

To show how phasor diagrams are used, we can analyze the circuit of Fig. 4-15. It shows an 80 ohm resistor in series with a 0.16 henry coil. A 60 Hz voltage of 200 volts is applied. There are four steps:

56

1. Compute the inductive reactance of the coil.
X_L = 6.28 × f × L = 6.28 × 60 × 0.16
  = 60 Ω

2. Find the impedance, using a phasor diagram. See Fig. 4-16.

   The resistance vector is drawn horizontally. The arrow points toward 3 o'clock, which is zero degrees. The length is 80 mm, representing 80 ohms. The inductive reactance vector is drawn vertically upward which is 90 degrees. It is 60 mm long since the coil has 60 ohms of inductive reactance.

   The impedance is represented by the hypotenuse. This is the phasor sum of R and X_L. The angle formed by the hypotenuse and the horizontal line is called the phase angle. The phase angle is always called Θ (theta).

   There are several ways of determining the impedance value.
   a. The easiest way is to measure the hypotenuse and convert its length to ohms.
   b. You can also use the *sum of the squares method.* Here you square the resistance value, square the reactance value; then add these two squares together and extract the square root.
   c. Use trigonometry. The tangent of an angle is the ratio of the opposite side to the adjacent side. In our case:

   $$\tan \Theta = \frac{X_L}{R} = \frac{60}{80} = 0.75$$

   Next, from the trigonometric tables we find which angle has a tangent of 0.75. We pick the closest angle, 37 degrees. Therefore, the answer is: Θ = 37°.

   Knowing the angle, we can find the impedance. The cosine of theta is the ratio of the adjacent side to the hypotenuse. In our case the answer is:

   $$\cos \Theta = \frac{R}{Z}, \text{ therefore } Z = \frac{R}{\cos 37°}$$

The trigonometric tables show that the cosine of 37 degrees is 0.7986 but 0.8 is close enough. Therefore,

$$Z = \frac{80}{.8} = 100 \ \Omega$$

3. Compute the current. In a simple series circuit like Fig. 4-15, there is only one current. The same current passes through the resistor and the coil. The resistance wants to keep this current in phase with the voltage. The inductive reactance wants to make the current lag the voltage by 90 degrees. They must compromise. The result is a current that lags the voltage by 37 degrees.

$$I = \frac{E}{Z} = \frac{200}{100} = 2.0 \ \underline{/37°} \ A$$

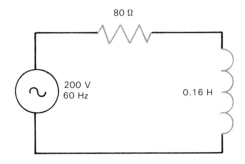

Fig. 4-15. Circuit containing a resistance and an inductance must be solved with a phasor diagram.

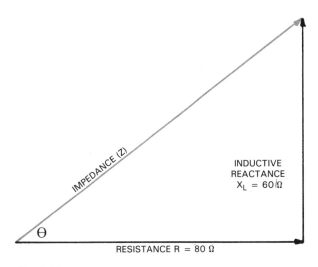

Fig. 4-16. Phasor diagram for circuit in Fig. 4-15. Impedance is represented by the hypotenuse.

Notice how the 37 degrees is written following the current value. This is called POLAR NOTATION. It can help you keep phase angles in mind when solving ac circuits. Your ammeter simply reads 2 A. It tells you nothing about the phase angle.

4. Compute the voltage drop across the resistor and the coil. Now that we know the current through the resistor we can use Ohm's Law to find the voltage drop.

$$E = IR = 2 \times 80 = 160 \; \underline{/0°} \; V$$

Likewise, we can use Ohm's Law to find the voltage drop across the coil:

$$E = IX_L = 2 \times 60 = 120 \; \underline{/90°} \; V$$

In a series ac circuit, the total voltage is the phasor sum of the individual voltage drops. Fig. 4-17 shows the voltage phasor diagram. It looks just like the impedance diagram. If you compute applied voltage by any of the three methods, it will come out to be 200 volts.

Actually testing a circuit like this might confuse you. Your voltmeter reads 160 V across the resistor. Then it reads 120 V across the coil. You could be a little startled to find only 200 volts across both. Do not feel bad. Many people are confused until they learn what happens in phase relationships.

You can get an idea of why the voltage drops "do not seem to add up" from Fig. 4-18. The wave form of the voltage drop across the coil is 90 degrees out of phase with the drop across the resistor. At every instant, the total drop is the actual sum of the instantaneous drops. The result, however, is the phasor sum.

PHASOR DIAGRAMS

Phasor diagrams are important in ac work. That does not mean that you have to draw one for every circuit. It is simply that they allow you to form a mental picture of ac current and voltage relationships. There are a few simple rules to remember when constructing and reading phasor diagrams:

1. There must be a reference. There are four kinds of phasor diagrams: voltage, current, power and impedance. Phase angles for all four are measured from a horizontal line, which is the reference. For current and power diagrams, voltage is the reference. For voltage and impedance diagrams, current is the reference. To help you keep this straight, the phasor diagrams in this chapter will show the reference line extended in a dashed horizontal line.

2. Positive angles (being above the reference line) are measured in a counterclockwise

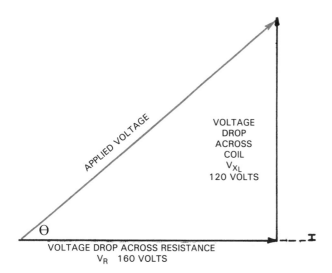

Fig. 4-17. This diagram will solve voltage for circuit shown in Fig. 4-15.

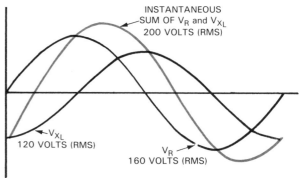

Fig. 4-18. At every instant of time, the total voltage is the sum of the individual voltage drops.

direction from the horizontal. Negative angles (below the reference line) are measured in a clockwise direction from the horizontal. Take, for example, the phasor diagram in Fig. 4-17. Current is the reference. Voltage across the resistor is in phase with current. You will see that the V_R vector is on top of the reference line. Voltage across the inductance, on the other hand, leads current by 90 degrees. Wherever current is in its cycle, V_{X_L} was there 90 degrees before. To show V_{X_L} leading, it is drawn at a positive 90 degree angle (rising vertically).

POWER IN AC CIRCUITS

It takes electrical power to force current through a resistance. This is true for ac circuits as well as dc.

However, inductance is different. If direct current is applied to a coil, a little power is needed to set up the magnetic field. After that, there is no inductance until the switch is opened. Then the coil gives up the energy stored in its field. But when alternating current is applied to a coil, inductance is present all of the time. During half of each cycle the magnetic field is expanding. This takes power from the source. The other half-cycle the field is collapsing. This returns power to the source.

However, you cannot talk about the power *used* by an inductance. *There is no power actually used.* There is, however, a definite amount of power shuttling back and forth between the source and the inductance. You can think of this as "shuttle power."

The term used for this "shuttle power" is REACTIVE POWER. Its symbol is P_q. The unit of reactive power is the VAR (for volt-ampere-reactive). *Reactive power does not do any useful work.* Even so, it must be transmitted to the user along with the ACTIVE POWER.

Active power is sometimes called true power or real power. As in direct current, active power has the symbol P and is measured in watts (W). Active power is the power that gets converted into work (such as producing light, heat and motion).

Both active power and reactive power go out from the power source. This combination is known as APPARENT POWER. *Apparent power is the phasor sum of active and reactive power.* The symbol for apparent power is Ps.

You can compute apparent power by multiplying voltage times current: $P_s = E \times I$. The unit of apparent power is volt-amperes, abbreviated VA.

Fig. 4-19 is the power phasor diagram for the circuit of Fig. 4-15. There is an 80 ohm resistor in series with an inductive reactance of 60 ohms, with 200 volts applied. We already know that current is 2 ampere and that the phase angle is 37 degrees. The best way to compute the power is as follows:
1. Compute the apparent power.
   $P_s = E \times I = 200 \times 2$
   $= 400$ volt-amperes (VA)
2. Compute the reactive power represented by the vertical line. This line is the side opposite the angle Θ. Current is lagging voltage.

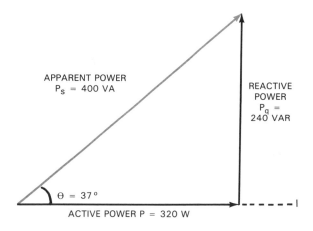

Fig. 4-19. Apparent power is the phasor sum of the active power and the reactive power.

Therefore, Θ is shown as a positive angle.

$$\sin \Theta = \frac{\text{opposite side}}{\text{hypotenuse}} = \frac{P_q}{P_s}$$

Therefore: $P_q = P_s \times \sin \Theta$

$P_q = 400 \times 0.6 = 240$ VAR

3. Compute active power. The active power used by the resistance is represented by the horizonal line. This line forms the side adjacent to the angle Θ.

$$\cos \Theta = \frac{\text{adjacent side}}{\text{hypotenuse}} = \frac{P}{P_s}$$

Therefore: $P = P_s \times \cos \Theta$

$P = 400 \times .8 = 320$ W

POWER RATING

Active power is expressed in watts or kilowatts. This is the actual power used. The electric meter in homes is called a kilowatt-hour meter. It measures power being used and multiplies it by time.

A 100 watt light bulb, for example, burning for 10 hours uses 1 kilowatt-hour of power. Many electrical devices, like lights and heaters, are rated in watts. When the power involved is small, there is no reason to be concerned with the difference between active and apparent power.

Where there may be large inductance in the circuit, apparent power becomes important. A power source must supply reactive power current as well as active power current. That is why you will find many devices rated in volt-amperes instead of watts. A pure inductance would not use any watts at all. Even so, reactive power current still must flow. The wattage rating gives you an idea of the work that can be done. The volt-ampere rating gives you an idea of the current capacity of a source, or the current drawn by the device.

POWER FACTOR

Of all the terms used in power work, one of the most important is POWER FACTOR. As you know, not all of the power supplied by a power source is used. Some of it is returned to the source. Very often you want to know how much of the power supplied is actually used. This is the ratio of power used to the power supplied. In other words, power factor is the ratio of active power to apparent power.

$$\text{Power Factor} = \frac{\text{active power}}{\text{apparent power}}$$

If we look at our power phasor diagram again, we see that this is the ratio of the adjacent side to the hypotenuse. Therefore, power factor is the cosine of the phase angle, Θ.

The power factor of the circuit of Fig. 4-15 is:

$$\text{PF} = \cos \Theta = \cos 37° = 0.8$$

This tells us that eight-tenths (80 percent) of the power supplied is actually used. The rest is returned to the source.

The highest number a power factor can have is 1. This level is called UNITY POWER FACTOR. All of the power supplied is used.

Unity power factor is best. For example, with 0.8 power factor, we realize 320 watts of active power from 2 amperes. In a unity power factor circuit with the same voltage, we can get 320 watts from 1.6 amperes.

$$I = \frac{P}{E} = \frac{320}{200} = 1.6 \text{ A.}$$

We lose in two ways when power factor is less than unity. Power lines, circuit breakers, and other devices have to be large to handle the extra current. Then too, a larger current means more lost power in the transmission lines ($P = I^2R$).

In inductive circuits, current lags voltage. Therefore, inductive circuits are said to have a lagging power factor. Power factors can also be leading, but this is rare. *When power factor is given, it is understood to be lagging.* If it does happen to be leading, it is described as a leading power factor.

REALISTIC COILS

We have been talking about the inductive reactance of coils. You could easily get the idea that they have no resistance. However, coils are wound with wire and wire does have resistance. In some coils this resistance is very small compared to the inductive reactance. Such coils act almost like pure inductance. Others, the majority, have both resistance and inductive reactance.

The drawing symbol for a coil is exactly the same as an inductance. When you see a coil in a circuit, remember that its resistance may be important. Fig. 4-20 shows one way of representing a coil. Both its resistance and inductance are given. Sometimes two symbols are used, like Fig. 4-21. In this case, the inductance symbol is taken to mean pure inductance.

Example 3

The coil shown in Fig. 4-22 has a resistance of 15 ohm and an inductance of 79.6 milli-henrys. Find the power factor when a 60 Hertz voltage is applied.
a. Convert millihenrys to henrys.
$$\text{henrys} = \text{millihenry} \times .001$$
$$= 79.6 \times .001 = 0.0796 \text{ H}$$

R = 15 Ω
L = 79.6 mH

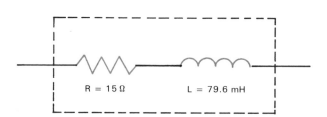

Fig. 4-20. When a coil has a significant amount of resistance, the value in both resistance and inductance can be marked on the coil symbol.

R = 15 Ω   L = 79.6 mH

Fig. 4-21. To emphasize a coil's resistance, a separate resistance symbol can be used.

b. Compute the inductive reactance.
$$X_L = 2\pi \times f \times L = 6.28 \times 60 \times 0.0796$$
$$= 30 \ \Omega$$
c. Find the tangent of Θ.
$$\tan \Theta = \frac{X_L}{R} = \frac{30}{15} = 2$$
d. From the trigonometric tables find the angle whose tangent is 2:
$$\Theta = 63.5°$$
e. From the trigonometric tables find the power factor.
$$PF = \cos \Theta = 0.45$$

IN-PHASE AND QUADRATURE CURRENT

The current flowing through the coil in *example 3* is lagging applied voltage by 63.5 degrees. There is only one current but it is made up of two parts. Now, these are not two separate currents. There is no way to measure either one of them alone. One part of the current is due to the resistance and is called the IN-PHASE component. The other part, due to the inductive reactance, is 90 degrees out of phase with the voltage. It is called the QUADRATURE component.

Sometimes we have to break a current down into its parts. Assume we find the current through the coil in Fig. 4-22 to be 3 A. This current is lagging the voltage by 63.5 degrees. We start our phasor diagram as shown in Fig. 4-23. A line representing 3 amperes of current is drawn at an angle of 63.5 degrees with the horizontal.

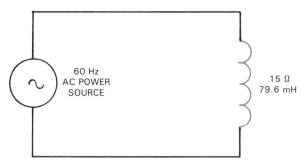

Fig. 4-22. Coil circuit for Example 3. Problem is to determine how much of the apparent power is active power.

The solution is shown in Fig. 4-24. The in-phase part of the current is represented by the adjacent side. Therefore:

$$I_{\text{IN-PHASE}} = I \times \cos \Theta$$
$$= 3 \times 0.45$$
$$= 1.35 \underline{/0°} \text{ A}$$

The quadrature part of the current is represented by the opposite side. Therefore:

$$I_{\text{QUADRATURE}} = I \times \sin \Theta$$
$$= 3 \times 0.9$$
$$= 2.7 \underline{/90°} \text{ A}$$

COILS IN PARALLEL

In a parallel circuit, the current splits. Part of it goes through one leg and the rest goes through the other. Each of these two currents has its own phase angle. The phase angle of the current coming from the source is a combination of both of these.

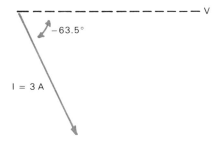

Fig. 4-23. Start of phasor diagram for determining in-phase and quadrature current. Vector marked "I" can be thought of as the hypotenuse of a right triangle.

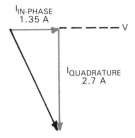

Fig. 4-24. Solution to Example 3. Horizontal vector of phasor diagram represents in-phase current and the vertical vector the quadrature current.

Example 4

Two coils are connected in parallel as shown in Fig. 4-25. Coil 1 has a resistance of 80 ohms and an inductive reactance of 60 ohms. Coil 2 has a resistance of 15 ohms and an inductive reactance of 30 ohms. Applied voltage is 120 V ac. Find the current through each coil, total current, apparent power, active power, reactive power and power factor.

a. Compute the impedance of coil 1. The impedance phasor diagram is shown in Fig. 4-26.

$$\tan \Theta_1 = \frac{X_L}{R} = \frac{60}{80}$$
$$= 0.75. \text{ Therefore, } \Theta_1 = 37°$$

$$\cos \Theta_1 = \frac{R}{Z}; \text{ therefore, } Z = \frac{R}{\cos \Theta_1}.$$
$$(\cos \Theta_1 = 0.8)$$
$$Z_1 = \frac{80}{0.8} = 100 \underline{/37°} \ \Omega$$

b. Compute current through coil 1.
$$I_1 = \frac{E}{Z_1} = \frac{120\underline{/0°}}{100\underline{/37°}} = 1.2\underline{/-37°} \text{ A}$$

c. Compute component parts of I_1.
$$I_1 \text{ (IN-PHASE)} = I_1 \cos \Theta_1 = 1.2 \times 0.8$$
$$= 0.96 \underline{/0°} \text{ A}$$
$$I_1 \text{ (QUADRATURE)} = I_1 \sin \Theta_1$$
$$= 1.2 \times 0.6$$
$$= 0.72 \underline{/90°} \text{ A}$$

d. Compute the impedance of coil 2. The impedance phasor diagram is shown in Fig. 4-27.

$$\tan \Theta_2 = \frac{X_L}{R} = \frac{30}{15} = 2. \text{ Therefore}$$
$$\Theta_2 = 63.5°$$
$$Z_2 = \frac{R}{\cos \Theta} = \frac{15}{.45} = 33.3 \underline{/63.5°} \ \Omega$$

e. Compute the current through coil 2.
$$I_2 = \frac{E}{Z_2} = \frac{120\underline{/0°}}{33\underline{/63.5°}} = 3.6\underline{/-63.5°} \text{ A}$$

f. Compute the component parts of I_2.
$$I_2 \text{ (IN-PHASE)} = I_2 \times \cos \Theta_2$$
$$= 3.6 \times .45 = 1.62 \underline{/0°} \text{ A}$$

$$I_2 \text{ (QUADRATURE)} = I_2 \times \sin \Theta_2$$
$$= 3.6 \times .9 = 3.24 \underline{/90°} \text{ A}$$

Inductance

g. Add the current through coil 1 to the current through coil 2. *Note: the in-phase components are added and the quadrature components are added.*

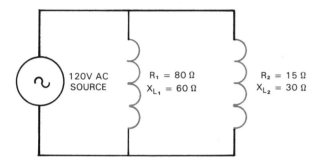

Fig. 4-25. Two coils connected in parallel in ac circuit. Currents must be combined (added together) to figure actual current.

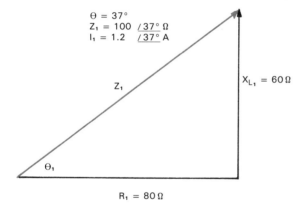

Fig. 4-26. Impedance diagram for coil 1 of Fig. 4-25.

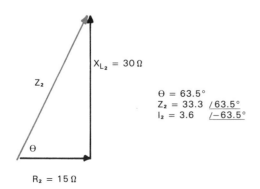

Fig. 4-27. Impedance phasor diagram for coil 2 of Fig. 4-25.

$$I_{\text{IN-PHASE}} = I_{1\ (\text{IN-PHASE})} + I_{2\ (\text{IN-PHASE})}$$
$$= 0.96 + 1.62 = 2.58\ \underline{/0°}\ A$$
$$I_{\text{QUADRATURE}} = I_{1\ (\text{QUADRATURE})} +$$
$$I_{2\ (\text{QUADRATURE})}$$
$$= 0.72 + 3.24$$
$$= 3.96\ \underline{/90°}\ A$$

The phasor diagram of this addition is shown in Fig. 4-28.

h. Compute the total current.

$$\text{Tan}\ \Theta = \frac{I_{\text{QUADRATURE}}}{I_{\text{IN-PHASE}}}$$
$$= 1.54;\ \text{therefore},\ \Theta = 57°$$
$$I = \frac{I_{\text{IN-PHASE}}}{\cos 57°} = \frac{2.58}{.544} = 4.75\ \text{amps}$$

i. Compute apparent power.
$$P = E \times I = 120 \times 4.75$$
$$= 570\ \text{volt amperes (VA)}$$

j. Compute power factor.
$$PF = \cos 57° = 0.544$$

k. Compute active power.
$$P = P \cos \Theta = 570 \times .544 = 310\ \text{W}$$

l. Compute reactive power: $P_q = P \sin \Theta$
$$= 570 \times .84 = 478\ \text{VAR}$$

SUMMARY

Inductance is the ability of a conductor to have a voltage induced into it when linked to a changing magnetic field.

One henry of inductance means that one volt results from a rate of change of current of one ampere per second.

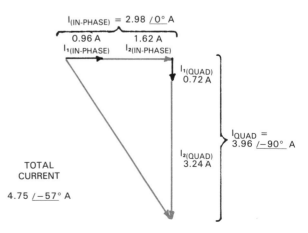

Fig. 4-28. Phasor addition of currents in coil 1 and coil 2.

...ctors in series is the
...inductances.

...tance of inductors in par-
...ed the same as resistors in

1. ...ge self-induced into a coil is 180 de-
g... out of phase with the voltage across it.
This has the effect of reducing coil current.

Inductance causes current to lag voltage by 90 degrees.

The inductive reactance of a coil is the ratio of ac voltage across it to the current through it.

Inductive reactance is computed from the equation: $X_L = 6.28 \times f \times L$. The unit of inductive reactance is the ohm.

Impedance is the total opposition to ac current caused by resistance and reactance.

The impedance of resistance and reactance in series is the phasor sum of the two.

The impedance of resistance and reactance in parallel is the applied voltage divided by total current, which is the phasor sum of the currents in each parallel lag.

Apparent power is the product of voltage and current. The unit of apparent power is the volt-ampere.

Active power is the power actually used by an electrical circuit or device. The unit of active power is the watt.

Reactive power is transmitted to a circuit or device, stored for half a cycle, then returned to the source. The unit of reactive power is the VAR.

TEST YOUR KNOWLEDGE

1. Voltage is induced into a conductor when there is a change in the magnetic flux linking the conductor. List two of the ways in which flux linkage can change.

2. What effect does inductance have on the phase relationship between current and voltage?

3. The unit of inductance is the _____.

4. Inductive reactance is (select the correct answer[s]):
   a. A quality similar to resistance.
   b. Proportional to both inductance and the frequency of the applied voltage.
   c. Measured in teslas.
   d. All of the above.
   e. None of the above.

5. A resistance and inductance in series causes current to lag voltage by an angle between _____ and _____ electrical degrees.

6. Describe the procedure for finding the impedance of a resistance and inductive reactance in parallel.

7. In a parallel circuit, current in each leg has its own phase angle. True or false?

8. Define and give the units for active power, reactive power, and apparent power.

PROBLEMS

1. Given: A voltage having a frequency of 100 hertz is applied to a 0.32 henry inductance. Find the inductive reactance.

2. Given: A 0.8 henry coil is connected in series with a 300 ohm resistance. Find the value and phase angle of current when 120 volts, 60 hertz is applied.

3. Given: A 0.8 henry coil is connected in parallel with a 300 ohm resistance. Find the value and phase angle of the total current when 120 volts, 60 hertz is applied.

5 CAPACITANCE

After studying this chapter, you will be able to:

☐ Describe capacitors and explain how they work.
☐ Compute the equivalent capacitance of capacitors in series and parallel.
☐ State the equation for computing capacitive reactance.
☐ Describe the effect that capacitance has on voltage with respect to current.
☐ Compute impedance of circuits containing resistance, inductive reactance, and capacitive reactance.
☐ Explain how capacitors are used to correct power factor.

In Chapter 1, we said that electricity is like wind. Wind is air in motion. Electricity is electrical charges in motion.

You cannot store wind. However, you can store air under pressure. When you release the pressure, air will flow once more.

Similarly, there is a way to store electrical charges. They can be collected on the plates of a CAPACITOR. To do this, there must be an ELECTROSTATIC FIELD set up in the insulator between the plates.

ELECTROSTATIC FIELDS

Electrostatic fields are very common. Lightning is the result of one. So is the shock you sometimes get from touching metal after walking across a carpet. We usually just call it "static electricity."

Electrostatic fields occur only in insulators. You never find them in conductors.

To get an electrostatic field, you must first have a voltage across the insulator. Voltage, remember, is a difference in electrical pressure. There are more electrons on one side than the other. However, no current flows. The electrons of an insulator cannot pass from one atom to the next. They are held tightly in their orbits. Even so, the electrical pressure has an effect on them.

ENERGY STORED IN AN ELECTROSTATIC FIELD

As shown in Fig. 5-1, the excess electrons on the surface exert a mechanical force on the electrons inside the insulator. Negative charges tend to repel other negative charges. This stretches the electrons' orbits like rubber bands. The energy it takes to distort the orbits is really the energy of the electrostatic field. It can be released later as current flow.

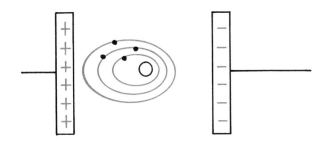

Fig. 5-1. Electrons in an insulator experience a force of attraction from positive charges on one plate. At the same time they are repelled by the negative charges on the other plate.

from the source push
_trons into a lopsided or-
_ one side of each atom more
_is is known as CHARGING.
_s are held on one surface until the
_ed voltage drops and there is an exter-
nal path between the plates.
2. Then they rush through this path to balance
each side. This movement is known as
DISCHARGING.

It is easy to see why an electrostatic field cannot be set up in a conductor. Electrons would
simply move from one atom to another. There
would be no distortion of orbits, no stored
energy.

FRANKLIN'S KITE

Two hundred years ago, there was no way to
generate a steady current. All experiments were
done with static electricity. They had to rub silk
cloth over spheres made of sulphur or glass. A
model of Benjamin Franklin's hand-driven
"generator" is shown in Fig. 5-2. Thin wires,
resting lightly on the sphere, collected the "electric fire," as they called it. Franklin had a good
reason for trying to attract lightning. He wanted
another source of electricity for his experiments.

Fig. 5-3 represents Franklin flying his historic
kite. It was a dangerous thing to do. Franklin,
however, was well aware of its dangers. He knew
the damage lightning can do. A description of
the event appeared in the first edition of the En-
cyclopedia Britannica published 1768-71:

"To demonstrate, in the completest manner
possible, the sameness of the electric fluid
with the matter of lightning, Dr. Franklin
contrived to bring lightning from the heavens,
by means of an electrical kite, which he raised
when a storm of thunder was perceived to be
coming on. This kite had a pointed wire fixed
upon it, by which it drew the lightning from
the clouds. This lightning descended by the
hempen string, and was received by a key tied

Fig. 5-2. Scale model of Benjamin Franklin's original electric
generator. Rotating glass ball was made to rub against a cloth
covered block.   (Franklin Institute)

to the extremity of it; that part of the string
which was held in the hand being of silk, that
the electric virtue might stop when it came to
the key. He found that the string would conduct electricity even when nearly dry, but that
when it was wet, it would conduct it quite
freely; so that it would stream out plentifully
from the key. . ."

THE LEYDEN JAR

Getting electricity from the clouds seemed
easier than rubbing a glass ball. Still, Franklin
could not depend on a thunderstorm every time
he wanted to experiment. He and others at that
time stored their electrical "fluid" or "fire" in
LEYDEN JARS. A sketch of the jar is shown in
Fig. 5-4.

Capacitance

Fig. 5-3. Benjamin Franklin flew a kite in a thunderstorm to prove that lightning was really electricity.

long time. However, if you touch the contact terminal and outer foil at the same time, you can get quite a shock. The stored electricity discharges very rapidly.

CAPACITORS

Today, Leyden jars are used only for demonstration. There are better ways to produce electricity. The Leyden jar turned out to be very important to electrical progress. It was the first capacitor.

Nowadays, capacitors come in all shapes and sizes. The ones used in electronics are usually very small. Power capacitors can be as small as a pencil box or larger than a filing cabinet. Like the Leyden jar, they all have two conducting surfaces, called PLATES. Between the plates is an insulating material called DIELECTRIC. Fig. 5-5 shows a typical power capacitor.

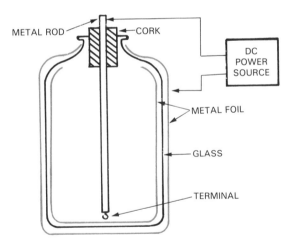

Fig. 5-4. Leyden jar. Early experimenters thought they had a ''jar full of electricity.'' Actually, the glass, itself, held an electrostatic charge.

Fig. 5-5. Cutaway view of power capacitor assembly. Like the Leyden jar, it stores up electric power. (Sprague Electric Co.)

The jar, invented in Leyden, England, is simply a glass jar lined with metal foil inside and out. A stopper insulates the contact terminal. A static electrical charge can be applied to the inside foil through the terminal. This sets up an electrostatic field in the glass. The jar becomes charged. It will hold the charge for a

Never pick up a capacitor without first checking that it is discharged. A charge may be held for days. Contact could result in a severe shock.

TYPES OF DIELECTRIC

There are four main types of dielectric:
1. Air or other gases.
2. Mineral oil or other liquid.
3. Glass, paper or other solid.
4. Combination of solid and liquid.

DIELECTRIC BREAKDOWN

In dielectric breakdown, the insulating material changes and conducts electricity across to other plate. This action may be compared to what happens when lightning strikes, Fig. 5-6.

A cloud acts as one capacitor plate. It becomes charged with an excess of electrons from friction as it moves through the sky. The other plate may be the earth or another cloud. The air between is the dielectric. Normally, air does not conduct electricity. But when the voltage in the charged cloud gets high enough, it causes the air to IONIZE. That is, its molecules become charged particles or IONS. This is the air's breakdown point. Lightning occurs at this instant. The lightning is the electrical charge in the cloud being conducted through the air to the other cloud or to the earth.

BREAKDOWN POINT

Like the air which ionized to become a conductor, every kind of dielectric has a point where it becomes a conductor. This is called its breakdown point. All capacitors are rated for a certain breakdown voltage. Be careful not to exceed it. When the dielectric begins to conduct, the capacitor is shorted out.

Even when the dielectric is in perfect condition, there is a tiny current flow through it. This is called LEAKAGE CURRENT. It is usually ignored in electrical power computations.

Fig. 5-6. The cloud acts as one plate of a capacitor. The other plate may be the earth. Air in between is the dielectric.

CAPACITANCE

The ability to store electricity in an electrostatic field is called CAPACITANCE. Capacitors are specifically made to do this. However, capacitance is not limited to capacitors. Current-carrying wires that are close together also set up an electrostatic field. This happens in power line cables, for example. The wires' insulation can break down if the voltage across it is too high. Power companies must be careful not to exceed the rated breakdown voltage of the insulation.

The capacitance within a power line cable also creates a second problem. Capacitance causes power losses by producing REACTIVE CURRENT. Reactive current is covered in detail later in this chapter. Power plant engineers often show the capacitance of power lines on a drawing like the one in Fig. 5-7.

CAPACITOR CHARGING

What happens to the current when a voltage is applied to a capacitor? Current cannot flow through the dielectric. However, this does not mean that there is no current.

The energy that sets up the electrostatic field comes from current. The current that flows is not the same as current from the source. Its intensity does not depend on applied voltage.

Capacitance

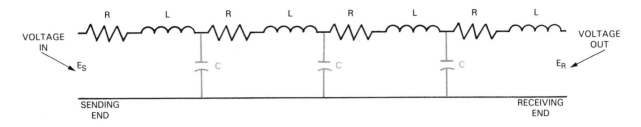

Fig. 5-7. Diagram shows capacitance in long transmission lines. Electrostatic field is set up in the insulation.

While the current is flowing, charges are stacking up on one plate and leaving the other. This is called CHARGING CURRENT.

Assume you apply dc to a capacitor. At first, there is a rush of charging current. Then, as the charge on the capacitor increases, current drops. Fig. 5-8 shows that charging current flows immediately after voltage is applied. Charging current stops when the voltage across the capacitor equals source voltage. The electrostatic field is established. The capacitor, itself, is now a voltage source. This is illustrated by Fig. 5-9.

CAPACITOR DISCHARGING

If you suddenly remove the source voltage, the capacitor remains charged. It can hold this charge for a very long time. That is why you must take extra care working on circuits having capacitors. Even with power off, you could get a bad shock from a capacitor. Experienced electricians always discharge them first. They use DISCHARGE RESISTORS to prevent too large a current.

If you gradually reduce the voltage applied to a capacitor, current will, at some point, flow back toward the source. This is illustrated by Fig. 5-10. It is still called charging current, even though the capacitor is discharging.

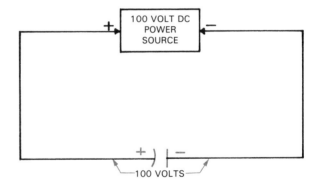

Fig. 5-9. Current stops flowing when the capacitor is charged to the value of the source voltage.

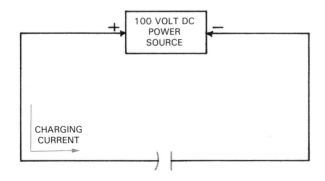

Fig. 5-8. Charging current flows briefly after a dc voltage is applied to a circuit containing a capacitor.

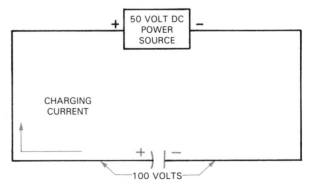

Fig. 5-10. When source voltage drops, the capacitor begins discharging and current flows back to power source.

Again, the intensity of this current does not depend on the value of the voltage. It depends on how fast the voltage is changing.

Of course, if you connect a low resistance between the terminals of a charged capacitor, the voltage changes very fast. This produces a large charging current. Devices like electronic strobes use this principle. They charge a capacitor slowly, then discharge it rapidly.

UNITS OF CAPACITANCE

The intensity of charging current depends on the rate at which voltage is changing. Capacitors are rated by the amount of current that results from the changing voltage. The unit of capacitance is the farad (symbol: F). *A capacitor is said to have a capacity of one farad when a current of one ampere is produced by a rate of change of one volt per second.*

In practice, a one farad capacitor would be large. The ones we will be discussing have values in microfarads. *A microfarad is one millionth of a farad.* Often, the Greek letter mu (μ) is used for micro. A 10 microfarad capacitor, for example, is shown as 10 μF.

The way a capacitor is made determines its capacitance. Capacitance depends on the area of its plates, as well as the material and thickness of the dielectric. The equation is:

$$C = \frac{K \times A}{T}$$

C is capacitance
A is area of the plates
T is thickness of the dielectric
K is dielectric constant

Some typical dielectric constants for common materials are: air, 1; paper, 3; and aluminum oxide, 8. From the equation you can see that the larger the plate area, the greater the capacitance. Also, the thinner the dielectric, the greater the capacitance.

The problem with thin dielectric is the danger of breakdown. The thinner the dielectric, the less voltage it takes to produce an arc-over between the plates.

For example, the thickness of the dielectric of a paper capacitor rated at 220 volts is 0.0015 centimetres. If it is rated at 110 volts, however, its dielectric is only 0.00075 centimetres thick.

For continuous duty, the voltage rating should not be exceeded. Most capacitors are conservatively rated, however. They can usually stand many times their rated voltage for a brief time.

Some ac motors have capacitors in their circuits. One type is connected for only a few seconds when starting. A second type is connected all the while the motor is running. The construction of a capacitor determines how it can be used.

PAPER CAPACITORS

Motor-running capacitors are made by rolling two or more layers of paper between two layers of aluminum foil. See Fig. 5-11, view A. The roll is then inserted into a metal case, and terminals

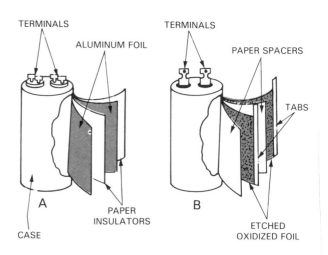

Fig. 5-11. Construction of capacitors varies according to their use. A—Paper capacitor used with running motors. B—Electrolytic capacitor used for starting ac motors. (Sprague Electric Co.)

are attached to the foil. The cover is sealed on the case. A small hole in the cover allows the paper to dry thoroughly. Insulating oil is added through the hole; then the hole is sealed.

The oil also increases the dielectric constant of the paper. The result is a capacitor with a high voltage rating in a small case.

ELECTROLYTIC CAPACITORS

Motor-starting capacitors may have 20 times the capacitance of motor-running capacitors. Yet, both are about the same size. Those used in motor starting are usually ELECTROLYTIC capacitors. They contain more capacitance in a smaller package but cannot be operated continuously. They are called "electrolytic" because they are made with electrolyte. In fact, one of the "plates" is an electrolyte—in this case, a thick chemical paste with electrical properties. The second plate is a strip of aluminum. The dielectric is a thin coat of aluminum oxide formed on the aluminum surface.

The aluminum foil sheets, with paper spacers, are rolled into a cylinder as shown in Fig. 5-11, view B. The spacers also act as wicks, to soak up the electrolyte. This cylinder is put into a housing and sealed.

Electrolytic capacitors for dc circuits are made the same way, except that only one of the foil sheets is treated. Aluminum oxide, you see, has a high resistance to current flow in one direction but a low resistance in the other. Direct current electrolytic capacitor terminals are marked positive (+) and negative (−). Care must be taken to connect them correctly in dc circuits. They must not be used at all in ac circuits.

NOTE: Electrolytic capacitors must be connected into circuits positive to positive and negative to negative. Otherwise the oxide film will be ruined. Sometimes the capacitor even explodes.

CAPACITORS IN SERIES

If two or more capacitors are in series, not as many charges can be stored. The equivalent capacitance of capacitors in series is computed just like resistors in parallel. If all the capacitances are the same, divide the value of one of them by the number in series. For example, the equivalent capacitance of three 15 microfarad capacitors in series is 5 microfarad ($\frac{15}{3} = 5$ μF). See Fig. 5-12.

If there are only two capacitors in series, use the *product over sum* method. For example, the equivalent capacitance of a 5 microfarad capacitor in series with a 15 microfarad capacitor is 3.75 microfarad. $\frac{5 \times 15}{5 + 15} = 3.75$ μF. See Fig. 5-13.

Three or more series capacitors having different values must be solved by the *reciprocals* method. For example, the equivalent capacitance of a 20 microfarad, a 25 microfarad and a 100 microfarad capacitor in series is:

10 μF   ($\frac{1}{20} + \frac{1}{25} + \frac{1}{100} = 10$ μF).

Fig. 5-12. If all capacitances in series are the same, divide the value of one by the number in the series. Equivalent capacitance of the series above equals 15 divided by 3 equals 5 microfarads.

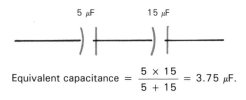

Equivalent capacitance = $\frac{5 \times 15}{5 + 15} = 3.75$ μF.

Fig. 5-13. Where two capacitances in series are not the same, equivalent capacitance is figured by the "product over sum" method.

CAPACITORS IN PARALLEL

Parallel connected capacitors are like storage tanks. Three tanks are shown in Fig. 5-14. The total amount they can store is the sum of the amount each can hold. In the same way, you can find the total capacitance of capacitors in parallel: simply add them up.

For example, the total capacitance of a 40 microfarad and a 50 microfarad capacitor in parallel is 90 microfarad (40 + 50 = 90 μF). See Fig. 5-15.

CAPACITORS IN AC CIRCUITS

Charging a capacitor sets up an electrostatic field. This field produces a back voltage that opposes the applied voltage. Current flows while the applied voltage is increasing. The capacitor is fully charged when the back voltage equals the applied voltage. Then current stops. As applied voltage decreases, current flows the other way. At every instant, current depends on the rate of change of the applied voltage. Current flow follows a sine wave pattern like the applied voltage.

CAPACITIVE REACTANCE

It is not possible to measure capacitor back voltage while a capacitor is charging or discharging. You must, therefore, concentrate on current. Pretend that capacitors have a quality like resistance. This quality is known as capacitive reactance (X_c). *Capacitive reactance, which is measured in ohms, is the ratio between the applied voltage and charging current.* This can be given as an equation: $X_c = \dfrac{E_c}{I_c}$.

Charging current, remember, is proportional to the *rate of change* of applied voltage. Therefore, capacitive reactance must depend on frequency, as well as capacitance. The larger the capacitance, the smaller the capacitive reactance. Also, the higher the frequency, the smaller the capacitive reactance. It might help to remember that capacitors block the flow of direct current. There, frequency is 0 hertz and only a tiny leakage current flows. In other words, capacitive reactance is infinitely high at 0 hertz. If you know both the capacitance and the frequency, you can compute capacitive reactance from the equation:

$$X_c = (\frac{1}{6.28 \times f \times C}) \qquad \begin{array}{l} X_c \text{ is in ohms} \\ f \text{ is in hertz} \\ C \text{ is in farads} \end{array}$$

Example 1

A 120 V, 60 Hz voltage is applied to a 19 microfarad capacitor. See Fig. 5-16. Find the capacitive reactance and charging current.

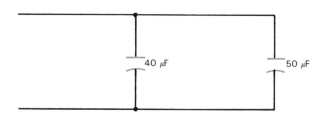

Fig. 5-15. Schematic of parallel-wired capacitors. Equivalent capacity is the sum of the two capacities (40 + 50 = 90 μF).

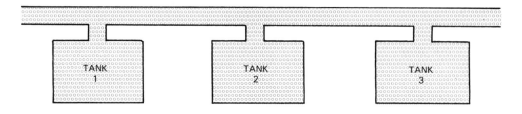

Fig. 5-14. Equivalent capacity of these three tanks is the sum of the individual capacities.

a. Convert microfarads to farads.

$19 \ \mu F = 0.000019 \ F$

b. Compute capacitive reactance.

$$X_c = (\frac{1}{6.28 \times f \times C})$$

$$= \frac{1}{6.28 \times 60 \times 0.000019}$$

$$X_c = \frac{1}{.007173} = 140 \ \Omega$$

c. Compute the charging current.

$$I_c = \frac{E}{X_c} = \frac{120}{140} = 0.86 \ A$$

Example 2

The frequency of the voltage in *Example 1* is 50 hertz. All other values are the same. Find capacitive reactance and charging current.

a. Compute the capacitive reactance.

$$X_c = (\frac{1}{6.28 \times 50 \times 0.000019})$$

$$X_c = \frac{1}{0.006} = 167.6 \ \Omega$$

b. Compute the charging current.

$$I_c = \frac{E_c}{X_c} = \frac{120}{167.6} = 0.72 \ A$$

PHASE SHIFT CAUSED BY CAPACITANCE

A capacitor in a circuit is like having a rubber disc across the inside of a tube full of marbles. Refer to Fig. 5-17. Pressure will force marbles against the disc. It stretches. At first, marbles on the other side move as though there were no disc. As it continues to stretch, the disc exerts a growing back pressure. Before long, back pressure equals the pressure being applied. Then the marble flow stops.

Now, assume that the pressure forcing the marbles against the disc is reduced. This makes the disc back pressure greater than the source pressure. Marbles begin to flow in the opposite direction. They will continue to flow in that direction until pressure drops on the back flow. This is shown by Fig. 5-18.

Fig. 5-19 has two curves. One shows how source pressure changes with time. The other shows how marble flow changes during the same time. Notice that the motion (flow) is out of step with pressure. This is the same effect that capacitance has in ac circuits. It makes current lead the applied voltage. Refer to Fig. 5-20.

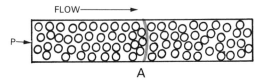

Fig. 5-16. Capacitive reactance can be computed from capacitance and frequency. Current can be computed from capacitive reactance and voltage.

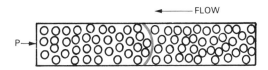

Fig. 5-18. Demonstrating why capacitors cause current flow to be out of phase with pressure. Stretched rubber disc moves marbles back toward pressure source as the pressure drops.

A

Pressure, P, is greater than back pressure from disc.

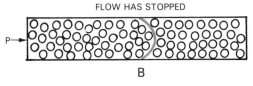

B

Pressure, P, is equal to back pressure from disc.

Fig. 5-17. Energy stored by a capacitor is like mechanical energy stored in this stretched rubber disc.

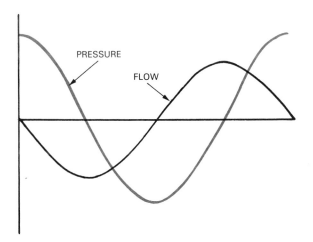

Fig. 5-19. Expressing pressure and marble flow of Fig. 5-18 in a sine wave. Flow leads the pressure.

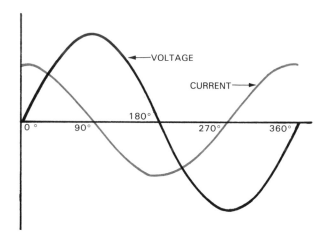

Fig. 5-20. A capacitor causes current to lead voltage by 90 degrees.

RESISTANCE AND CAPACITIVE REACTANCE IN AC CIRCUITS

Coils always contain some resistance along with inductive reactance. This is not true of capacitors. There may be a very small leakage current through a capacitor but it does not concern us. In our computations, the capacitors are assumed to be perfect.

RESISTOR AND CAPACITOR IN SERIES

Phasor diagrams are used to solve circuits containing capacitance. Refer to the circuit of Fig. 5-21. It has a 176.8 μF capacitor in series with a 40 ohm resistor. Applied voltage is 100 volts alternating current at 60 hertz. We will determine the capacitive reactance, the impedance, the phase angle, and the active, reactive and apparent power.

1. Convert microfarads to farads.
   $176.8 \times 10^{-6} = 0.0001768$ F
2. Compute the capacitive reactance.
   $$X_c = (\frac{1}{6.28 \times f \times c})$$
   $$= (\frac{1}{6.28 \times 60 \times 0.0001768})$$
   $$X_c = \frac{1}{0.067} = 15 \ \Omega$$
3. Now construct the phasor diagram. See Fig. 5-22.

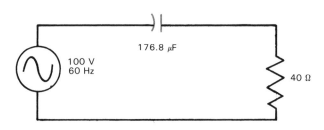

Fig. 5-21. Current through resistor is in phase with voltage across it. But current through capacitor is 90 degrees ahead of current through it.

a. First, draw a horizontal line 40 units long. This represents the resistance.
b. Capacitive reactance is represented by a line 15 units long. It is drawn straight downward.
c. The impedance is the hypotenuse. We can find the phase angle Θ first. It is the angle whose tangent equals the inductive reactance divided by the resistance.
   $$\tan \Theta = \frac{X_c}{R} = \frac{-15}{40} = -.375$$
   Therefore, $\Theta = -20.5$ degrees.
4. Since $\cos \Theta = \frac{R}{Z}$, you can compute the impedance from the equation $Z = \frac{R}{\cos \Theta}$. From the trigonometric tables, you can determine that $\cos \Theta = 0.94$.

   Therefore, $Z = \frac{40}{.94} = 42.6 \ \Omega$

Note that power factor equals cos Θ. Therefore, the power factor is 0.94.

5. Now that you know the impedance, you can use Ohm's Law to compute total current.

$$I = \frac{E}{Z} = \frac{100\underline{/0°}}{42.6\underline{/-20.5°}} = 2.35\ \underline{/\,20.5°}\ A$$

Notice that a minus sign is used to show that this is a leading current.

6. At this point, you have enough information to compute apparent, active and reactive power.

Apparent power P_s = E × I = 100 × 2.35
= 235 volt amperes (VA)

Active power P = P_s cos Θ = 235 × 0.94
= 221 W

also P = I^2R = 2.35 × 2.35 × 40 = 221 W

Reactive power P_q = P_s sin Θ = 235 × 0.35
= 82.3 VAR

also P_q = I^2X_c = 2.35 × 2.35 × 15
= 83 VAR

The slight difference in the answer comes from rounding off values.

RESISTANCE AND CAPACITANCE IN PARALLEL

A different approach solves ac parallel circuits. The impedance is not the phasor sum of the resistance and reactance. First you must find the total current.

Total current in an ac parallel circuit is the phasor sum of the currents in each leg. Refer to the circuit in Fig. 5-23. An 88.4 microfarad capacitor is in parallel with a 40 ohm resistor. A 100-volt ac source applies a voltage at a frequency of 60 Hz. Determine the total current impedance,

power factor and the apparent, active and reactive power.

1. Convert microfarads to farads.
88.4 μF × 10^{-6} = 0.0000884 F

2. Compute capacitive reactance.

$$X_c = \left(\frac{1}{6.28 \times 60 \times 0.0000884}\right)$$
$$= \frac{1}{.03333}$$
$$X_c = 30\ \Omega$$

3. Find the in-phase current ($I_{IN\text{-}PH}$).

$$I_{IN\text{-}PH} = \frac{E}{R} = \frac{100}{40} = 2.5\ \underline{/0°}\ A$$

4. Find the quadrature current (I_{QUAD}).

$$I_{QUAD} = \frac{E}{X_c} = \frac{100\underline{/0°}}{30\underline{/-90°}} = 3.33\ \underline{/\,90°}\ A$$

5. Construct the current phasor diagram shown in Fig. 5-24. Find Θ.

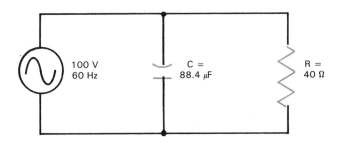

Fig. 5-23. In reactive ac parallel circuits, currents must be combined.

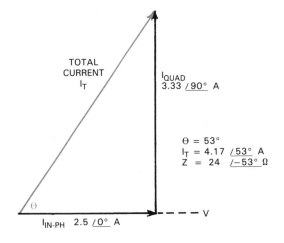

Fig. 5-24. Current phasor diagram is set up to solve the circuit shown in Fig. 5-23.

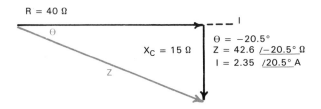

Fig. 5-22. This phasor diagram solves the circuit in Fig. 5-21.

$$\tan \Theta = \frac{I_{QUAD}}{I_{IN\text{-}PH}} = \frac{3.33}{2.5} = 1.33$$

Therefore, $\Theta = 53°$

6. Find the power factor, $\cos \Theta$.

   PF $= \cos 53° = 0.6$

7. Compute the total current.

   $$I_t = \frac{I_{IN\text{-}PH}}{\cos \Theta} = \frac{2.5}{.6} = 4.17 \underline{/53°} \text{ A}$$

8. Compute the impedance.

   $$Z = \frac{E}{I} = \frac{100 \underline{/0°}}{4.17 \underline{/53°}} = 24 \,\Omega \underline{/-53°}$$

9. At this point there is enough information to compute apparent, active and reactive power.

   Apparent power $(P_s) = E \times I = 100 \times 4.17$
   $= 417$ volt-amperes

   Active power $(P) = P_s \times \cos \Theta$
   $= 417 \times 0.6 = 250$ W

   Also, $P = I_{IN\text{-}PH}{}^2 \times R = 2.5^2 \times 40 = 250$ W

   Reactive power $P_q = P_s \times \sin \Theta$
   $= .417 \times 0.8 = 333$ VAR

   Also, $P_q = I_{QUAD}{}^2 \times X_c = 3.33^2 \times 30$
   $= 333$ VAR

COIL AND CAPACITOR IN SERIES

You will now see what happens when coils and capacitors turn up in the same circuit, Fig. 5-25. Assume there is a capacitor having 15 ohms of capacitive reactance in series with a coil having 15 ohms of resistance and 30 ohms of inductive reactance. Applied voltage of the circuit is 100 volts ac. Compute the current, active, reactive and apparent power and power factor.

1. Since this is a series circuit, construct an impedance phasor diagram. See Fig. 5-26. A horizontal vector 15 units long represents the resistance of the coil. A vertical vector 30 units long is drawn upward to represent the inductive reactance. A vertical vector 15 units long is drawn downward to represent the capacitive reactance. Notice that the capacitive reactance is 180 degrees out of phase with the inductive reactance.

2. Compute the equivalent reactance of the circuit. The equivalent reactance of inductance and capacitance in series is the algebraic difference. The smaller is simply subtracted from the larger.

   $X_{EQ} = X_L - X_c = 30 - 15 = 15 \,\Omega$

   This 15 ohms is inductive reactance. The 15 ohms of capacitive reactance cancel out 15 ohms of inductive reactance. See the phasor diagram in Fig. 5-27.

3. Find the phase angle.

   $$\tan \Theta = \frac{X_{EQ}}{R} = \frac{15}{15} = 1$$

   Therefore $\Theta = 45°$

4. Find the power factor.

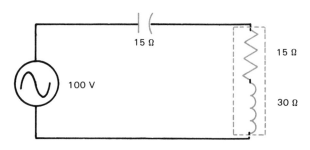

Fig. 5-25. To solve circuits containing capacitors and coils in series, first combine the reactances with algebra. Then perform phasor addition on the resistance and resulting reactance.

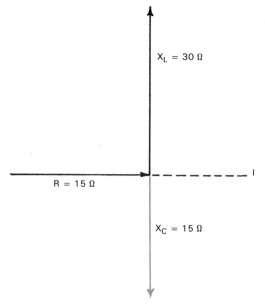

Fig. 5-26. Phasor diagram for circuit shown in Fig. 5-25. Capacitive reactance is drawn downward because it is 180 degrees out of phase with inductive reactance.

Capacitance

PF = cos 45° = 0.707
5. Find the impedance.
$$Z = \frac{R}{\cos \Theta} = \frac{15}{0.707} = 21.2 \; \Omega$$
6. Find the current.
$$I = \frac{E}{Z} = \frac{100\underline{/0°}}{21.2\underline{/45°}} = 4.7 \; A \; \underline{/-45°}$$
7. Compute the apparent power.
$$P_s = E \times I = 100 \times 4.7 = 470 \; VA$$
8. Compute the active power.
$$P = P_s \times \cos \Theta = 470 \times 0.707 = 332.3 \; W$$
$$\text{Also, } P = I^2 \times R = 4.7^2 \times 15 = 331.4 \; W$$
9. Compute the reactive power.
$$P_q = P_s \times \sin \Theta = 470 \times 0.707$$
$$= 332.3 \; VAR$$
$$\text{Also, } P_q = I^2 X = 4.7^2 \times 15 = 331.4 \; VAR$$

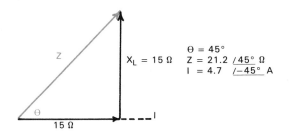

Fig. 5-27. Completed phasor diagram solves circuit described in Fig. 5-25.

COIL AND CAPACITOR IN PARALLEL

When solving parallel ac circuits you have to work with currents. The total current is the phasor sum of the current in each leg.

It is easy to combine these currents if they are either in phase or 90 degrees out of phase. However, suppose they are at some other angle. Then you must compute the in-phase and quadrature components.

Let us take a specific example. Refer to Fig. 5-28. Assume that 100 volts is applied to a capacitor and coil in parallel. The capacitor has a capacitive reactance of 50 ohms. The coil has a resistance of 15 ohms and an inductive reactance of 30 ohms. We will compute the current in each leg, the total current, the impedance, the power

factor and the apparent, active, and reactive power.
1. Compute the current through the capacitance leg.
$$I_{X_c} = \frac{E}{X_c} = \frac{100\underline{/0°}}{50\underline{/-90°}} = 2 \; \underline{/90°} \; A$$
2. Compute the current through the inductive leg. To do this, first compute the impedance of this leg. Fig. 5-29 shows the coil's impedance phasor diagram. The phase angle, Θ_{COIL}, is determined as follows:
$$\tan \Theta_{COIL} = \frac{X_L}{R} = \frac{30}{15} = 2$$
Therefore, $\Theta_{COIL} = 63.5°$
The impedance of the coil, Z_{COIL}, can be computed.
$$Z_{COIL} = \frac{R}{\cos \Theta_{COIL}} = \frac{15}{0.446} = 33.6 \; \Omega$$
$$I_{COIL} = \frac{E}{Z_{COIL}} = \frac{100\underline{/0°}}{33.6\underline{/63.5°}} = 3 \; \underline{/-63.5°} \; A$$
3. Now separate the 3 amperes into its in-phase and quadrature components.
$$I_{IN\text{-}PH} = I_{COIL} \times \cos \Theta = 3 \times 0.446$$
$$= 1.34 \; \underline{/0°} \; A$$

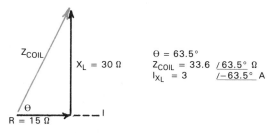

Fig. 5-28. To combine currents that are out of phase by different amounts, first determine the in-phase components and quadrature components.

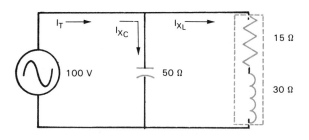

Fig. 5-29. Phasor diagram for inductive leg of parallel circuit in Fig. 5-28.

$$I_{QUAD} = I_{COIL} \times \sin \Theta = 3 \times 0.895$$
$$= 2.68 \angle -90° \text{ A}$$

4. The current phasor diagram for the parallel circuit is shown in Fig. 5-30. Notice that the 2 amperes of quadrature leading current is subtracted directly from the 2.68 amperes of lagging current. This leaves 0.68 amperes of lagging current. Now we can find the phase angle Θ for the entire circuit.

$$\tan \Theta = \frac{I_{QUAD}}{I_{IN\text{-}PH}} = \frac{0.68}{1.34} = 0.5$$

Therefore, $\Theta = 26.6°$

The power factor, PF $= \cos \Theta = 0.894$

$$\text{Total current, } I_t = \frac{I_{IN\text{-}PH}}{\cos \Theta} = \frac{1.34}{.894}$$
$$I_t = 1.5 \angle -26.6° \text{ A}$$

Notice that coil current is almost 3 amperes and capacitor current is 2 amperes. Yet the total current is only 1.5 amperes. This can only mean that some current is shuttling between the coil and capacitor.

5. Compute the equivalent impedance.

$$Z = \frac{E}{I_t} = \frac{100\angle 0°}{1.5\angle -26.6°} = 66.67 \angle 26.6° \text{ } \Omega$$

6. It is interesting that any ac circuit can be represented by an equivalent circuit. An ac equivalent circuit is a resistance and reactance in series. This is shown in Fig. 5-31. The equivalent resistance and reactance are computed as follows:

$$R_{EQ} = Z \cos 26.6° = 66.67 \times .894 = 60 \text{ } \Omega$$
$$X_{EQ} = Z \sin 26.6° = 66.67 \times .448 = 30 \text{ } \Omega$$

7. Compute the various powers.

Apparent power (P_s) $= E \times I_t = 100 \times 1.5$
$= 150$ VA

Active power (P) $= P_s \cos \Theta = 150 \times .894$
$= 134$ W

Also, $P = I_t{}^2 R_{EQ} = 1.5^2 \times 60 = 135$ W

Also, $P = I_{COIL}{}^2 R = 3^2 \times 15 = 135$ W

Reactive power $P_q = P_s \sin \Theta = 150 \times .448$
$= 67.2$ VAR

Also $P_q = I_t{}^2 X_{EQ} = 1.5^2 \times 30 = 67.5$ VAR

USE OF CAPACITORS

Capacitors cause current to lead voltage by 90 degrees. This is the opposite of coils, which cause current to lag voltage. There are two main uses for capacitors in ac power work. Both are related to this shift in phase.

CAPACITORS USED WITH MOTORS

Some types of ac motors require what looks like two-phase power. That is, the current through one coil must lead the current through a second coil by about 90 degrees. You must start with single-phase since two-phase ac is not produced. A capacitor is then added to create this phase shift. Fig. 5-32 shows a typical capacitor-start motor. This application will be covered in detail in Chapter 9, *Single-Phase AC Motors*.

POWER FACTOR CORRECTION

Power factor is the ratio of the active power being used to the total power supplied. The rest is called reactive power. It is stored for half a cycle, then returned to the source. Most loads are inductive. This means that current lags volt-

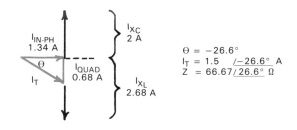

Fig. 5-30. Phasor solution for the parallel circuit shown in Fig. 5-28.

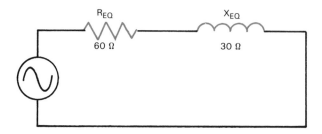

Fig. 5-31. This circuit is the equivalent of the parallel circuit shown in Fig. 5-28.

Fig. 5-32. Motors like this one are used in many home appliances. Motor-starting capacitor is mounted on top of case.

age. The effect of a low power factor is shown in Fig. 5-33, view B. The shaded portion of the power curve above zero is the active power. Below zero is the reactive power.

What we want is unity power factor, as represented by Fig. 5-33, view A. That is when all the power supplied is used. Low power factor costs money. For one thing, useless current is being transmitted. Power companies charge for this by including power factor in the rate schedule. All power handling devices must be

rated to handle this useless current. Then too, power losses are larger than they should be.

We cannot eliminate the inductance. Motors will not operate without it. What we can do is add a capacitor. This is called POWER FACTOR CORRECTION. Fig. 5-34, view A, shows a motor operating without power factor correction. The feeder line must supply both useful and reactive current. The reactive current merely shuttles back and forth between the generator (power source) and the motor. Fig. 5-34, view B, shows a capacitor installed near the motor. Now the reactive current flows only between the capacitor and the motor.

Current from the collapsing magnetic field of the motor charges the capacitor. The next half cycle the capacitor discharges. This provides the magnetizing current to the motor.

Power factor correcting capacitors are rated in vars or kilovars (KVAR). The rating shows how much reactive power the capacitor will supply. Since this is leading reactive power, it cancels out the lagging reactive power of the inductance. Therefore, each kilovar of capacitor value decreases the total reactive power by the same amount. A 15 KVAR capacitor on a motor, for example, will cancel 15 KVAR of inductive reactive power.

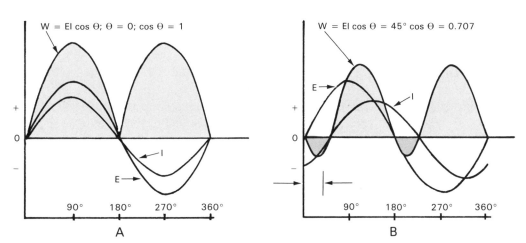

Fig. 5-33. Curves showing effect of low power factor with the same values of voltage and current. A—There is no reactive power when voltage and current are in phase. B—Active power is reduced to 70 percent when current lags voltage 45 degrees.

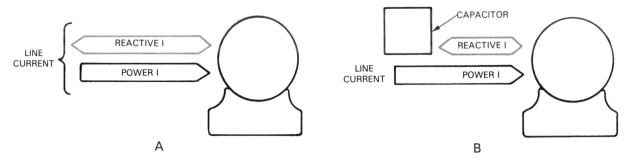

Fig. 5-34. Effect of a capacitor on power usage. A—Reactive power must come from the feeder line when there is no capacitor. B—Reactive power is received from the power factor correction capacitor installed near the motor.

SUMMARY

Capacitors have the ability to store energy in an electrostatic field.

The electrostatic field is set up in insulating material called dielectric which is placed between metal plates.

Capacitors become charged to the value of the voltage applied to their terminals.

A charged capacitor is a voltage source.

Current does not actually flow through a capacitor. When dc is applied, charging current flows briefly. When ac is applied, the continuous charging and discharging makes alternating current appear to flow through it.

The unit of capacitance is the farad (F). However, the value of most capacitors is in microfarads (μF).

The total capacitance of capacitors in series is computed like resistors in parallel.

The total capacitance of capacitors in parallel is the sum of the individual capacitances.

The capacitive reactance is the ratio of applied voltage to changing current.

Capacitive reactance is measured in ohms and may be computed from the equation $X_c = \dfrac{1}{6.28fC}$.

Capacitance causes current to lead the applied voltage by 90 degrees.

When a coil and capacitor are in series, the total reactance is the difference between capacitive reactance and inductive reactance.

When a coil and capacitor are in parallel, the total reactive (quadrature) current is the difference between the leading current and the lagging current.

TEST YOUR KNOWLEDGE

1. What is capacitance? What is the unit of capacitance?
2. What is the total capacitance of capacitors in series?
3. (Select best answers.) Capacitors are said to (pass, block) dc and (pass, block) ac.
4. Capacitive reactance is inversely proportional to both _____ and _____.
5. What is the phase relationship between the ac voltage applied to a capacitor and the resulting charging current?
6. A capacitor affects lagging power factor in the following way:
   a. It cancels out lagging power.
   b. It has no effect whatever.

PROBLEMS

1. Given: A 60 Hz voltage is applied to a 53 microfarad capacitor. Find the capacitive reactance in ohms.
2. Given: A capacitive reactance of 50 ohms is in series with 100 ohms of resistance. Find the impedance and the angle by which current leads voltage.
3. Given: A capacitance and resistance are in parallel. Total current is 5 amperes when 120 volts at 60 hertz is applied. Active power is 360 watts. Find the apparent power, reactive power and power factor.

Capacitance

4. Given: A coil having a resistance of 5 ohms and an inductive reactance of 25 ohms is in series with a capacitor having a capacitive reactance of 30 ohms. Find the impedance.

5. Given: A coil whose resistance is so small that it can be ignored, having an inductive reactance of 20 ohms, is in parallel with a capacitor having a capacitive reactance of 30 ohms. When 120 volts, 60 hertz is applied, what is the current in each leg? What is the total current? What is the angle of lead or lag?

6. Given: A voltage of 120 volts, 60 hertz is applied to a coil having a resistance of 20 ohms and inductive reactance of 35 ohms. Find the power factor. A 120.5 microfarad capacitor is connected in series. Find the improved power factor and angle of lead or lag.

Industry photo. Design, construction and maintenance of installations such as this substation provide careers for persons with a proper training in electricity.

6

INTRODUCTION TO TRANSFORMERS

After studying this chapter, you will be able to:

☐ List two main purposes of transformers.
☐ Describe how core losses are determined from open circuit tests.
☐ Compute secondary voltage from primary voltage and turns ratio.
☐ Compute primary current from secondary current and turns ratio.
☐ Describe how the equivalent resistance and reactance of transformers are determined from short circuit tests.
☐ Compute the voltage regulation of a transformer.

Transformers come in many sizes. Some power transformers are big as a house. Electronic transformers, on the other hand, can be as small as a cube of sugar. All transformers have at least one coil; most have two although they may have many more.

The usual purpose of transformers is to change the level of voltage. But sometimes they are used to isolate a load from the power source.

TYPES OF TRANSFORMERS

Standard power transformers have two coils. These coils are labeled PRIMARY and SECONDARY. The primary coil is the one connected to the source. The secondary coil is the one connected to the load. There is no electrical connection between the primary and secondary. The secondary gets its voltage by induction.

The number of volts induced into the secondary is usually different than that applied to the primary. For example, Fig. 6-1 shows pad-mounted STEP-DOWN transformers designed to be used for reducing high voltage and distributing it to commercial buildings at a lower voltage. (A step-down transformer lowers the voltage.) Units like the ones pictured may have as much as 34,500 volts in the primary and as little as 5000 volts in the secondary.

About the only place you will see a STEP-UP transformer is at the generating station. Typically, electricity is generated at 13,800 volts. It is stepped up to 345,000 volts for transmission. The next stop is the substation where it is stepped down to distribution levels, around 15,000 volts. Large substation transformers,

Fig. 6-1. Substation transformers at a manufacturing plant await shipment to customers. (Standard Transformer Co.)

Fig. 6-2. Bank of transformers at an electrical substation step voltage down to distribution levels of 15,000 volts. (Westinghouse Electric Corp., Large Power Transformer Div.)

Fig. 6-3. Step-down transformers of this type are seen mounted on power line poles. (Standard Transformer Co.)

like the ones shown in Fig. 6-2, have cooling fins to keep them from overheating. Other transformers are located near points where the electric power is used, Fig. 6-3.

TRANSFORMER CONSTRUCTION

The coils of a transformer are electrically insulated from each other. There is a magnetic link, however. The two coils are wound on the same core. Current in the primary magnetizes the core. This produces a magnetic field in the core. The core field then affects current in both primary and secondary.

There are two main designs of cores:
1. The CORE type, shown in Fig. 6-4, view A,

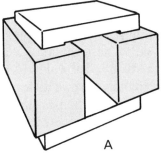

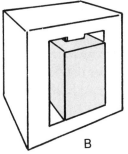

A

B

Fig. 6-4. Two basic types of transformer. A—Core type has the core inside the coils. B—Shell type has the core on the outside or surrounding the coil.

has the core inside the windings.

2. The SHELL type, shown in Fig. 6-4, view B, has the core outside.

Smaller power transformers are usually the core type. The very large transformers are the shell type. There is no difference in their operation, however.

Coils are wound with copper wire. The resistance is kept as low as possible to keep losses low.

IDEALIZED TRANSFORMERS

Transformers are very efficient. The losses are often less than 3 percent. This allows us to assume that they are perfect in many computations.

Perfect means that the wire has no resistance. It also means that there are no power losses in the core.

Further, we assume that there is no flux leakage. That is, all of the magnetic flux links all of the turns on each coil.

EXCITATION CURRENT

To get an idea of just how small the losses are, we can take a look at the EXCITATION CURRENT. Assume that nothing is connected to the secondary. If you apply rated voltage to the primary, a small current flows. Typically, this excitation current is less than 3 percent of rated current.

Excitation current is made up of two parts. One part is in phase with the voltage. This is the current that supplies the power lost in the core. Core losses are due to EDDY CURRENTS and HYSTERESIS.

Eddy currents circulating in the core result from induction. The core is, after all, a conductor within a changing magnetic field.

Hysteresis loss is caused by the energy used in lining up magnetic domains in the core. The alignment goes on continuously, first in one direction, then in the other.

The other part of the excitation current magnetizes the core. It is this magnetizing current that supplies the "shuttle power." *Shuttle power is power stored in the magnetic field and returned to the source twice each cycle.* Magnetizing current is quadrature (90 degrees out of phase) with the applied voltage. The phasor diagram of excitation current is shown in Fig. 6-5.

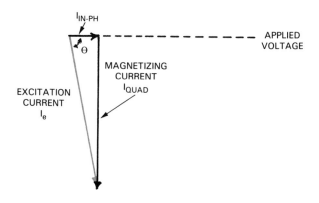

Fig. 6-5. Phasor diagram for excitation current. Magnetizing current is 90 degrees out of phase with applied voltage.

OPEN CIRCUIT TEST

Core losses are a very small part of the total power handled by a transformer. They would be hard to measure accurately while a transformer is in use. Core losses can be determined, however, by what is called an "open circuit test."

The amount of core loss is proportional to the magnetic field. The magnetic field, in turn, is proportional to applied voltage. If you apply full voltage to the primary and leave the secondary open, you set up the field in the core. This magnetic field is the same as it would be if you loaded the secondary. Therefore, the eddy current and hysteresis losses are the same as in a loaded transformer.

The current you measure, however, is all excitation current. Fig. 6-6 shows a diagram for hooking up an open circuit test. Power used by the wire's resistance is ignored. Current is so small and resistance so low that calculations are not affected.

Example 1

A voltage of (V_{PRI}) of 120 volts is applied for the open circuit test shown in Fig. 6-6. The ammeter reads 0.335 amperes of excitation current (I_e). The wattmeter reads 8.0 watts of active power. The rated current of the transformer (I_{FL}) is 10 amperes. Note that excitation current is 3.35 percent of rated current. Find the in-phase current (I_{IN-PH}), the magnetizing current (I_M), the phase angle (Θ) and the power factor of the exciting current.

a. Compute the equivalent resistance of the core losses.

$$P = \frac{V_{PRI}^2}{R_{EQ}}, \text{ therefore } R_{EQ} = \frac{V_{PRI}^2}{P}$$

$$R_{EQ} = \frac{120^2}{8} = \frac{14,400}{8} = 1800 \ \Omega$$

b. Compute the in-phase part of the exciting current.

$$P = I_{IN-PH}^2 R_{EQ}, \text{ therefore } I_{IN-PH} = \sqrt{P/R_{EQ}}$$

$$I_{IN-PH} = \sqrt{8/1800} = \sqrt{.0044}$$
$$= 0.067 \ \underline{/0°} \text{ A}$$

NOTE: The equivalent circuit for the core losses and magnetizing current is shown in Fig. 6-7. Remember, there is really no physical junction inside the transformer. This is just a way of accounting for the exciting current. We can now put numbers on the exciting current phasor diagram of Fig. 6-5. The hypotenuse is the total excitation current. The base is the in-phase current.

c. Find the power factor.

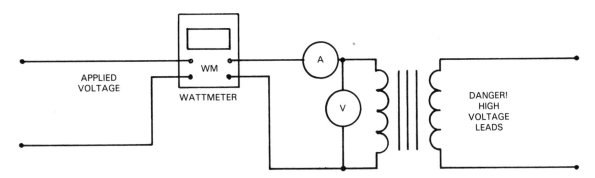

Fig. 6-6. During the open circuit test, the ammeter and voltmeter should be removed when measuring watts. Do you know why?

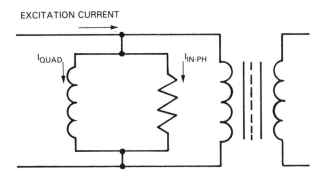

Fig. 6-7. Equivalent circuit of an unloaded transformer.

$$PF = \cos \Theta = \frac{I_{\text{IN-PH}}}{I_e} = \frac{0.067}{0.335}$$

$$PF = 0.2$$

d. Find the phase angle.
From the "trig" tables: $\Theta = 78.5°$

e. Compute the quadrature magnetizing current.

$$I_M = I_e \times \sin \Theta = 0.335 \times .98$$
$$= 0.33 \underline{/-90°} \text{ A}$$

SECONDARY WINDING

Magnetizing current is very important to the operation of a transformer. It causes a magnetic field in the core, Fig. 6-8, which expands and collapses in a sine wave pattern. The flux of this field links the secondary coil. The changing flux linkages cause electron motion in the wire. A voltage appears across the secondary terminals.

It is important to note that the movement of electrons merely produces a voltage. There is no current flow in the secondary as long as its terminals are open. The secondary is, in fact, a voltage source. Current will flow only if there is an external path between the secondary terminals. This is the LOAD CURRENT.

VOLTAGE INDUCED IN THE SECONDARY

The core flux links the primary coil as well as the secondary. In the primary this is simply self-inductance. In our computation, we assume a perfect coil. That is, the induced back voltage exactly equals the applied voltage. For example,

an applied voltage of 120 volts will self-induce a back voltage in the primary of 120 volts.

The induced back voltage is the sum of the voltages induced in each turn, or loop, of the coil. If a 12-turn coil has a back voltage of 120 volts, there must be 10 volts induced into each turn.

The secondary is linked by this same flux. Therefore, it must have the same voltage induced in each turn. If 10 volts are induced in each turn of the primary, there must be 10 volts induced in each turn of the secondary. A secondary of six turns, then, has 60 volts across its terminals.

TURNS RATIO

The voltage applied to the primary coil divided by the number of turns gives you volts per turn. This is equal to the volts per turn of the secondary. Mathematically, the equation is:

$$\frac{E_{\text{PRI}}}{N_{\text{PRI}}} = \frac{E_{\text{SEC}}}{N_{\text{SEC}}}$$

E_{PRI} is the voltage applied to the primary.
N_{PRI} is the number of turns on the primary.
E_{SEC} is the voltage across the secondary.
N_{SEC} is the number of turns on the secondary.

This equation assumes that the induced back voltage in the primary equals the applied voltage. The equation can be rewritten as:

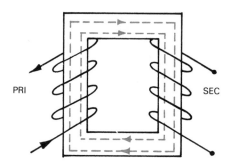 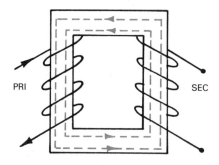

Fig. 6-8. Expanding and collapsing magnetic field (caused by primary current) induces voltage into the secondary. Magnetic field is shown in color.

$$E_{PRI} = E_{SEC} \times (\frac{N_{PRI}}{N_{SEC}})$$

The ratio between the primary and secondary turns $(\frac{N_{PRI}}{N_{SEC}})$ is called the TURNS RATIO.

Normally, you do not know the number of turns of the coils. The manufacturer gives you the turns ratio. To simplify equations, $\frac{N_{PRI}}{N_{SEC}}$ is represented by the letter "a." The equation then becomes:

$$E_{PRI} = E_{SEC} \times a \quad \text{or}$$
$$E_{SEC} = \frac{E_{PRI}}{a}$$

Example 2

The step-down transformer shown in Fig. 6-9 has a turns ratio of 2. Primary voltage is 120 volts. Find the secondary voltage.

$$E_{SEC} = \frac{E_{PRI}}{a} = \frac{120}{2} = 60 \text{ V}$$

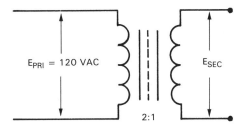

Fig. 6-9. A step-down transformer reduces voltage in proportion to the turns ratio.

Example 3

The transformer of *Example 2* is turned around to become a step-up transformer. Now the turns ratio is 0.5. Find the secondary voltage when there is 120 volts applied to the transformer.

$$E_{SEC} = \frac{E_{PRI}}{a} = \frac{120}{0.5} = 240 \text{ V}$$

As you can see from these examples, either coil can be the primary. It depends on which coil is connected to the source. Therefore, transformers are marked to show HIGH SIDE and LOW SIDE. *High side terminals are marked H. Low side terminals are marked X.*

CURRENT FLOW IN SECONDARY

There is no current in the secondary if there is no electrical path between its terminals. The only current in the primary is the excitation current. Now, let us connect a resistance load to the secondary and see what happens, Fig. 6-10.

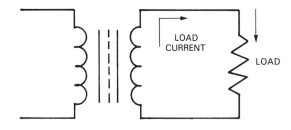

Fig. 6-10. Load current passes through the secondary winding of a transformer.

The load is in series with the secondary coil. Therefore, load current passes through the wire of this coil. Whenever current flows through a wire, a magnetic field is set up. The secondary field is 180 degrees out of phase with the primary field. It opposes the primary field head-on.

When these two fields combine, several things happen all at once.
1. The primary field is weakened. This reduces primary back voltage.
2. Additional current is allowed to flow through the primary. This strengthens the primary field.
3. Now the secondary field is canceled by the extra primary field. The core is left with the original field from the magnetizing current.

CURRENT DRAWN FROM THE SOURCE

As the load changes, there is no change in core flux. The secondary field is proportional to the current times the number of turns ($N_{SEC} \times$

I_{SEC}). The primary field, which equals the secondary field, is proportional to its number of turns and current ($N_{PRI} \times I_{PRI}$). The transformer primary automatically draws the right amount of current from the source. At all times, $N_{PRI} \times I_{PRI} = N_{SEC} \times I_{SEC}$. The excitation current is ignored since it is such a small part of total current.

CURRENT TRANSFORMATION

The current transformer equation,

$$N_{PRI} \times I_{PRI} = N_{SEC} \times I_{SEC}$$

can be rewritten as:

$$I_{SEC} = I_{PRI} \times \left(\frac{N_{PRI}}{N_{SEC}}\right)$$

The expression, $\frac{N_{PRI}}{N_{SEC}}$, is the turns ratio, "a." Therefore the equation can be written as:

$$I_{SEC} = I_{PRI} \times a \text{ or}$$
$$I_{PRI} = \frac{I_{SEC}}{a}$$

Example 4

The power transformer shown in Fig. 6-11 has 14,560 volts applied to the primary. Secondary voltage is 208 volts. The load is drawing 210 amperes. Find the primary current.

a. Compute the turns ratio.

$$a = \frac{N_{PRI}}{N_{SEC}} = \frac{E_{PRI}}{E_{SEC}}$$
$$a = \frac{14,560}{208} = 70$$

b. Compute primary current.

$$I_{PRI} = \frac{I_{SEC}}{a} = \frac{210}{70} = 3 \text{ A}$$

From this example, you can see why power companies step up their voltage so high. The higher the voltage, the lower the current.

EQUIVALENT CIRCUIT OF A TRANSFORMER

Transformers can change voltage and current levels. The thing they cannot change is power. *Except for small losses, power out always equals power in.* What is more, this is just as true for apparent and reactive power as it is for active power.

RESISTANCE TRANSFORMATION

To show how a transformer supplies power, we will take a simple example. Refer to Fig. 6-12. A 2 to 1 step-down transformer has 120 volts applied to primary. There is, therefore, 60 volts across the secondary. A 5 ohm resistance on the secondary draws 12 amperes. Primary current is, therefore, 6 amperes.

There are no reactances involved. We can easily compute power by multiplying voltage

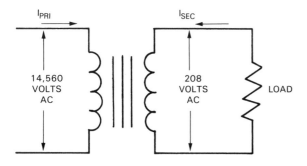

Fig. 6-11. In a step-down transformer, the secondary has low voltage and high current, while the primary has high voltage and low current.

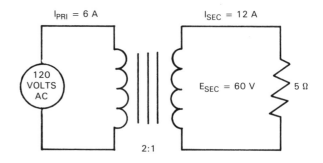

Fig. 6-12. Except for small losses, output power equals input power.

times current. Power used by the load is (60 × 12) 720 watts. The power supplied by the source is (120 × 6) 720 watts.

Power drawn from the secondary causes less back voltage in the primary. Less back voltage means larger primary current. The power source does not know or care why it has to supply the current. As long as voltage and current are in phase, the opposition from back voltage looks the same as opposition from resistance.

What we have in the primary circuit is an opposition that permits a current of 6 amperes when 120 volts is applied. The ratio between voltage and current gives us the equivalent resistance in the primary circuit. In this case it is ($\frac{120}{6}$) or 20 ohms.

The equivalent circuit is shown in Fig. 6-13. The power source does not even know the transformer is there. All it "sees" is a resistance. Notice that the value of this resistance is different from the load resistance. It has been *transformed* by the turns ratio of the transformer. Often we need to know what a load looks like to the power source. This is called *referring* the resistance (or reactance) to the primary.

To refer any resistance, reactance, or impedance to the primary, simply multiply it by the turns ratio squared. We saw that the 5 ohm load on the secondary of the 2:1 transformer looked like 20 ohms to the power source:

$$R_{EQ} = R_{LOAD} \times a^2 = 5 \times 2^2 = 5 \times 4$$
$$R_{EQ} = 20 \ \Omega.$$

Similarly, a 15 ohm load referred to the primary has an equivalent resistance of (15 × 4) 60 ohms.

Resistance can also be referred from the primary to the secondary. To refer a resistance (or reactance) to the secondary, divide by the turns ratio squared.

$$R_{EQ(SEC)} = \frac{R_{PRI}}{a^2}$$

REACTANCE TRANSFORMATION

If the load contains a capacitor or coil, secondary current is out of phase with secondary voltage. This does not make any difference inside the core, however. The secondary's magnetic field still changes the back voltage in the primary. This, in turn, varies the current drawn from the source. The secondary's phase angle is reflected in the primary without change.

A transformer with reactive load is shown in Fig. 6-14, view A. Its equivalent circuit is shown in Fig. 6-14, view B. We will show the transformation results in different voltage, current, and impedance without changing the power or phase angle.

The secondary circuit consists of 4 ohms of resistance connected in series with 6 ohms of inductive reactance and 3 ohms of capacitive reactance. Applied voltage is 120 volts across the primary. This is a step-down transformer with a turns ratio of 4 to 1.

First, let us take a look at the secondary circuit. Output voltage is ($\frac{120}{4}$) 30 volts. The impedance triangle is shown in Fig. 6-15. If you construct a vector from a resistance of 4 ohms and a reactance of 3 ohms (6-3), the result is an impedance of 5 ohms.

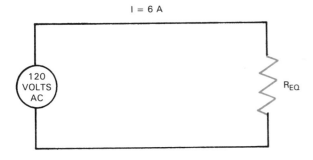

I = 6 A

120 VOLTS AC

R_{EQ}

Fig. 6-13. Simplified equivalent circuit of an idealized transformer with resistive load.

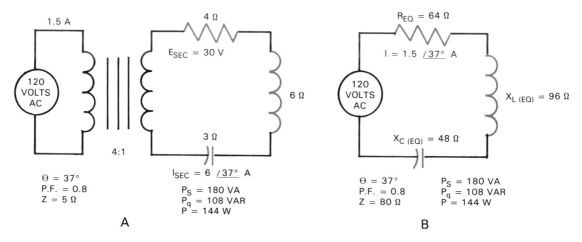

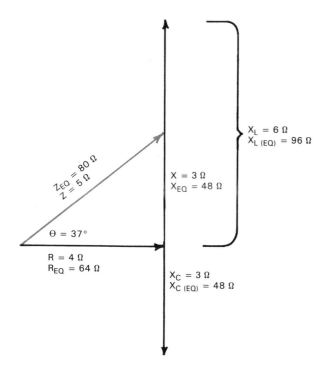

Fig. 6-14. Transformer with reactive load. A—Actual circuit. B—Equivalent circuit with resistance and reactances referred to the primary.

Fig. 6-15. Impedance phasor diagram for the circuit of Fig. 6-14.

Now let us see what all this looks like to the power source. Refer to Fig. 6-14, view B. Following are the equivalent values when referred to the primary. Each secondary value is multiplied by a^2 ($4^2 = 16$):

Resistance (4×16) = 64 Ω
Inductive reactance (6×16) = 96 Ω
Capacitive reactance (3×16) = 48 Ω
Net reactance ($96 - 48$) = 48 Ω

THE IMPEDANCE DIAGRAM

We can find the phase angle by computing its tangent:

$$\tan \Theta = \frac{X_{(EQ)}}{R_{(EQ)}} = \frac{48}{64} = 0.75$$

The angle whose tangent is 0.75 is 37 degrees. This is the same as the secondary phase angle. Therefore, we know that the power factor (cos 37 degrees) is 0.8.

The referred impedance is a^2 times load impedance (5×16). It is also the referred resistance divided by cos 37 degrees ($\frac{64}{0.8}$). Either way, it is 80 ohms. 120 volts across 80 ohms produces a primary current of 1.5 $\underline{/37°}$ amperes.

Also, $I_{PRI} = \dfrac{I_{SEC}}{a} = \dfrac{6}{4} = 1.5 \ \underline{/37°}$ A.

The power factor, which is the cosine of the phase angle, is $\dfrac{4}{5} = 0.8$. The phase angle, therefore, is 37 degrees; current is $(\dfrac{30}{5})$ 6 $\underline{/-37°}$ A. Apparent power is (30×6) 180 volt-amperes. Active power is ($180 \times .8$) 144 watts. Reactive power is (180×0.6) 108 VAR.

Apparent power is 180 volt-amperes (120 × 1.5). Active power is 144 watts (180 × 0.8). Reactive power is 108 VAR (180 × 0.6). So you see, power drawn from the source is the same as that supplied to the load. The back voltage in the primary determines how much current will be drawn from the source. The value of this back voltage is, in turn, determined by secondary current.

REALISTIC TRANSFORMERS

Most transformer calculations assume a perfect transformer. There are times, though, when you need to know how much power is really lost. From open circuit tests, you can compute core losses. Core losses are the same at all loads. Two other classes of losses, however, vary with load. These are *resistance* and *leakage reactance*.

RESISTANCE

Wire resistance is kept as low as possible. Still, you cannot eliminate it entirely. It is like having a resistor in series with the coil.

The resistance of the primary wire is shown in Fig. 6-16. Its effect is to reduce the value of the voltage that gets transformed. The voltage drop across this resistance plus the back voltage equals the applied voltage. This voltage drop gets larger as more current flows. This makes secondary voltage decrease as load is applied.

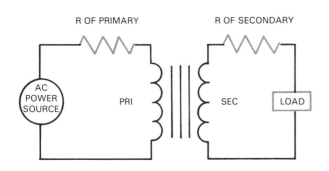

Fig. 6-16. Resistance of the conductors of each winding produce copper losses.

Fig. 6-16 also shows the resistance of the secondary wire. It, too, produces a voltage drop that increases with load. Power losses that are produced by primary and secondary resistances are called COPPER LOSSES.

LEAKAGE REACTANCE

We have been assuming that all of primary flux linked the secondary. Actually, not all of the flux stays in the core. Some of it leaks off into the air. This flux links the primary but not the secondary. In the primary it helps produce the back voltage. It does not, however, help to produce secondary voltage. It is like having an inductor in series with the primary.

The leakage reactance is shown in Fig. 6-17. Leakage flux gets larger as load increases. The increase in flux leakage makes the reactance increase as more current passes through the primary. Some of the secondary flux also leaks off into the air. It does not link the primary so it does not get in on the transformer action. It also produces a voltage drop that increases with the load.

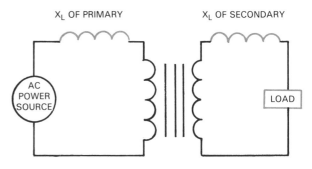

Fig. 6-17. Leakage reactance is like having an inductance in series with each winding.

EQUIVALENT CIRCUIT OF A REALISTIC TRANSFORMER

What the power source actually "sees" as it looks at a loaded transformer is shown in Fig. 6-18. First, there is the parallel branch composed of a resistance and an inductance.

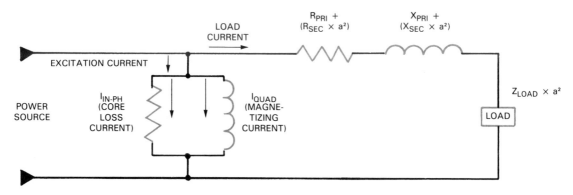

Fig. 6-18. Complete equivalent circuit of a realistic transformer, including losses, coil resistances, and leakage reactances.

There is not really a branch like this in a transformer. However, core loss current and magnetizing current act as though there were.

The resistance represents core losses. Magnetizing the core requires a quadrature current, like an inductance. Current from the source is made up of load current and excitation current.

Next, the power source "sees" a resistance. This is the actual value of the primary wire resistance plus the equivalent secondary wire resistance. You see, the resistance of the wire in the secondary has been "referred" to the primary. The source sees secondary resistance times the turns ratio squared.

The inductance is the leakage inductance of the primary plus the "referred" secondary leakage inductance.

Finally, there is the load. The load also has been referred to the primary. Therefore, the source sees load impedance times the turns ratio squared.

PREDICTING SECONDARY VOLTAGE

It is easy to predict no-load secondary voltage. Since excitation current is so small, there are almost no copper or reactance losses. It is more difficult when the transformer is loaded. The resistance produces a voltage drop which increases with load. Also, the value of the leak-

age reactance affects both value and phase angle of the secondary voltage.

There would be no problem if the rating of a transformer were low enough. You could simply connect a load and measure secondary voltage. This would not be practical, however, with large power transformers. You might need to borrow an entire city for the test. Engineers need to know the equivalent resistance and equivalent reactance. Then they can compute secondary, or load, voltage. The way to find these values is to perform a short circuit test.

SHORT-CIRCUIT TEST

Short-circuit tests must be done carefully. First, the transformer low side is shorted out. A variable voltage is applied to the high side. Starting at zero you bring the applied voltage up very slowly. The diagram of a short circuit test is shown in Fig. 6-19. Around 5 percent of rated high-side voltage, rated currents will flow in both primary and secondary coil.

You do not have to worry about core losses. The secondary field cancels the primary field, as it does in a loaded transformer. What is left is the core field which is proportioned to applied voltage. That means you have only 5 percent of normal core flux. This is not enough to affect the wattmeter reading. The wattmeter measures only active power. This is the power used by the copper losses. You can compute apparent power from the voltage and amperage. With this information, you can compute the values of equivalent resistance and equivalent reactance.

GENERAL SAFETY

Great care must be used in working around or on transformers. To prevent damage to equipment or injury to persons, the secondary winding of current transformers must be shorted out before connecting or disconnecting equipment to it. Most transformers have a shorting device for this purpose.

If transformers are being inspected or examined they must be removed from all sources of electric power at both primary and secondary terminals. Windings should be grounded. Fuses should be removed at power source. Circuit breakers must be tripped and protected against accidental turning on of power.

Example 5

The equivalent circuit of a short circuit test on a 1000 VA transformer is shown in Fig. 6-20. This is a 2:1 step-down transformer. Rated primary current is 4.17 amperes. The voltage, at rated current, is 9.3 volts. Active power is 36 watts. Find equivalent resistance, impedance, and reactance as seen by the power source.

NOTE: Since this is a step-down trans-

former, the high side is the primary. Measurements have been made on the high side. Therefore, the computations give us the values as seen by the power source.

a. Compute the equivalent primary resistance, $R_{EQ(P)}$.

$$R_{EQ(P)} = \frac{P}{I^2} = \frac{36}{(4.17)^2} = 2.07 \ \Omega$$

b. Compute the equivalent primary impedance, $Z_{EQ(P)}$.

$$Z_{EQ(P)} = \frac{E}{I} = \frac{9.3}{4.17} = 2.23 \ \Omega$$

c. From the phasor diagram of Fig. 6-21, find the phase angle, Θ.

$\Theta =$ angle whose cosine is $\dfrac{R}{Z}$

$$\cos \Theta = \frac{2.07}{2.23} = .928 \therefore \Theta = 21.8°$$

d. Compute the equivalent primary reactance $X_{EQ(P)}$.

$$X_{EQ(P)} = Z \sin 21.8° = 2.23 \times .37$$
$$= 0.83 \ \Omega$$

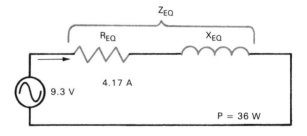

Fig. 6-20. Circuit showing the equivalent resistance and the leakage reactance of both coils as determined by the short-circuit test.

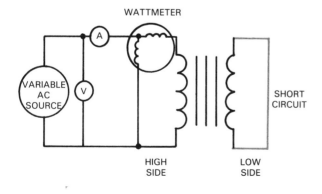

Fig. 6-19. Circuit diagram for a short-circuit test of a transformer.

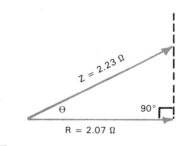

Fig. 6-21. Phasor diagram for the circuit of Fig. 6-19.

VOLTAGE REGULATION

The output voltage of a transformer will be different at different loads. This is because load current also flows through the secondary coil. The larger this current, the larger the IR drop caused by the equivalent resistance. In *Example 5,* we compute the equivalent primary resistance. However, the output voltage sees equivalent secondary resistance. Therefore, we have to refer this resistance to the secondary. Then we can see the effect this resistance has on output voltage.

The equivalent primary resistance must be referred to the secondary. Fig. 6-22 shows the equivalent circuit as seen by the load. *To refer any resistance or reactance to the secondary, you must divide the primary value by the turns ratio squared.*

DEFINITION OF VOLTAGE REGULATION

The idealized transformer had no resistance or leakage reactance. Its full load voltage is the same as its no-load voltage. It, therefore, would have zero *voltage regulation. Voltage regulation tells us how much secondary voltage changes. It is the ratio of change to full load voltage.* This ratio is then multiplied by 100 and given as a percentage. The percentage voltage regulation equation is:

$$\% \text{ V.R.} = \frac{\text{no load V} - \text{full load V}}{\text{full load V}} \times 100$$

The equivalent resistance and reactance of a transformer do not change. You might, therefore, expect the voltage regulation to depend entirely on the value of the load current. The voltage regulation, however, also depends on the power factor of the load. A unity power factor load current produces one value. The same load current with a lagging and leading power factors would produce different values.

REGULATION WITH RESISTIVE LOAD

Always start with full load conditions when computing voltage regulation. You can use phasor addition to compute no-load voltage. Current is in phase with voltage when the load is resistive. The phasor diagram of secondary voltage is shown in Fig. 6-23. First, there is the horizontal vector for full load voltage. Then add the IR voltage drop due to the equivalent secondary resistance. Draw a vector at 90 degrees to the IR vector. This is the back voltage (IX) caused by leakage reactance.

This creates a right triangle whose hypotenuse represents no-load voltage.

Example 6

Consider the 1000 volt-ampere transformer of *Example 5.* Full load voltage is 120 volts. A unity power factor load is applied. Full load current is $(\frac{1000}{120})$ 8.34 amperes. Find the percent voltage regulation.

a. Refer primary resistance and reactance to the secondary.

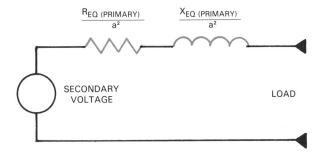

Fig. 6-22. The equivalent circuit of the resistance and leakage reactance as seen by the load.

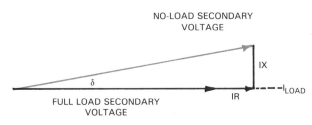

Fig. 6-23. Phasor diagram of the secondary voltage when a unity power factor (resistive) load is connected.

$$R_{EQ(S)} = R_{EQ(P)} \times \frac{1}{a^2} = \frac{2.07}{4} = 0.52 \ \Omega$$

$$X_{EQ(S)} = X_{EQ(P)} \times \frac{1}{a^2} = \frac{0.83}{4} = 0.21 \ \Omega$$

b. Compute IR voltage.
   $8.34 \times 0.52 = 4.34$ V
c. Compute the total in-phase voltage.
   $E_{IN\text{-}PH} = 120 + 4.34 = 124.34$ V
d. Compute the IX voltage.
   $8.34 \times 0.21 = 1.75$ V
e. Find the δ angle.
   $$\tan \delta = \frac{1.75}{124.34} = 0.014$$
   therefore $\delta = 0.8°$
f. Compute the no-load voltage.
   $$E_{NL} = \frac{E_{IN\text{-}PH}}{\cos 0.8} = \frac{124.34}{0.9999} = 124.35 \text{ V}$$
g. Compute percent of voltage regulation.
   $$VR = \frac{E_{NL} - E_{FL}}{E_{FL}} \times 100$$
   $$= \frac{124.35 - 120}{120} \times 100$$

$$VR = \frac{4.35}{120} \times 100 = 3.6\%$$

REGULATION WITH INDUCTIVE LOAD

When load current lags secondary voltage, both resistance voltage drop and inductance voltage are at a different phase angle with secondary voltage. Refer to Fig. 6-24. Again, start with the same horizontal vector for terminal voltage. A dotted line represents current, which is lagging voltage. A vector, representing the IR drop, is drawn at the end of the no-load vector. Note the direction of the IR vector. It is parallel with the current line. Also note the IX vector. It is vectorially added to the no-load and IR voltages. Its angle is 90 degrees to the current line.

The phasor sum of IX voltage and the in-phase voltage is no-load voltage. It is represented by a vector drawn from the beginning of the first vector to the end of the last.

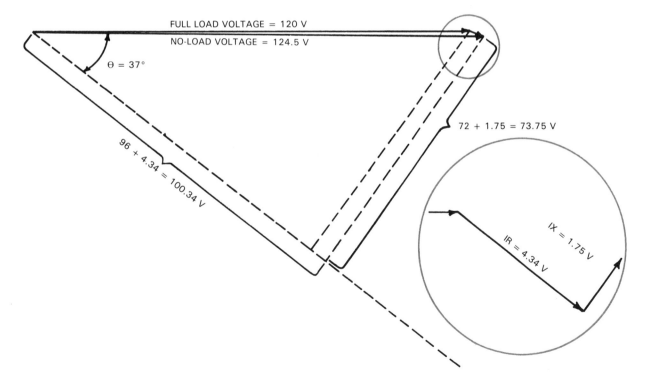

FULL LOAD VOLTAGE = 120 V
NO-LOAD VOLTAGE = 124.5 V
$\theta = 37°$
96 + 4.34 = 100.34 V
72 + 1.75 = 73.75 V
IR = 4.34 V
IX = 1.75 V

Fig. 6-24. Phasor diagram of the secondary voltage when a lagging power factor (inductive) load is connected.

Example 7

In the transformer of *Example 6*, secondary equivalent resistance is 2.07 ohms. Secondary equivalent reactance is 0.83 ohms. Full load voltage and current are 120 volts and 8.34 amperes. An inductive load having 0.8 lagging power factor is connected to the secondary. Find the percentage of voltage regulation.

a. Refer to Fig. 6-25. First, compute E_{FL} cos Θ 120 × 0.8 = 96 V.

b. IR voltage was computed in *Example 6* as 4.34 volts. This is added directly to the voltage. 96 + 4.34 = 100.34 V. This is the adjacent side of the right triangle.

c. The angle that has the cosine of 0.8 is 37 degrees. The sine of 37 degrees is 0.6. Compute E_{FL} sin Θ 120 × 0.6 = 72 V.

d. IX voltage was computed in *Example 6* as 1.75 volts. This is added directly to the previous voltage: 72 + 1.75 = 73.75 V. This is the opposite side of the right

triangle (hypotenuse).

e. Compute the tangent of the angle.
$$\frac{73.75}{100.34} = 0.735$$

f. The cosine of the angle whose tangent is 0.735 is 0.806.

g. Compute the no-load voltage.
$$E_{NL} = \frac{100.34}{0.806} = 124.5 \text{ V}$$

h. Compute the percent voltage regulation.
$$\text{V.R.} = \frac{E_{NL} - E_{FL}}{E_{FL}} \times 100$$
$$= \frac{124.5 - 120}{120} \times 100$$
$$= \frac{4.5}{120} \times 100 = 3.75\%$$

REGULATION WITH CAPACITIVE LOAD

You seldom run into a capacitive load. Still, it is interesting to see what happens to voltage regulation.

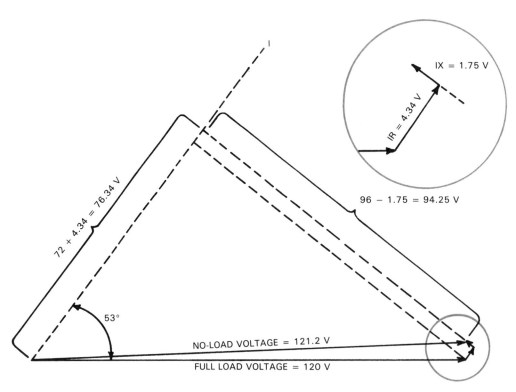

Fig. 6-25. Phasor diagram of the secondary voltage when a leading power factor (capacitive) load is connected.

Current in the secondary leads the voltage. The induced voltage of the inductive reactance may not oppose secondary voltage. It can acutally make up for some of the IR voltage drop. A transformer would have better voltage regulation with a leading power factor load.

Example 8

The transformer of *Example 7*, is given a capacitive load having 0.6 leading power factor. Find percent voltage regulation.

a. Refer to the phasor diagram that is shown in Fig. 6-25.

NOTE: The adjacent side is the sum of E_{FL} cos Θ and the IR drop.

b. Compute E_{FL} cos Θ.
$120 \times 0.6 = 72$ V.

c. The total adjacent side therefore is:
$72 + 4.34 = 76.34$ V
NOTE: The opposite side is the difference between E_{FL} sin Θ and $I \times X$. Since cos Θ is 0.6 then Θ is 53 degrees and sin Θ is 0.8.

d. Compute E_{FL} sin Θ.
$120 \times 0.8 = 96$ V

e. The total opposite side therefore is:
$96 - 1.75 = 94.25$ V

f. Compute the tangent of the angle.
$$\frac{94.25}{76.34} = 1.235$$

g. The cosine of the angle whose tangent is 1.235 is 0.63.

h. Compute the no-load voltage.
$$E_{NL} = \frac{76.34}{0.63} = 121.2 \text{ V}$$

i. Compute the percent voltage regulation.
$$VR = \frac{121.2 - 120}{120} \times 100$$
$$= \frac{1.2}{120} \times 100 = 1\%$$

TRANSFORMER EFFICIENCY

Efficiency is the ratio between output and input. It might seem easy to compute efficiency. Just measure input power and output power.

The problem is that transformers are so efficient. Your meter might be 3 percent off. A transformer might have only 3 percent losses (97 percent efficiency). That means your computation could be 100 percent wrong.

A better way is to use the losses you computed with the open and shorted tests. You measure the output power. Then you add the losses to the output power to give you input power. The equation is:

$$\% \text{ Efficiency} = \frac{\text{output}}{\text{output} + \text{losses}} \times 100$$

What are the losses? First, there is the core loss. This does not change with load. Second, there is the copper loss. This is the loss caused by the primary equivalent resistance. It depends on primary current ($P = I^2R$). Copper loss is different for each value of current. Therefore, when you compute efficiency, it must be for a specific load with a specific power factor.

Example 9

A voltage of 240 volts is applied to the primary of a 2:1 step-down 1 kVA transformer. A unity power factor load is drawing full load rated current of 8.33 amperes. Core losses are 10 watts. Equivalent primary resistance is 2.07 ohms. Find the efficiency.

a. Compute output power.
$P = E \times I = 120 \times 8.33 = 1000$ W

b. Compute copper losses at full load.
$P = I_{PRI}^2 R_{EQ} = (4.17)^2 \times 2.07$
$= 17.34 \times 2.07 = 36$ W

c. Compute total losses.
$36 + 10 = 46$ W

d. Compute input power (output plus losses).
$1000 + 46 = 1046$ W

e. Compute transformer efficiency.
$$\% \text{ Eff.} = \frac{\text{Output power}}{\text{Input power}} \times 100$$
$$= \frac{1000}{1046} \times 100 = 95\%$$

SUMMARY

Transformers consist of two or more coils, insulated electrically but linked magnetically.

The coil connected to the power source is called the primary; the one to which the load is connected is called the secondary.

The ratio between the number of turns on the primary and the number of turns on the secondary is called the turns ratio.

Secondary voltage equals primary voltage times the turns ratio.

Primary current equals secondary current times the turns ratio.

Primary ampere-turns equals secondary ampere-turns.

Exciting current and core losses are determined from open circuit tests.

Equivalent resistance, reactance, impedance, and copper losses are determined from short circuit tests.

A load impedance is referred to the primary by multiplying it by the turns ratio squared.

Voltage regulation of a transformer is the percentage increase in voltage from full load to no load.

TEST YOUR KNOWLEDGE

1. (Circle the correct answers.) In a two coil transformer, the coil with the greater number of turns is the (high, low) side. The one with the fewer number of turns is the (high, low) side.
2. Explain the function of a step-up transformer and of a step-down transformer.
3. Describe the process by which the secondary of a transformer is able to draw current from the power source.
4. (Circle the correct answer.) The current in the primary of a step-down transformer is (greater, less) than the secondary current.
5. What is meant by "voltage regulation"? How is it computed?
6. Power loss in a transformer is caused by (check all correct answers):
   a. Resistance of the conductor (copper loss).
   b. Core loss.
   c. Leakage reactance.
   d. All of the above.
   e. None of the above.

PROBLEMS

1. Given: A transformer has 100 turns on the high side and 50 turns on the low side. Find the turns ratio.
2. Given: A step-down transformer has a turns ratio of 10:1. Secondary voltage is 120 volts and load current is 5 amperes. Find the primary current.
3. Given: The transformer of *Problem 2* has a 24 ohm load connected. What is the value of the load referred to the primary?
4. Given: A 1000 volt-ampere transformer rated at 480 volts, 2.08 amperes on the high side. An open circuit test produced a power reading of 9.6 watts and 0.1 amperes of exciting current. Find the in-phase and quadrature components of the exciting current.
5. Given: The transformer of *Problem 4* with a low side rating of 60 volts, 16.68 amperes. A short circuit test taken with the low side shorted produced the following readings: 21 volts; 2.08 amps; 40.6 watts. Find the equivalent resistance, impedance, and reactance.
6. Given: The no-load voltages of a transformer secondary is 64 volts. When loaded, the voltage drops to 60 volts. Find the voltage regulation.

Construction of transformer. Huge coils are carefully placed over cores. Coils rest on yoke pads already in position.   (McGraw-Edison Co.)

Besides transformers, substations also contain huge circuit breakers like these.

7

TRANSFORMER APPLICATIONS

After studying this chapter, you will be able to:

☐ Define the terms additive polarity and subtractive polarity.
☐ Explain the principle of three-phase power.
☐ Compute line and phase voltages and currents in delta and wye connections.
☐ Describe the operation of an autotransformer.

Most transformers are used to step voltage up or down in alternating current circuits. The terminals, therefore, cannot be marked plus (+) and minus (−) like dc terminals. Their polarity changes every half cycle. The INSTANTANEOUS POLARITY of the secondary coil, however, is determined by the direction of the magnetic field which the primary coil produces. There must, then, be a definite relationship between the instantaneous polarity of the primary and secondary terminals.

TERMINAL MARKINGS

Assume a transformer has one primary coil and one secondary coil. There would be four terminals. The high-side terminals are marked H1 and H2. The low-side terminals are marked X1 and X2. The code is this: terminal X1 is positive at the instant H1 is positive.

Such a transformer is illustrated by Fig. 7-1. Note that the wire of both coils is wrapped around the core in the same direction. Voltage is induced with the same polarity in each coil. The terminals, however, may be arranged in either of two ways. The X1 (low-side) terminal may be directly across from terminal H1 (high-side) or diagonally opposite it.

Transformers are said to have either ADDITIVE or SUBTRACTIVE POLARITY. Subtractive polarity, which is standard, means that the X1 terminal is directly opposite the H1 terminal.

POLARITY TESTS

Some older transformers may have additive polarity but their terminals may not be marked. However, you must know the polarity of the terminals if it is to be connected with other transformers. The following test will tell you if they have a subtractive or an additive polarity:

1. Connect a jumper between one of the high-side terminals (call it H1) and the terminal directly opposite it. This is shown in Fig. 7-2.
2. Apply a voltage to the high-side winding. Any voltage will do, just so it is less than the rated voltage of the transformer.

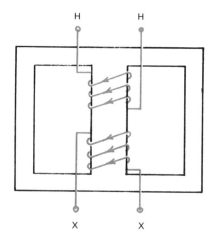

Fig. 7-1. There is a definite relationship between instantaneous polarity of the high-side terminals and the low-side terminals.

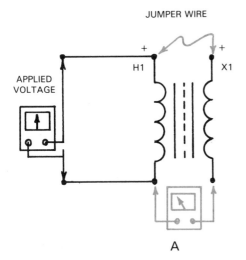

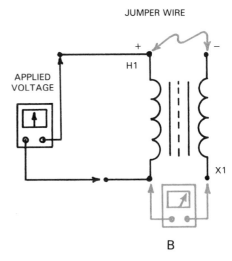

Fig. 7-2. Using the voltmeter to find the polarity of transformers. A—Subtractive polarity. The voltmeter reads the difference between applied high-side voltage and the induced low-side voltage. B—Additive polarity. The voltmeter reads the sum of the applied high-side voltage and the induced low-side voltage.

3. Measure the voltage between H2 and the other low-side terminal.

Subtractive polarity

If the voltmeter reads less than the voltage you applied, the polarity is subtractive. See Fig. 7-2, view A. The terminal directly across from H1 is X1. Terminal X1, remember, always has the same polarity as H1. Therefore, the jumper is between like polarities (plus to plus). The voltages subtract.

Additive polarity

If the voltmeter reads more than the voltage you applied, the polarity is additive. See Fig. 7-2, view B. The terminal directly across from H1 is X2. The jumper connects the two coils series-aiding (plus to minus). The voltages add.

MULTI-COIL TRANSFORMERS

Power transformers often have two or more coils on their secondaries. This allows you to get different voltage or current values. Fig. 7-3 shows a distribution transformer with two identical secondary coils. The terminals of the first coil are X1 and X2. The terminals of the second coil are X3 and X4. Terminal X3 is positive when X1 is positive.

For single-phase connections, it might help to

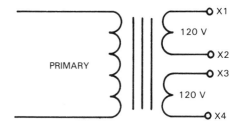

Fig. 7-3. Secondary coils can be connected in different ways to produce different voltage and current ratings.

think of flashlight cells. For double the voltage, you connect minus to plus (series-aiding). For double the current you connect plus to plus (parallel). Assume the primary in Fig. 7-3 has 2400 volts applied. The turns ratio is 20:1 for each of the secondary coils. The transformer is rated for 24 kVA (24,000 volt-amperes). With full load connected, the primary carries 10 amperes.

Fig. 7-4 shows one of the ways this transformer can be used. Separate 100 ampere loads are connected to each coil. Each load has 120 volts across it.

Fig. 7-5 shows terminals X2 and X3 joined. Now you can get 120 volts from X1 and the common and another 120 volts from X4 and the common. Each may have up to 100 amperes of load. Notice, however, that these loads are not independent. The current through one secondary coil affects the current through the other one. Heavy current in coil 1, for example, causes greater primary current. This, in turn, causes an increase in the secondary voltage of the second coil.

Note also, in Fig. 7-5, that you get 240 volts from X1 to X4. All three voltages (two 120 volts and one 240 volts) can be used, as long as the total load is no more than 100 amperes.

Fig. 7-6 shows a parallel connection. Now

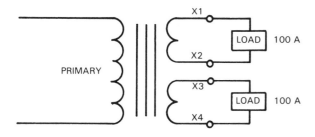

Fig. 7-4. Each secondary coil can be loaded separately.

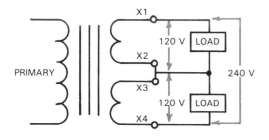

Fig. 7-5. When connected series-aiding, the total voltage is the sum of individual coil voltages.

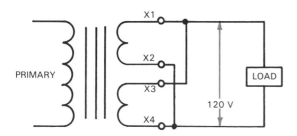

Fig. 7-6. When connected in parallel, the total current capacity is the sum of current from each coil.

there is only one secondary voltage available — 120 volts. However, each coil is capable of supplying 100 amperes. It is just like paralleling two dry cells. A 200-ampere load may be safely connected.

DOT NOTATION

Sometimes, the terminal numbers are shown on drawings. It is more likely, however, that the DOT NOTATION will be used to show relative polarity. The drafter places a dot at the ends of two coils. The terminals with the dots have the same polarity. Use of the dot notation is shown in Fig. 7-7.

TRANSFORMER TAPS

The output voltage of a transformer depends on the number of turns on the secondary. A certain voltage is induced into each turn. You could say the turns are connected series-aiding. The voltage across the secondary is the sum of the voltages across each turn. You can pick off other voltages by making connections in the middle of the secondary coil. These connections are called TAPS.

The transformer shown in Fig. 7-8 has a 50 percent tap. This means the tap is connected midway between the ends. This is a 10:1 step-down transformer with 2400 volts on the primary. Secondary voltages are as follows: X1 to X2 = 120 volts; X2 to X3 = 120 volts; X1 to X3 = 240 volts.

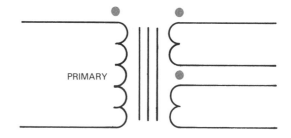

Fig. 7-7. The dots indicate which terminals have the same instantaneous polarity.

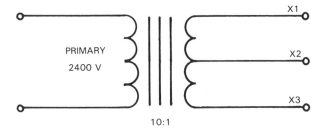

Fig. 7-8. This transformer has a center-tapped secondary.

TAP CHANGERS

As we learned in Chapter 6, the secondary voltage goes down when the load goes up. To make up for this voltage loss, power transformers often have TAP CHANGERS. Fig. 7-9 shows the terminal board used for changing taps. A diagram of the tap-changing principle is shown in Fig. 7-10. Taps are usually placed on the primary winding. With them you change the transformer's turns ratio. A 5 percent tap cuts out 5 percent of the primary turns.

For example, the transformer shown in Fig. 7-10 has 2400 volts on the primary. When it was installed, secondary voltage was 240 volts. However, because of an increase in load, the voltage has dropped to 228 volts. To bring the voltage back up to 240 volts, you use the 5 percent tap.

By cutting out 5 percent of the primary turns, you change the turns ratio from 10:1 to 9.5:1. Now the no-load secondary voltage is 252.6 volts (2400 ÷ 9.5). When loaded, the output voltage is again 240 volts.

THREE-PHASE POWER

Between any two ac power lines there is a voltage which is changing in a sine wave pattern. This is illustrated in Fig. 7-11. Normally, you think of this as a single-phase voltage. It might, however, be one phase of a POLYPHASE voltage system. Polyphase means more than one phase. In polyphase systems, there are several

Fig. 7-9. Terminal board of tap-changing transformer. This one will handle 69,000 volts.

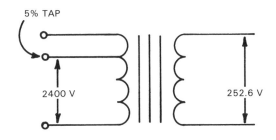

Fig. 7-10. The use of transformer taps changes the effective turns ratio.

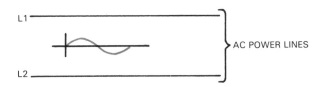

Fig. 7-11. A voltage supply of one phase is represented by one wave form.

voltages which change direction and amplitude at different times.

Electricity supplied by the power company is three-phase. In most homes, only single-phase is used. The three-wire service entrance has two "hot" wires and a "neutral" wire. This is illustrated by Fig. 7-12. The two voltages from the neutral to each of the "hot" lines are in phase.

Three-phase power (usually supplied to factories and commercial buildings) may also enter over three wires. The difference is in the voltage between each pair of wires. There are three separate voltages which have a definite phase relationship.

PHASE RELATIONSHIP

In Fig. 7-13, the three-phase power lines are called L1, L2 and L3. There is a sine wave voltage with a 240-volt RMS value between lines L1 and L2. There are also 240 volts between lines L2 and L3. Further, the other pair of lines, L1 and L3, also have 240 volts between them. The wave forms of these three voltages are drawn on the same time base in Fig. 7-14. Notice that each voltage is 120 electrical degrees from the others.

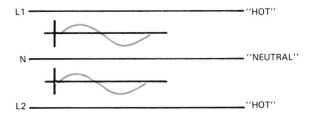

Fig. 7-12. Two in-phase voltages represented by wave forms.

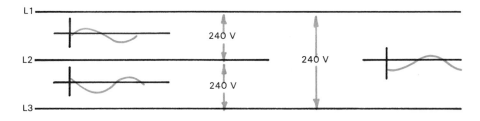

Fig. 7-13. A three-phase voltage consists of three separate voltages. The sine wave shows each voltage reaching its peak 120 electrical degrees apart from the other voltages.

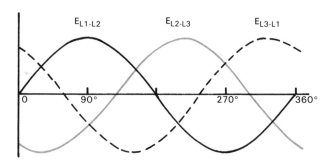

Fig. 7-14. Wave forms of the three voltages of a three-phase system are 120 degrees apart.

Coil polarity has a little different meaning in three-phase circuits. For example, we may talk about the positive terminal of the three coils. This does not mean that they are positive at the same instant. It means that these three terminals are positive 120 electrical degrees after each other. Fig. 7-15 shows the three voltages as vectors. *When we combine these vectors in phasor diagrams we must observe polarities.*

THREE-PHASE CONNECTIONS

There are two principal ways of connecting transformer coils (either primaries or secondaries) in three-phase circuits:
1. One is called DELTA.
2. The other is called WYE, or STAR.

The delta connection is shown in Fig. 7-16. Note that the coils are drawn to form the Greek letter, delta (Δ). This is a series-aiding connection with coils connected minus to plus.

The wye connection is shown in Fig. 7-17. Some people prefer the term "star," but wye (for the letter Y) is used more often. The junction, N (for neutral) is formed by the negative ends of all three coils.

Fig. 7-18 shows a typical three-phase transformer used to supply a large office building. Notice the flat wire used for the coils. This is known as a "pancake winding." The coils are wound on three legs of one core.

Usually, three separate single-phase transformers are used for three-phase service. Then, if one transformer fails, the other two are not affected and will still function. If one coil of a three-phase transformer fails, however, the entire transformer is out of service.

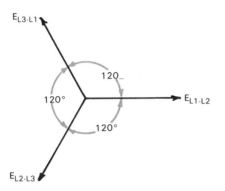

Fig. 7-15 The three voltages can be represented by vectors 120 degrees apart.

Fig. 7-18. A three-phase transformer with "pancake" windings. Units like this are used in large office buildings. (Standard Transformer Co.)

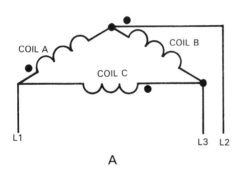

A

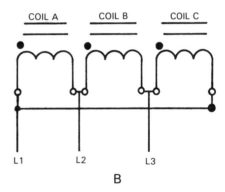

B

Fig. 7-16. Delta connection. A—Symbolic way of showing the connection. B—The way "delta" connections are usually shown on schematic diagrams.

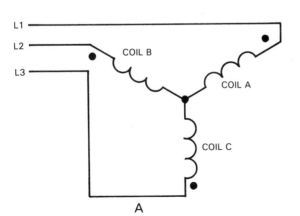

A

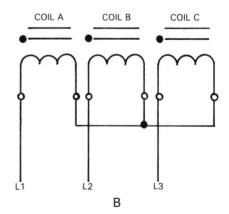

B

Fig. 7-17. Wye connection. A—Symbolic way of showing the connection. B—The way the "wye" connection is usually shown on schematic diagrams.

There is no difference in the connections or operation, whether one three-phase or three single-phase transformers are used.

TERMS USED IN THREE-PHASE POWER

We have already discussed line voltage. This is the voltage between lines L1 and L2, L2 and L3, and L1 and L3. The voltage across a coil is called PHASE VOLTAGE. We identify the coils as coil A, coil B and coil C. Phase A voltage, then, is the voltage across coil A. Phase B voltage is the voltage across coil B and phase C voltage is voltage across coil C. (Note: "voltage across the coil" is the voltage measured between its terminals. The voltage may have been applied to the coil, generated in it or it may be a combination of the two.)

In our three-phase computations, we assume the load is balanced. That means that the current in each coil is the same. You could measure this coil current, known as PHASE CURRENT, by connecting an ammeter in series with the coil. In practice, however, the ammeter goes in one of the lines. It measures what is called LINE CURRENT. This current is the same in each of the three lines when a balanced load is connected.

DELTA VOLTAGES

The relationship between line voltage and phase voltage in a delta connection is the easiest to see. Fig. 7-19 shows a delta-connected primary with 2400 volts applied. Line L1 is connected to one terminal of coil A. The other terminal of coil A is connected to line L2. The voltage between L1 and L2 must, therefore, be the same as the voltage across coil A.

The same relationship exists for the other two line and phase voltages. We can, therefore, make the general statement: *In a delta connection, phase voltage equals line voltage*. The mathematical equation is:

$$E_{PH} = E_{LINE}$$

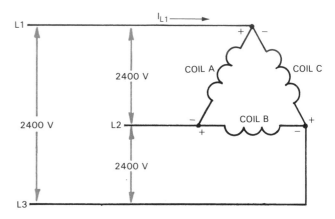

Fig. 7-19. Each line-to-line voltage of a delta connection is the voltage across one of the coils.

DELTA CURRENTS

The relationship between line and phase currents in delta circuits is not as simple as the voltages. We have to use a phasor diagram. Current in line L1 flows into the junction formed by the positive terminal of coil A and the negative terminal of coil C (refer to Fig. 7-19 once more).

From Kirchhoff's law, we know that the current entering a junction equals the current leaving it. Line 1 current must, therefore, be the vectorial sum of phase A current and phase C current.

Assume that Line 1 current is 17.3 amperes. The values of the phase currents are not yet known. All we are sure of is that they are equal and that there are 120 degrees between them. We can show the currents by the vectors in Fig. 7-20, view A.

The solution to the problem is shown in view B, Fig. 7-20. At the junction, coil A has the opposite polarity of coil C. At every instant, line 1 current equals coil A current *minus* coil C current. To subtract vectorially, you reverse the direction of the vector, then add in the normal manner.

Drawing the dotted line marked $-I_C$ in Fig. 7-20, view B, is the first step in the solution.

Next, the $-I_C$ vector's starting point is moved over to the end of line I_A. Their vector sum is the solid line marked I_{LI}. The diagram is proportioned so as to make I_{LI} represent 17.3 amperes. If we measure either I_A or I_C, we find them to be 10 amperes. In other words, the line current turns out to be 1.73 times the phase current. We can, therefore, make the general statement: *In a delta connection, phase current equals line current divided by 1.73.* The mathematical equation is:

$$I_{PH} = \frac{I_{LINE}}{1.73}$$
$$\text{or } I_{LINE} = I_{PH} \times 1.73$$

COMPUTING 1.73

You may be interested in how the factor of 1.73 is computed. Fig. 7-21, view A, shows the triangle formed by I_A, $-I_C$ and I_{LINE}. This is an isosceles triangle. The angle formed by I_A and $-I_C$ is 120 degrees. The two equal angles must therefore be 30 degrees ($180° - 120° = 60°$; $60° \div 2 = 30°$). In Fig. 7-21, view B, we draw a line from the apex (top) perpendicular to the base. Now there are two right triangles. Each has a hypotenuse equal to phase current. Also, each has an adjacent side equal to one-half I_{LINE}.

Therefore:

$$\cos 30° = \frac{1/2\ I_{LINE}}{I_{PH}} \qquad \cos 30° = 0.866$$
$$I_{PH} \times 0.866 = \frac{I_{LINE}}{2}$$
$$I_{PH} = \frac{I_{LINE}}{2 \times 0.866} = \frac{I_{LINE}}{1.73}$$

DELTA-CONNECTED SECONDARY

A delta secondary connection is made the same as the primary. You have to connect minus

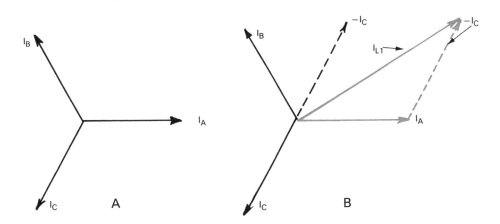

Fig. 7-20. In a delta connection, line current is the vectorial sum of the current in two coils. With a balanced load, line current is 1.73 times phase current. A—Vectors. B—Solution is reached by vector diagram.

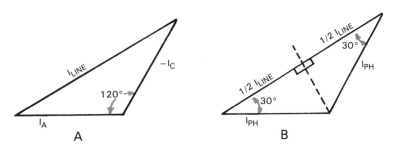

Fig. 7-21. Diagrams show how the constant, 1.73, is determined. A—Phasor diagram. B—Graphic representation of solution using trigonometry.

to plus. If your are not sure of the polarity of the coils, you should make voltage checks as you go. Assume the primaries of three transformers are connected correctly and your job is to connect the secondary into a delta.

First connect coil A to coil B as shown in Fig. 7-22, view A. Then, turn on the supply and measure the voltage between the open terminals. If they are correctly connected, the voltage across the series combination equals the voltage across one. This is shown by the phasor diagram of Fig. 7-22, view B. For example, if the voltage across coil A is 100 volts, the voltage across A and B in series is also 100 volts.

If one coil is backward, however, you would have the situation shown in Fig. 7-23. Instead of coil B voltage being 120 degrees behind coil A

voltage, it would be 60 degrees ahead. The combined voltage is 1.73 times coil A voltage. For example, the voltage across the open terminals is 173 volts.

If you have correctly connected coil B to coil A, all you need do to close the delta is connect coil C. Use extra care! Connect one terminal of coil C to coil B. Then apply voltage to the primary and measure the voltage across the open terminals of coil A and coil C. The phasor diagrams for the right and wrong connections are shown in Fig. 7-24 and Fig. 7-25. If the correct coil C terminal is connected to B, the voltage is zero. If incorrect, coil C voltage is exactly in phase with the A-B voltage. The voltage across the open terminals is DOUBLE the phase voltage. It would be very dangerous to connect that terminal of coil C to coil A.

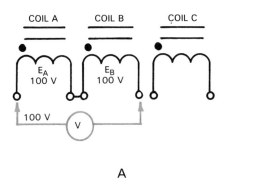
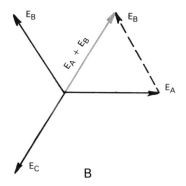

Fig. 7-22. Diagram of the first step in making a delta connection. A—With two coils correctly connected, the total voltage of two coils is the same as single coil voltage. B—Phasor diagram of delta connection.

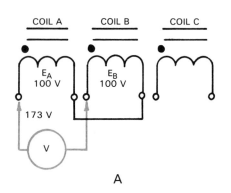
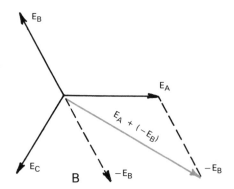

Fig. 7-23. Incorrectly connected coils. Total voltage is 1.73 times each coil voltage. A—Voltmeter connection for testing voltage. B—Resulting vector diagram show coil B voltage leading coil A by 60 degrees.

Remember, never attempt to close a delta connection until you make sure the voltage across the open terminals is zero.

DELTA POWER

Assume a balanced resistive load is connected to the delta-connected secondaries of three transformers. *The power drawn from each coil is equal to phase voltage times phase current* $(P = E_{PH} \times I_{PH})$. The total power delivered is three times the power supplied by one coil. Total apparent power is, therefore:

total $P_s = 3 \times E_{PH} \times I_{PH}$
where P_s is in volt-amperes.

Normally, you will be measuring line, rather than phase values. We must, therefore, change the equation to show line voltage and current. For phase voltage, we can substitute line voltage. They are the same for a delta connection. Phase current, on the other hand, equals line current divided by 1.73. The equation becomes:

total $P_s = 3 \times \dfrac{E_{LINE} \times I_{LINE}}{1.73}$ or

total $P_s = \dfrac{3}{1.73} \times E_{LINE} \times I_{LINE}$

total $P_s = 1.73 \times E_{LINE} \times I_{LINE}$

It should be noted that if a balanced reactive load is connected, the active power is:

total $P = 1.73 \times E_{LINE} \times I_{LINE} \times$ P.F.
where P is in watts.

Example 1
Three delta-connected transformers are supplying a 0.8 lagging power factor load.

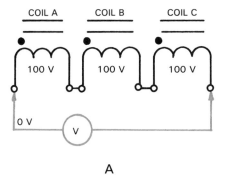

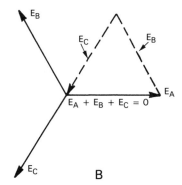

Fig. 7-24. Final step in making a delta connection. Total voltage across the three coils is O. A—Diagram of correct connection being tested. B—Phasor diagram of resulting voltage.

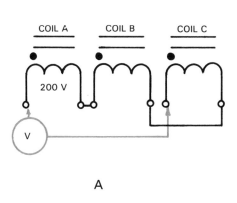

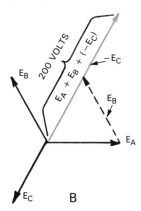

Fig. 7-25. Third coil is incorrectly connected. Total voltage is twice coil voltage. The delta must not be closed. A—Voltmeter connection. B—Phasor diagram of resulting voltage.

Each secondary supplies 10 amperes. Line voltage is 240 volts. Find the total apparent power and total active power.

a. Compute line current.

$$I_{LINE} = 1.73 \times I_{PH} = 1.73 \times 10$$
$$= 17.3 \text{ A}$$

b. Compute total apparent power.

$$\text{total } P_s = 1.73 \times E_{LINE} \times I_{LINE}$$
$$= 1.73 \times 240 \times 17.3$$
$$= 7,200 \text{ volt-amperes}$$
$$= 7.2 \text{ kVA (approx.)}$$

c. Compute total active power.

$$P = P_s \times PF = 7.2 \times .8 = 5.76 \text{ kW}$$

OPEN DELTA

When making a delta connection, you start by connecting coil B to coil A. This is shown in Fig. 7-26. At that point, you had three equal voltages. This connection is known as the OPEN DELTA, or the "VEE."

Assume that coil A voltage is 100 volts. Coil B voltage is also 100 volts. Likewise, the voltage across both coils is 100 volts. If you consider each pair of lines, you see that in each case, line voltage equals phase voltage and line current equals phase current. The power delivered by each coil therefore is:

$$P = E_{PH} \times I_{PH} = E_{LINE} \times I_{LINE}$$

The total power is twice the power per coil.

This is not two-thirds the power delivered by a delta connection, however. It works out to 58 percent.

Open delta connections can be used to supply reduced power for a new installation. For example, a new section of town where a third transformer can be added when the area is built up. It can also be used in an emergency. If one transformer in a delta fails, it can be removed from the circuit. You can get by on the lower power until a new transformer can be installed.

Example 2

Two transformers connected open delta are supplying a 0.8 lagging power factor load. Line current is 10 amperes. Line voltage is 240 volts. Find the total apparent power and the total active power. Compare these values with those computed for a delta connection in *Example 1*.

a. Compute the total apparent power.

$$\text{total } P_s = 1.73 \times E_{LINE} \times I_{LINE}$$
$$= 1.73 \times 240 \times 10$$
$$= 4.152 \text{ kVA}$$

b. Compare with *Example 1* ($P_s = 7.2$)

$$\frac{4.152}{7.2} \times 100 = 58\%$$

c. Compute the total active power.

$$\text{total } P = 4.152 \times .8 = 3.3 \text{ kW}$$

d. Compare with *Example 1* ($P = 5.76$ kW)

$$\frac{3.3}{5.76} \times 100 = 57\%$$

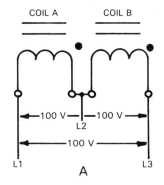
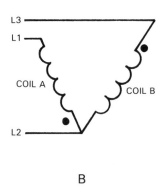

Fig. 7-26. The "open delta" is sometimes called the "Vee" connection. A—How to connect the open delta. B—Schematic diagram.

WYE CONNECTION

A wye connected secondary is shown in Fig. 7-27. There are four terminals. Three connect to the lines L1, L2 and L3. The fourth connects to a neutral wire. Line voltage is obtained from any pair of lines. Phase voltage can also be obtained between neutral and any one of the lines. For example, a factory may use a 208 three-phase line voltage for machinery and the 120 single-phase voltage for lighting.

The neutral junction is formed by connecting together the ends with the same polarity. Primaries, secondaries, or loads can be wye connected.

WYE VOLTAGES

The voltage between the neutral wire and any of the lines is the phase voltage. This can be seen from Fig. 7-27. The voltage between two lines is the phasor sum of two phase voltages. The voltage phasor diagram is shown in Fig. 7-28. The voltage between line L1 and L2 is the voltage across the series combination of coil A and coil B. Coil B has the opposite polarity of coil A, however. Therefore, its phasor must be reversed before adding. The solution shows L1 to L2 voltage to be 1.73 times the voltage across one coil. We can, therefore, make the general statement:

In a wye connection, line voltage equals 1.73 times phase voltage.

The mathematical equation is:

$$E_{LINE} = 1.73 \times E_{PH}; \text{ or } E_{PH} = \frac{E_{LINE}}{1.73}$$

WYE CURRENTS

Again refer to Fig. 7-27. There is no junction between line L1 and coil A. The current through line L1 also passes through coil A. The same is true for the other two lines and coils. We can, therefore, make the general statement:

In a wye connection, line current equals phase current.

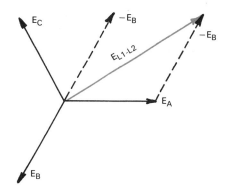

Fig. 7-28. Line voltage is 1.73 times phase voltage in a wye connection.

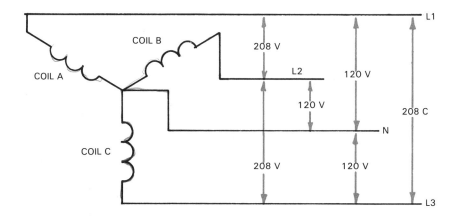

Fig. 7-27. With a wye connection, one voltage value can be obtained from line to line (line voltage), and another from line to neutral (phase voltage).

The mathematical equation is:

$$I_{LINE} = I_{PH}$$

WYE-CONNECTED SECONDARY

Before applying a load, you must be sure the secondary is connected correctly. You must connect plus to plus. If you are not sure of the polarity of the coils, you can make the connection, then measure the voltage.

For example, assume the phase voltage of each coil is 100 volts. Connect coil A to coil B. Then measure the voltage across the open ends. If they are correctly connected, the voltage across the combination will be 173 volts, as shown in Fig. 7-29. If incorrect, you will have 100 volts. This is shown in Fig. 7-30.

When you have coils A and B connected cor-

rectly, you can add coil C. The voltage across C and A (or B) should be 173 volts.

WYE POWER

Assume a balanced resistive load is connected to the wye-connected secondaries of three transformers. The power drawn from each coil is equal to phase voltage times phase current ($P = E_{PH} \times I_{PH}$). The total power delivered is three times the power supplied by one coil. Total apparent power is therefore:

$$\text{total } P_s = 3 \times E_{PH} \times I_{PH}$$

In wye circuits, phase current and line current are the same. Therefore, I_{LINE} can be substituted directly for I_{PH}. Phase voltage, on the other hand, equals line voltage divided by 1.73 ($E_{PH} = \dfrac{E_{LINE}}{1.73}$).

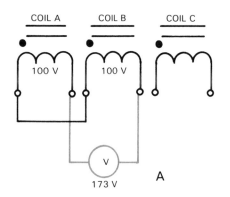
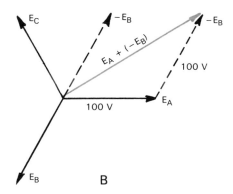

Fig. 7-29. The first step in making a wye connection. With two coils correctly connected, the total voltage is 1.73 times each coil voltage. A—Making the voltmeter connection. B—Drawing the phasor diagram for a wye connection.

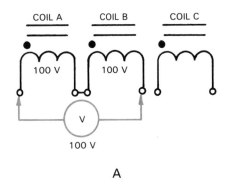
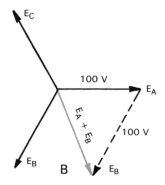

Fig. 7-30. Incorrectly connected coils. Total voltage is equal to each coil voltage. A—Incorrect hookup. B—Phasor diagram.

The equation, therefore, becomes:

$$\text{total } P_s = 3 \times \frac{E_{LINE}}{1.73} \times I_{LINE}$$

$$\text{total } P_s = \frac{3}{1.73} \times E_{LINE} \times I_{LINE}$$

$$\text{total } P_s = 1.73 \times E_{LINE} \times I_{LINE}$$

This is exactly the same equation we had for a delta connection. When you are measuring or computing three-phase power, it does not matter whether the connection is delta or wye.

MEASURING THREE-PHASE POWER

Active power is measured with a wattmeter. Wattmeters have two coils. One coil, called the POTENTIAL COIL, is connected like a voltmeter. The other, called the CURRENT COIL, is connected like an ammeter. The meter multiplies voltage times current times the power factor.

Fig. 7-31 shows a typical wattmeter used in school laboratories. The "dot notation" is used so the two coils may be connected with the correct polarity. A correctly connected wattmeter is shown in Fig. 7-32.

POWER IN EACH PHASE

Fig. 7-33 shows a three-phase balanced resistive load. Wattmeters are connected in each leg. Each one measures the power used by one phase. The value shown on a wattmeter is $E \times I \times \cos \Theta$. The angle, Θ, is zero degrees for this resistive load, making $\cos \Theta = 1$. The wattmeters all read the same. If phase voltage is 100 volts and phase current is 5 amperes, each wattmeter reads 500 watts. The total power used is the sum of the three wattmeter readings, or 1500 watts.

The big problem with this method of measuring three-phase power is that, for many loads, the neutral point is not accessible. Also, it would not work with delta-connected loads. Usually, the only values we can measure are line

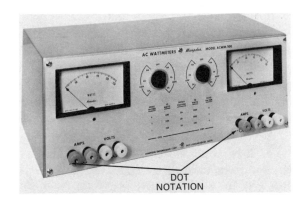

Fig. 7-31. Wattmeter is typical of those used in school laboratories.   (Hampden Engineering Corp.)

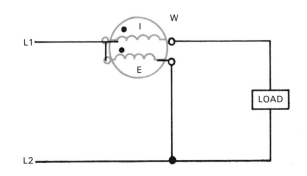

Fig. 7-32. Schematic diagram shows wattmeter connected in a circuit.

voltage and line current. Fortunately, there is a way to use these values with only two wattmeters to measure three-phase power.

TWO-WATTMETER METHOD

Fig. 7-34 shows the connections for the two-wattmeter method of measuring three-phase power. The load shown is wye-connected, but a delta connection works the same. The total three-phase power is the algebraic sum of the two wattmeter readings.

There are three facts you must keep in mind when using this method:

1. The polarity of the wattmeter coil connections is very important.
2. The only condition under which the wattmeters will have the same reading is unity

power factor (resistive) load.

3. When the load's power factor is below 0.5, one of the wattmeters will read downscale. You must reverse the connections to the potential coil to make it read upscale. Then you must subtract its reading from the other wattmeter reading.

To show what the two wattmeters read and why, we will take three examples. The first is with a unity power factor load. The second is with a 0.8 power factor. This is the typical power factor of a loaded induction motor. The third example is a 0.4 power factor, which is typical for an unloaded motor.

Example 3

Line current, I_L, to a three-phase resistive load is 2 amperes. Line voltage, V_L, is 208

volts. Find the reading of the two watt-meters and the total power consumed by the load.

a. For the first wattmeter, determine the phase angle, δ, between $E_{L1\text{-}L3}$ and I_{L1}. The phasor diagram in Fig. 7-35, view A, is for the first wattmeter. The phase voltages are shown 120 degrees apart. Note that the phase current (I_{L1}) is shown in phase with the phase voltage (E_A). The wattmeter's potential coil is responding to the voltage between lines L1 and L3. This is the phasor sum of the phase voltages E_A and $-E_C$. View A shows that $E_{L1\text{-}L3}$ leads E_L and I_{L1} by 30 degrees.

b. For the second wattmeter, determine the phase angle, δ, between $E_{L2\text{-}L3}$ and I_{L2}.

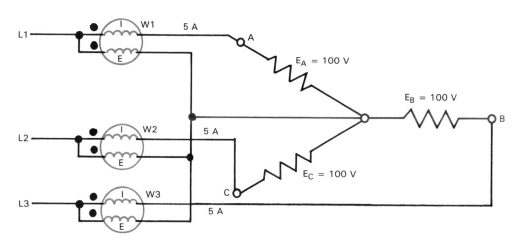

Fig. 7-33. The use of three wattmeters to measure three-phase power. Total power is the sum of the three wattmeter readings.

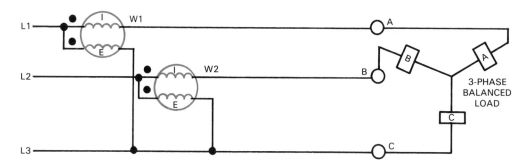

Fig. 7-34. The use of two wattmeters to measure three-phase power. The total power is the algebraic sum of the two watt-meter readings.

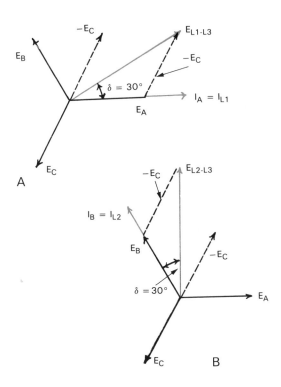

A

B

Fig. 7-35. Finding wattage with a unity power factor (resistive) load. A—Wattmeter 1 reads E_{L1-L3} x I_{L1} x cos 30°. B— Wattmeter 2 reads E_{L2-L3} x I_{L2} x cos 30°.

The phasor diagram of Fig. 7-35, view B, is for the second wattmeter. The phase current through coil B (which is also the line current, I_{L2}) is shown in phase with the phase voltage, E_B. The voltage coil of the second wattmeter is responding to the voltage between lines L2 and L3. This is the phasor sum of the phase voltages, E_B and $-E_C$. Fig. 7-35, view B, shows that E_{L2-L3} lags I_{L2} by 30 degrees.

c. Determine the reading of the first wattmeter (W1). The wattmeter reads line voltage times line current times the cosine of the phase angle, δ, between them.

$$W1 = E_{L1-L2} \times I_{L1} \times \cos \delta$$
$$= 208 \times 2 \times \cos 30°$$
$$W1 = 416 \times 0.866 = 360 \text{ W}$$

d. Determine the reading of the second wattmeter (W2).

$$W2 = E_{L2-L3} \times I_{L2} \times \cos \delta$$
$$= 208 \times 2 \times \cos 30°$$

$$W2 = 416 \times 0.866 = 360 \text{ W}$$

e. Compute total power, P_T.
The total power is the sum of the two wattmeter readings:
$$P_T = W1 + W2 = 360 + 360$$
$$= 720 \text{ W}$$
Note: This answer can be confirmed by computing power from the equation: $P_T = 1.73 \times E_L \times I_L \times \cos \Theta$ where cos Θ is the load power factor:
$$P_T = 1.73 \times 208 \times 2 \times 1 = 720 \text{ W}$$

Example 4

Line current, I_L, to a 208 volt three-phase induction motor having a 0.8 power factor is 2 amperes. Find the reading of two watt-meters and total power delivered to the motor.

a. For wattmeter 1, determine the phase angle, δ, between E_{L1-L2} and I_{L1}.
The phasor diagram of Fig. 7-36, view A, is for wattmeter 1. Phase current lags

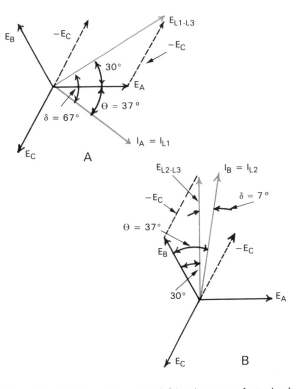

A

B

Fig. 7-36. Finding wattage with 0.8 lagging power factor load. A—Wattmeter 1 reads E_{L1-L3} x I_{L1} x cos 67°. B—Wattmeter 2 reads E_{L2-L3} x I_{L2} x cos 7°.

phase voltage by 37 degrees. (The angle whose cosine is 0.8 is 37 degrees.) Once again, line voltage, $E_{L1\text{-}L2}$, is 30 degrees ahead of E_A. This time, however, $E_{L1\text{-}L2}$ leads I_{L1} by 67 degrees (30 degrees + Θ).

b. Determine the reading of wattmeter W1.

$W1 = 208 \times 2 \times \cos 67°$

$W1 = 416 \times 0.39 = 162.5$ W

c. For wattmeter W2, determine the phase angle, δ between $E_{L2\text{-}L3}$ and I_{L2}.

The phasor diagram of Fig. 7-36, view B, is for wattmeter 2. Phase (line) current lags phase voltage by 37 degrees. However, line voltage leads phase voltage by 30 degrees. That means line voltage, $E_{L2\text{-}L3}$, lags line current, I_{L2}, by 7 degrees (30° − Θ).

d. Determine the reading of wattmeter 2.

$W2 = 208 \times 2 \times \cos 7°$

$W2 = 416 \times 0.99 = 412.9$ W

e. Compute total power, P_T.

$P_T = W1 + W2 = 162.5 + 412.9$
$ = 575.4$

Note: This answer can be confirmed by computing power from the equation: $P_T = 1.73 \times V_L \times I_L \times \cos \Theta$

$P_T = 1.73 \times 208 \times 2 \times 0.8$
$ = 575.7$ W

Example 5

A 208 volt three-phase induction motor is running unloaded. Line current is 2 am-

peres. Power factor is 0.4. Find the readings of the two wattmeters and the total power used.

a. For wattmeter 1, determine the phase angle, δ, between $E_{L1\text{-}L2}$ and I_{L2}.

The phasor diagram of Fig. 7-37, view A, is for wattmeter 1. Phase current lags phase voltage by 66.4 degrees. (The angle whose cosine is 0.4 is 66.4 degrees.) Once again, line voltage, $E_{L1\text{-}L2}$, is 30 degrees ahead of E_A. That makes line voltage lead I_L by 96.4 degrees (30° + Θ).

b. Determine the reading of wattmeter 1.

$W1 = 208 \times 2 \times \cos 96.4°$

$W1 = 416 \times -0.11 = -46.4$ W

Note: The cosine of angles between 90 degrees and 270 degrees have a minus sign. The reading of −46.4 watts represents power being returned to the source. This power must be subtracted from W2.

c. For wattmeter 2, determine the phase angle, δ, between $E_{L2\text{-}L3}$ and I_{L2}.

The phasor diagram of Fig. 7-37, view B, is for wattmeter 2. Phase (line) current lags phase voltage by 66.4 degrees. However, line voltage leads phase voltage by 30 degrees. That means line voltage, $E_{L2\text{-}L3}$, lags line current, I_{L2}, by 36.4 degrees (30° − Θ).

d. Determine the reading of wattmeter 2.

$W2 = 208 \times 2 \times \cos 36.4°$

$W2 = 416 \times 0.8 = 336.5$ W

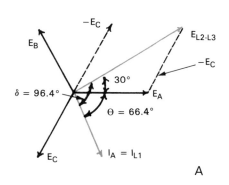

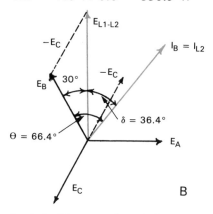

Fig. 7-37. Finding wattage and total power with 0.4 leading power factor load. A—Wattmeter 1 reads $E_{L1\text{-}L3} \times I_{L1} \times \cos 96.4°$. B—Wattmeter 2 reads $E_{L2\text{-}L3} \times I_{L2} \times \cos 36.4°$.

Turns ratio for some transformers can be varied by use of special devices. Pictured are a tap selector and reversing switches for a load tap changer. (McGraw-Edison Co.)

e. Compute total power, P_T.

$P_T = W1 + W2 = -46.4 + 336.5$

$P_T = 290.1$ W

Note: This answer can be confirmed by computing power from the equation (The answers will be slightly different because of rounding of figures.): $P_T = 1.73 \times 208 \times 2 \times 0.4 = 287.9$ W

From these three examples it is easy to see that for any power factor, wattmeter 1 reads $E_{L1\text{-}L3} \times I_{L1} \times \cos(30° - \Theta)$. Wattmeter 2 reads $E_{L2\text{-}L3} \times I_{L2} \times \cos(30° + \Theta)$.

At a power factor of 0.5 ($\Theta = 60°$), wattmeter 2 reads zero watts ($\cos 90° = 0$).

At power factors higher than 0.5, the sum of the two wattmeter readings is the total power.

At power factors below 0.5, the total power

used is the difference between the two wattmeter readings.

TRANSFORMER CONNECTIONS

Either the primary or secondary may be connected in a delta. Also, either may be connected in a wye. That gives us four types of three-phase connections.

1. Both the primary and secondary may be delta-connected. This is called a DELTA-DELTA connection.
2. The primary may be delta-connected, while the secondary is wye-connected. This is called a DELTA-WYE connection.
3. The primary may be wye-connected, while the secondary is delta-connected. This is called a WYE-DELTA connection.
4. Both the primary and secondary may be wye-connected. This is called a WYE-WYE connection.

To show the effect of these different connections, we will go through four examples. In each, we will use the same three transformers. All three have a turns ratio of 10:1. The primary has 2400 volts, three-phase applied. The load consists of three 100 ohm resistances, which are wye-connected.

Example 6

The connection shown in Fig. 7-38 is a delta-delta (Δ-Δ). Find the primary phase voltage, secondary phase voltage, secondary line voltage, load voltage, load current, secondary line current, secondary phase current, primary phase current and primary line current.

a. Compute delta primary phase voltage.
$$E_{PH(PRI)} = E_{LINE(PRI)} = 2400 \text{ V}$$

b. Compute secondary phase voltage.
$$E_{PH(SEC)} = \frac{E_{PH(PRI)}}{a} = \frac{2400}{10} = 240 \text{ V}$$

c. Compute delta secondary line voltage.
$$E_{LINE(SEC)} = E_{PH(SEC)} = 240 \text{ V}$$

d. Compute wye load voltage.
$$E_{LOAD} = \frac{E_{LINE}}{1.73} = \frac{240}{1.73} = 138.6 \text{ V}$$

e. Compute load current.
$$I_{LOAD} = \frac{E_{LOAD}}{R} = \frac{138.6}{100} = 1.4 \text{ A}$$

f. Compute wye secondary line current.
$$I_{LINE(SEC)} = I_{LOAD} = 1.4 \text{ A}$$

g. Compute delta secondary phase current.
$$I_{PH(SEC)} = \frac{I_{LINE(SEC)}}{1.73} = 0.8 \text{ A}$$

h. Compute primary phase current.
$$I_{PH(PRI)} = \frac{I_{PH(SEC)}}{a} = \frac{0.8}{10} = 0.08 \text{ A}$$

i. Compute delta primary line current.
$$I_{LINE(PRI)} = I_{PH(PRI)} \times 1.73 = .08 \times 1.73 = 0.14 \text{ A}$$

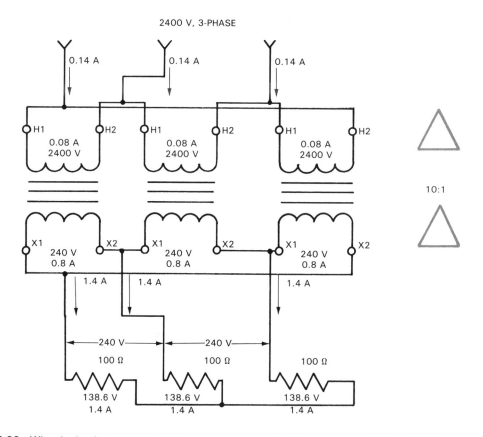

Fig. 7-38. When both primary and secondary are connected delta, it is known as a "delta-delta" connection.

Transformer Applications

Example 7

Fig. 7-39 shows three transformers in a delta-wye connection. Find the primary phase voltage, secondary phase voltage, secondary line voltage, load voltage, load current, secondary line current, secondary phase current, primary phase current and primary line current.

a. Compute delta primary phase voltage.
$$E_{PH(PRI)} = E_{LINE(PRI)} = 2400 \text{ V}$$

b. Compute secondary phase voltage.
$$E_{PH(SEC)} = \frac{E_{PH(PRI)}}{a} = \frac{2400}{10} = 240 \text{ V}$$

c. Compute wye secondary line voltage.
$$E_{LINE(SEC)} = E_{PH(SEC)} \times 1.73$$
$$= 240 \times 1.73 = 415.2 \text{ V}$$

d. Compute wye load voltage.
$$E_{LOAD} = \frac{E_{LINE}}{1.73} = \frac{415.2}{1.73} = 240 \text{ V}$$

e. Compute load current.

$$I_{LOAD} = \frac{E_{LOAD}}{R} = \frac{240}{100} = 2.4 \text{ A}$$

f. Compute wye secondary line current.
$$I_{LINE(SEC)} = I_{LOAD} = 2.4 \text{ A}$$

g. Compute wye secondary phase current.
$$I_{PH(SEC)} = I_{LINE(SEC)} = 2.4 \text{ A}$$

h. Compute primary phase current.
$$I_{PH(PRI)} = \frac{I_{PH(SEC)}}{a} = \frac{2.4}{10} = 0.24 \text{ A}$$

i. Compute delta primary line current.
$$I_{LINE(PRI)} = I_{PH(PRI)} \times 1.73$$
$$= 0.24 \times 1.73 = 0.42 \text{ A}$$

Example 8

In the wye-wye connection of Fig. 7-40, find primary phase voltage, secondary phase voltage, secondary line voltage, load voltage, load current, secondary line current, secondary phase current, primary phase current and primary line current.

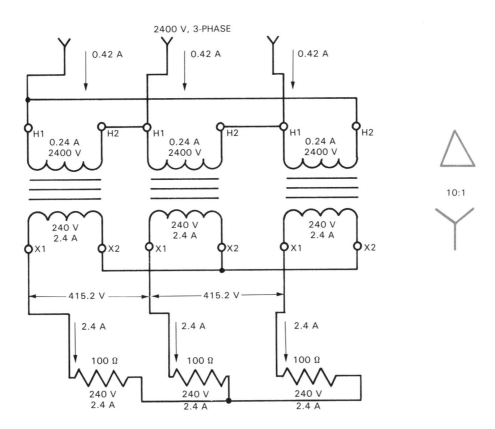

Fig. 7-39. When the primary is connected delta and the secondary is connected wye, it is known as a "delta-wye" connection.

a. Compute wye primary phase voltage.

$$E_{PH(PRI)} = \frac{E_{LINE(PRI)}}{1.73} = 1386 \text{ V}$$

b. Compute secondary phase voltage.

$$E_{PH(SEC)} = \frac{E_{PH(PRI)}}{a} = \frac{1386}{10} = 138.6 \text{ V}$$

c. Compute wye secondary line voltage.

$$E_{LINE(SEC)} = E_{PH(SEC)} \times 1.73 = 240 \text{ V}$$

d. Compute wye load voltage.

$$E_{LOAD} = \frac{E_{LINE}}{1.73} = 138.6 \text{ V}$$

e. Compute load current.

$$I_{LOAD} = \frac{E_{LOAD}}{R} = \frac{138.6}{100} = 1.4 \text{ A}$$

f. Compute wye secondary line current.

$$I_{LINE(SEC)} = I_{LOAD} = 1.4 \text{ A}$$

g. Compute wye secondary phase current.

$$I_{PH(SEC)} = I_{LINE(SEC)} = 1.4 \text{ A}$$

h. Compute primary phase current.

$$I_{PH(PRI)} = \frac{I_{PH(SEC)}}{a} = \frac{1.4}{10} = 0.14 \text{ A}$$

i. Compute wye primary line current.

$$I_{LINE} = I_{PH} = 0.14 \text{ A}$$

Example 9

Fig. 7-41 shows a wye-delta connection. Find the primary phase voltage, secondary phase voltage, secondary line voltage, load voltage, load current, secondary line current, secondary phase current, primary phase current and primary line current.

a. Compute wye primary phase voltage.

$$E_{PH(PRI)} = \frac{E_{LINE(PRI)}}{1.73} = \frac{2400}{1.73} = 1386 \text{ V}$$

b. Compute secondary phase voltage.

$$E_{PH(SEC)} = \frac{E_{PH(PRI)}}{a} = \frac{1386}{10} = 138.6 \text{ V}$$

c. Compute delta secondary line voltage.

$$E_{LINE(SEC)} = E_{PH(SEC)} = 138.6 \text{ V}$$

d. Compute wye load voltage.

$$E_{LOAD} = \frac{E_{LINE(SEC)}}{1.73} = \frac{138.6}{1.73} = 80 \text{ V}$$

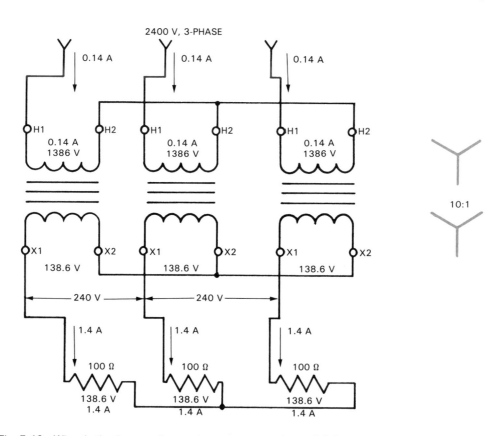

Fig. 7-40. When both primary and secondary are connected wye, it is known as a "wye-wye" connection.

e. Compute load current.

$$I_{LOAD} = \frac{E_{LOAD}}{R} = \frac{80}{100} = 0.8 \text{ A}$$

f. Compute secondary line current.

$$I_{LINE(SEC)} = I_{LOAD} = 0.8 \text{ A}$$

g. Compute delta secondary phase current.

$$I_{PH(SEC)} = \frac{I_{LINE}}{1.73} = \frac{0.8}{1.73} = 0.46 \text{ A}$$

h. Compute primary phase current.

$$I_{PH(PRI)} = \frac{I_{PH(SEC)}}{a} = \frac{0.46}{10} = 0.046 \text{ A}$$

i. Compute wye primary line current.

$$I_{LINE(PRI)} = I_{PH(PRI)} = 0.046 \text{ A}$$

AUTOTRANSFORMERS

Core flux induces a back voltage into the primary at the same time it induces voltage in the secondary. This voltage is self-induced in every coil turn. It would even be induced in turns that do not have voltage applied to them. This is the principle of the AUTOTRANSFORMER. Autotransformers have only one coil. Still, they are able to step up and step down voltages.

Fig. 7-42 shows a step-down autotransformer. The part of the coil that has voltage applied to it is the primary. In this case, the entire winding is the primary. One terminal is common to the primary and the load. The other load terminal is connected to a tap. Location of the tap determines if it is a step-up or step-down transformer. The part of the coil to which the load is connected is the secondary.

The ratio of primary to secondary voltages is the ratio of turns in the primary section to turns in the secondary section. In fact, all the ratios that apply to two-coil transformers apply to autotransformers.

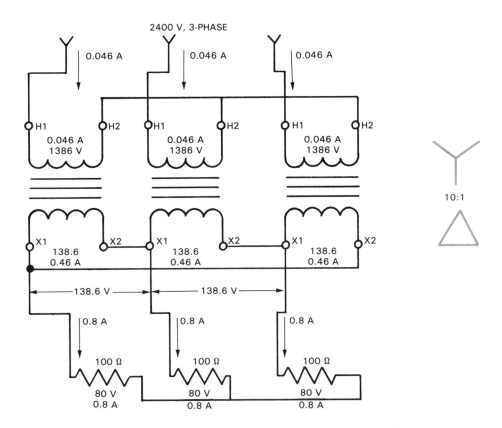

Fig. 7-41. When the primary is connected wye and the secondary is connected delta, it is known as a ''wye-delta'' connection.

STEP-DOWN AUTOTRANSFORMER

Assume that 208 volts is applied to the autotransformer in Fig. 7-42. The primary includes all of the turns but the secondary includes only half the turns. The turns ratio is 2:1. Secondary voltage is, therefore, 104 volts ($\frac{208}{2} = 104$). There is a 10.4 ohm resistive load. Load current is 10 amperes ($\frac{104}{10.4} = 10$).

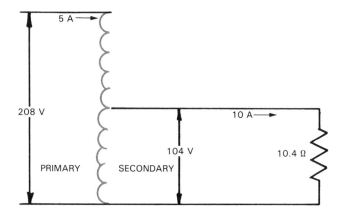

Fig. 7-42. Step-down autotransformer. Primary and secondary are on the same coil. Primary uses all of the turns.

Primary current is 5 amperes ($\frac{10}{2} = 5$). It may be hard at first to understand how the secondary can have twice the primary current. After all, the primary and secondary are the same coil. Secondary current, however, is the sum of two separate currents. One is primary current which also flows in the secondary circuit. The other is produced by back-voltage, which is self-induced in the secondary part of the coil.

The primary current of 5 amperes is *conducted* to the secondary current. The other 5 amperes is *transformed* in the autotransformer.

STEP-UP AUTOTRANSFORMER

Fig. 7-43 shows an autotransformer that steps up voltage. This time, we have applied voltage to some of the turns. This part of the coil acts as the primary. The load is connected to the entire coil, which acts as the secondary.

Assume the output is 140 volts when there is 120 volts applied to the primary. A resistive load of 14 ohms is connected. Again we have a secondary current of 10 amperes. Output power is therefore:

$$P_o = E_{SEC} \times I_{SEC} = 140 = 10 \times 1400 \, W$$

We can assume that input power equals output power. That allows us to compute primary current.

$$P_I = P_O = 1400 = 120 \times I_{PRI}$$

$$I_{PRI} = \frac{1400}{120} = 11.67 \, A$$

APPLICATIONS OF AUTOTRANSFORMERS

Autotransformers are often used where a small change in voltage is needed. For example, they can boost voltages on a large transmission line. Or, they can convert a 208-volt circuit to 240 volts. They also may reduce the voltage applied to three-phase motors while starting.

When the turns ratio is small, autotransformers use less steel and copper than two-winding

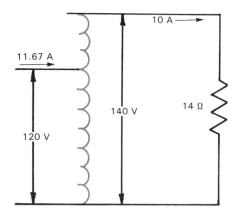

Fig. 7-43. Step-up autotransformer. Secondary uses all of the turns.

transformers. For turns ratio above 5:1, this is no longer true.

Autotransformers have a more serious limitation. They do not isolate the load from the source.

For example, Fig. 7-44 shows what might happen if a fault occurs. In Fig. 7-44, view A, everything is fine; 2400 volts is being stepped down to 480 volts. Now suppose one of the turns in the secondary part breaks. As you can see from view B of Fig. 7-44, the entire 2400 volts appears across the load!

Some autotransformers are made with variable outputs. They may have a number of taps like the one shown in Fig. 7-45. A lever makes contact with the taps to change the output from 5 percent to 110 percent of the primary voltage.

Variable autotransformers are very widely used in laboratory power supplies and for motor control. Fig. 7-46 shows a commercial unit. Known as Variac, it is a single layer coil wound on a doughnut-shaped core. Insulation is removed from each turn along a narrow band. A contact arm, called a brush, is always in contact with one or more of the wires. By turning a knob, you can move the brush from one end of the coil to the other. In this way, you can get any voltage from 0 volts up to the applied volt-

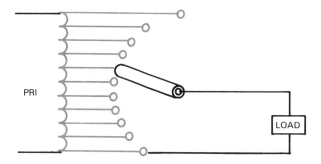

Fig. 7-45. Variable-output autotransformers can provide a number of voltages by use of a movable contact.

Fig. 7-46. One type of commercially made variable auto-transformer. This one is known as "Variac."

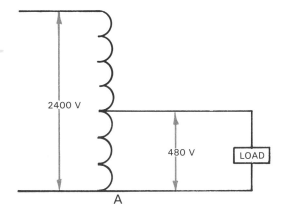

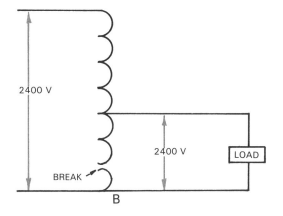

Fig. 7-44. Result of a breakdown in an autotransformer. A—Transformer is working properly. B—A break in the autotransformer secondary puts the full primary voltage across the load.

age. In some cases, voltage is applied to a tap on the primary. This widens the range to a value above the primary voltage. For example, a typical Variac operates from 120 volts. The output range is 0 to 140 volts.

SUMMARY

There is a definite relationship between the instantaneous polarities of coils that are magnetically linked.

In the marking of transformer terminals, low side terminal X1 is positive at the instant high side terminal H1 is positive.

Subtractive polarity means that the X1 terminal is directly opposite the H1 terminal.

Additive polarity means that the X1 terminal is diagonally opposite the H1 terminal.

Polarity must be observed when interconnecting coils of a multicoil secondary.

Dot notation is a graphic system for showing relative polarity in coil terminals.

Three-phase voltage consists of three voltages displaced 120 degrees from each other.

There are two principal three-phase connections, wye and delta.

In a wye connection, terminals with the same polarity are connected to a neutral point.

In a delta connection, coils are connected series-aiding, minus to plus.

The total apparent power of either wye or delta circuit is 1.73 times voltage times current.

A third type of three-phase connection is the vee, also known as an open delta.

The vee connection delivers 58 percent of the power delivered by either the wye or delta three-phase connection.

Autotransformers have a single coil. Some of its turns are common to both primary and secondary.

TEST YOUR KNOWLEDGE

1. What is meant by the polarity of magnetically linked coils?
2. Describe a way of telling whether a transformer has additive or subtractive polarity.
3. What is the purpose of tap changers on transformers?
4. Explain the use of the dot notation on coils.
5. Why should the voltage across the open terminals of a delta connection be measured before attempting to close a delta?
6. Give one advantage and one disadvantage of an autotransformer.

PROBLEMS

1. Given: A 10:1 step-down transformer, equipped for tap changing, has a primary voltage of 13,800 volts. Find the no-load secondary voltage on the 10 percent tap and on the 7 percent tap.
2. Given: The line-to-neutral voltage of the secondary of a wye-connected transformer is 120 volts. Find the line (line-to-line) voltage.
3. Given: The line voltage and line current of the primary of a wye-connected transformer is 173 volts and 5 amperes. Find the phase voltage and phase current.
4. Given: Transformer of *Problem 3* has delta-connected secondary. This is a step-down transformer with turns ratio of 10:1. Find secondary line voltage and line current.
5. Given: Load on transformer of *Problem 4* is a balanced resistive load. Find total power output and input (ignore losses).
6. Given: A 2:1 step-down autotransformer with 120 volts on the primary and a 10-ohm resistive load. Find the output power and primary current.

8

INTRODUCTION TO GENERATORS AND MOTORS

After studying this chapter, you will be able to:

☐ Define the terms "generator action" and "motor action."

☐ Compute mechanical power from torque and speed.

☐ List the factors that determine the voltage generated by a generator.

☐ List the factors that determine the frequency of the voltage generated by an alternator.

☐ Explain the effects of counter torque in a generator and cemf in a motor.

Generators change mechanical energy into electrical energy. Motors change electrical energy into mechanical energy.

Generators and motors are a lot alike. They are made in the same general way. Further, they both depend on the same electromagnetic principles for their operation.

The first principle is called GENERATOR ACTION. It is also called INDUCTION. *Voltage can be induced into a wire that is in a magnetic field.* This happens when the magnetic flux is cut by the wire.

In some cases, the wire moves. In other cases, the field moves. In still others, both are moving, but at different speeds.

It takes mechanical energy to cause the motion. The motion causes electricity to be generated.

The second principle is called MOTOR ACTION. This is simply the mechanical forces between magnets. When two magnets (or electromagnets) approach each other, one will be either pulled toward or pushed away from the other.

Some motors use one permanent magnet and one electromagnet. Others use two electromagnets. Either way, electrical energy creates at least one of the magnetic fields. Then, the forces between the two magnetic fields cause the motion.

CONSTRUCTION

Each type of motor and each type of generator is built to do its own special job. All of them have two main parts, Fig. 8-1.

One part is called the STATOR. The stator is stationary; it does not move. The other part, the ROTOR, is mounted on bearings so it can rotate. The rotor shaft sticks out beyond the housing.

Fig. 8-1. All generators have a rotor and a stator. The rotor rotates. The stator does not move.
(Westinghouse Electric Co.)

For a generator, a PRIME MOVER is coupled to this shaft. For a motor, the rotor shaft is coupled to a MECHANICAL LOAD.

GENERATORS

The input to a generator is mechanical power. If its output is direct current, it is called a DC GENERATOR. If its output is alternating current, it is known as an ALTERNATOR.

Prime movers apply mechanical power to turn the rotor. Water wheels, windmills, steam and hydroturbines, diesel and jet engines, and hand cranks are all examples of prime movers.

To make a rotor turn, the prime mover applies TORQUE. *Torque simply means turning force.* For example, when you tighten a nut on a bolt, you apply torque with a wrench, Fig. 8-2.

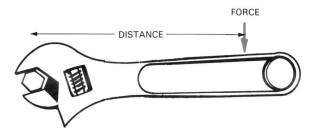

Fig. 8-2. Torque is a force which tends to cause rotational motion.

TORQUE

The amount of torque applied to a shaft is proportional to two things:

1. Amount of force applied.
2. The distance the force is applied away from the center of the shaft.

Fig. 8-3 shows a crank used to rotate a shaft. *The torque, or turning force, is directly proportional to both force and distance.* The more force you apply, the greater the torque. Also, the longer the crank handle, the greater the torque. The equation is:

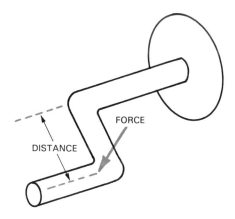

Fig. 8-3. The amount of torque produced by this crank is the product of force and distance.

Torque (T) = force (F) × distance (d)

In Conventional units of measure, force is expressed in pounds. The abbreviation is lbf, for POUNDS FORCE. One lbf is the amount of force that would be applied by a weight of one pound. Distance is measured in feet (ft.). Torque is expressed in pound force-feet (lbf-ft.). Torque may also be expressed in ounce force-inches (ozf-in.).

In SI metric, the unit of force is the newton (N). One newton is the amount of force that would be applied by 0.102 kilograms (kg) which is about 1/4 pound. Distance is measured in metres (m). Torque is expressed in newton metres (N·m).

Example 1

A force of 20 newtons (4.5 pounds) is applied to the crank handle shown in Fig. 8-3. The lever arm is 0.5 metres (1.64 ft.) long. Find the torque in both SI metric and Conventional units.

a. Compute torque in SI metric units.
   T = F × d = 20 × 0.5
   = 10 newton metres (N·m)
b. Compute torque in Conventional units.
   T = F × d = 4.5 × 1.64
   = 7.38 pound force-feet (lbf-ft.)

Ten newton metres equal 7.38 pound force-feet. Therefore, a newton metre equals 0.738 pound force-feet. Also a 1 pound force-foot equals 1.36 newton metres.

Prime movers are often rated by how much torque they can produce. You should not think of torque as a specific force applied at a specific distance. Rather, you should think of torque as mechanical energy.

MECHANICAL POWER

Assume the crank in *Example 1* is coupled to a generator rotor. Further, assume it takes a torque of 20 newton metres (14.76 foot-pounds) to start the rotor turning. The rotor will not turn because the turning force is not great enough. Even though there is torque, no MECHANICAL POWER is being delivered. To get mechanical power, you must have motion.

Power is the rate of using energy. You could turn the crank fast or slow. It takes more power to turn it fast than it does to turn it slow. This means that mechanical power depends on speed as well as torque.

In the SI metric system, mechanical power is measured in watts (W). In our computations we use rotor speed in revolutions per minute (rpm). This is the speed that results from the torque. The equation for finding power is:

$$P_m = \frac{S \times T}{9.55}$$

P_m is the mechanical power in watts
S is the speed in revolutions per minute
T is the torque in newton metres (N·m)
9.55 is the constant for metric measure

If we know the power and the speed, we can compute the torque from the equation:

$$T = \frac{9.55 \times P_m}{S}$$

In the Conventional system, mechanical power is measured in horsepower (hp). As with SI metric, we express rotor speed in revolutions per minute (rpm). The equation is:

$$P_m = \frac{S \times T}{5252}$$

P_m is the mechanical power in horsepower
S is the speed in revolutions per minute
T is the torque in pound force-feet
5252 is a constant for Conventional measure

If we know the power and the speed, we can compute torque from the equation:

$$T = \frac{5252 \times P_m}{S}$$

Example 2

A prime mover drives a generator as shown in Fig. 8-4. Torque applied to the generator shaft is 4.07 N·m (3 lbf-ft.). The generator is being driven at a speed of 1750 rpm. Find the mechanical power delivered to the generator in both SI metric units and Conventional units.

a. Compute power in SI metric units.

$$P_m \text{ (watts)} = \frac{T \text{ (N·m)} \times S \text{ (rpm)}}{9.55}$$

$$P_m = \frac{4.07 \times 1750}{9.55} = 746 \text{ W}$$

b. Compute power in Conventional units.

$$P_m \text{ (hp)} = \frac{T \text{ (lbf-ft.)} \times S \text{ (rpm)}}{5252}$$

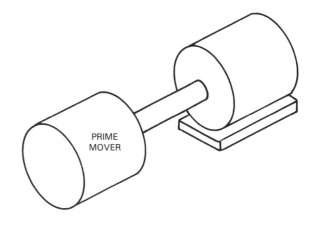

Fig. 8-4. The amount of power produced is proportional to torque and speed.

$$P_m = \frac{3 \times 1750}{5252} \cong 1 \text{ hp}$$

Remember that 746 watts equal 1 horsepower. This is very important. Do not forget that one watt of mechanical power equals one watt of electrical power. Any time you know the mechanical power in horsepower, you can convert it to watts. That way, you can compare mechanical and electrical power in the same units. If you want to convert watts to horsepower, the equation is:

1 watt = 0.00134 horsepower

GENERATOR FIELDS AND WINDINGS

Fig. 8-5 shows four types of generators. To generate electricity you have to start with a main magnetic field. Then, this field must be cut by conductors. The main field can be produced by a permanent magnet. This permanent magnet may be part of the stator as in view A, Fig. 8-5. Or, it may be the rotor as in view B, Fig. 8-5.

The main field can be an electromagnetic field instead of a permanent magnet. The coil that produces it is called the FIELD WINDING, or just the FIELD.

The field may be wound on the stator as in view C, Fig. 8-5. Or, it may be wound on the rotor as in view D, Fig. 8-5.

The conductors, into which electricity is induced, make up the armature winding. The armature winding of dc generators and most small alternators is on the rotor, view C, Fig. 8-5. However, in some ac applications the armature winding is on the stator, views B and D, Fig. 8-5.

HOW GENERATORS WORK

How do generators convert mechanical power to electrical power? Let us take a look at a simple alternator. Refer to Fig. 8-6.

The main magnetic field comes from a pair of permanent magnets. Note that the north pole of

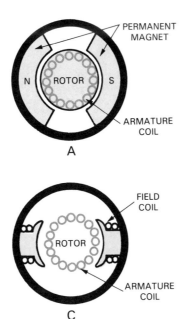

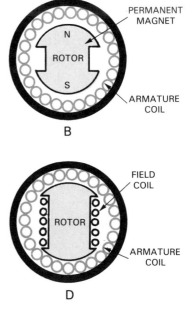

Fig. 8-5. Electricity is generated by changing flux linkages. A—Flux from permanent-magnet stator links conductors on rotor. B—Flux from permanent-magnet rotor links conductors on stator. C—Flux from electromagnetic stator links conductors on the rotor. D—Flux from electromagnetic rotor links conductors on stator.

one faces the south pole of the other. The special curved shape produces the strongest possible field. The armature coil is wound on the rotor. Each end of this coil is attached to its own metal band. These bands are called SLIP RINGS. This is where the generated voltage appears.

To pick up the generated voltage, you must have an electrical path from the slip rings to the generator's terminals. This is done with small pieces of carbon or metal, called BRUSHES. Brushes are held tightly against the slip rings by springs.

As the armature coil revolves, its conductors cut across the magnetic field. This causes a voltage to be induced into the coil. We can figure out the direction in which current will flow from the *right-hand rule for generators.*

RIGHT-HAND RULE FOR GENERATORS

To determine the polarity of a generator, you first must know two directions.

1. The direction (north to south) of the magnetic field.
2. The direction in which the conductor is moving, as it cuts through the field.

You can always determine the directions through use of the right-hand rule for generators. See Fig. 8-7.

The thumb is pointed upward, the index or forefinger is pointing to the left. The middle finger is pointing toward your body.

The forefinger points in the direction of the field flux. (To help you remember, note that *forefinger* starts with the same letter as *field flux.*)

The thumb is pointing in the direction that the wire is moving. (Both *thumb* and *moving* contain an "m.")

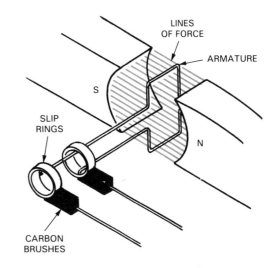

Fig. 8-6. In small alternators, ac voltage is generated in the armature coils as they cut through the magnetic field provided by permanent magnets.

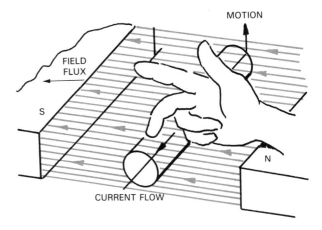

Fig. 8-7. Right-hand rule for generators. Position of thumb and fingers give directions for current, flux and conducting wire.

Now, your middle finger is parallel to the wire. It is pointing in the direction of conventional current flow. Conventional current flows minus-to-plus inside a voltage source. Therefore, the middle finger is pointed toward the plus terminal.

Remember that current does not really begin to flow until there is a closed circuit between the generator terminals. However, we describe the polarity by the direction current will flow when a load is connected.

POLARITY OF GENERATOR TERMINALS

Fig. 8-8 is a simplified generator. It shows how the terminal polarities change as the loop of wire rotates. To help you keep track of the motion, the wire has been marked A-B-C-D. In view A, Fig. 8-8, the section of wire marked A-B is moving downward through the field. The section marked C-D is moving upward through the field.

Now, let us take one section at a time. We need not worry about section B-C. It is outside the field and not cutting through any magnetic flux.

Using the right-hand rule, we can see that point A will be positive, compared to B. In the C-D section, point C is made positive compared to D. A-B is connected series-aiding to C-D. This makes terminal D negative and terminal A positive.

In view B, Fig. 8-8, the loop has gone through half a revolution. Applying the right-hand rule again, we see that the polarity has reversed. Now, terminal D is positive and terminal A is negative.

THE AMOUNT OF VOLTAGE INDUCED

As each loop of the armature coil moves from one part of the field to another, it links a different number of flux lines. It is this changing flux linkage that induces voltage into the wire. The most voltage is induced at the instant this change is the greatest. That is the instant the wire is cutting through the field at right angles, as in Fig. 8-8.

The other extreme is shown in Fig. 8-9. Twice during each revolution, the wire travels parallel to the flux lines for an instant. At that instant, the wire does not cut through any flux at all. Induced voltage is zero, then. At all other times, however, the wire cuts through at an angle. Therefore, as the wire goes around, the voltage

in it rises to a peak, falls to zero, reverses direction, then rises to a peak in the opposite direction. One revolution is shown in Fig. 8-10. Direction of current flow is shown by the "x" and dot in the wire ends. Think of the dot as the arrow tip and the "x" as the feathered end. When you see the tip current is moving toward you. When you see the feather it is moving away.

As long as the rotor turns at a constant speed, a sine wave voltage is induced. The value of that voltage, however, depends on rotor speed. The faster the speed, the greater the voltage.

The value of the voltage also depends on the strength of the magnetic field. The stronger the field, the greater the voltage. Generated voltage (E_G) is therefore proportional to both magnetic flux, ϕ, and rotor speed (S). The equation is:

$$E_G = K_G \times \phi \times S$$

E_G is the generated voltage
ϕ is flux
S is the speed in rpm

In this equation, the constant, K_G, depends on the design of the generator. In permanent-magnet generators, the flux, ϕ, is always the same value, too. Therefore, these two factors can be combined into what is called the voltage constant, K_E. This makes it easier to compute voltage at different speeds.

$$K_G \times \phi = K_E \text{ volts per rpm;}$$
therefore, $E_G = K_E \times S$

Example 3
A permanent-magnet generator has a voltage constant (K_E) of 0.002 volts per rpm. Find the generated voltage at 2000 rpm and 4000 rpm.
a. Compute the generated voltage at 2000 rpm.
  $E_G = K_E \times S = 0.002 \times 2000 = 4 \text{ V}$
b. Compute the generated voltage at 4000 rpm.
  $E_G = K_E \times S = 0.002 \times 4000 = 8 \text{ V}$

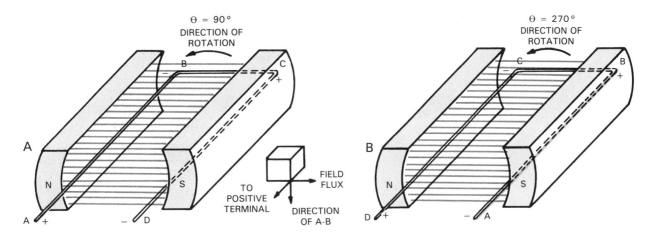

Fig. 8-8. Maximum voltage is induced at the instant the conductor cuts through the magnetic field at right angles. A—Segment A-B has current moving from B to A. B—Current reversed moving from point A to point B.

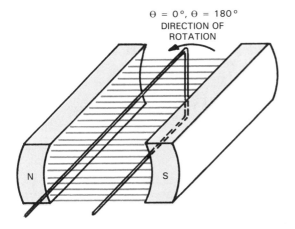

Fig. 8-9. No voltage is induced at the instant the conductor moves parallel to the magnetic field.

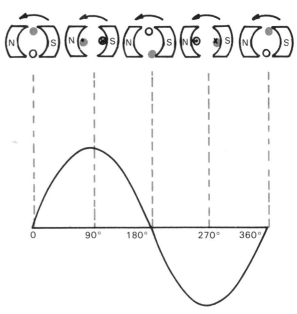

Fig. 8-10. The instantaneous value of the generated voltage follows a sine wave pattern as the conductor rotates. Reversal of current flow is shown on wire ends. Segment A-B is shown in blue.

FREQUENCY OF GENERATED VOLTAGE

The generator pictured at the top of Fig. 8-10 has two poles. One is a north pole; the other is a south pole. Every time the armature coil (single loop of wire) goes around one turn, one cycle of ac is generated. One revolution a second produces one cycle of electric current a second. One cycle per second is a frequency of 1 hertz.

In Fig. 8-11 we have added another pair of poles. (Poles always come in pairs.) Now, one revolution of the rotor produces two cycles of ac. The frequency, in hertz, is twice the number of revolutions per second.

The frequency of the generated voltage, in hertz, is always equal to the speed in revolutions per second times the number of pairs of poles. Speed, however, is usually measured in revolutions per minute. Therefore we have to divide rpm by 60. The equation is:

$$f \text{ (hertz)} = \frac{S \text{ (rpm)}}{60} \times \text{pairs of poles}$$

131

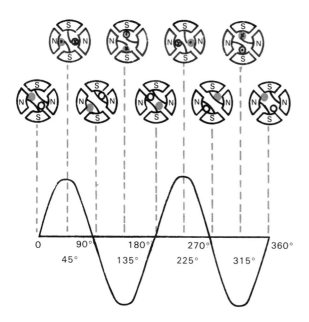

Fig. 8-11. With two pairs of poles, one revolution produces two cycles of alternating current.

Example 4

A two-pole generator is driven at 1800 rpm. Find the frequency of the generated voltage. Find the frequency of a four-pole generator driven at the same speed.

a. Compute the frequency of the two-pole generator.

$$f = \frac{S}{60} \times \text{pairs of poles}$$

$$= \frac{1800}{60} \times 1 = 30 \text{ Hz}$$

b. Compute the frequency of the four-pole generator.

$$f = \frac{S}{60} \times \text{pairs of poles}$$

$$= \frac{1800}{60} \times 2 = 60 \text{ Hz}$$

ROTATIONAL LOSSES

You have seen how a prime mover causes a rotor to turn. Some mechanical power is needed just to do this. It supplies the rotational losses of WINDAGE and friction. This mechanical power is not converted to electrical power. Additional mechanical power is needed when an electrical load is connected to the generator.

GENERATOR LOAD

Assume you have a load (such as a headlamp) connected between the generator's terminals. A certain value of current will flow. This current depends on the terminal voltage and the load resistance ($I = \frac{E}{R}$). From Fig. 8-12 you can see that this current also flows through the armature coil. Electrical power ($P = E \times I$) is drawn from the generator by the load. At the same time, the generator draws mechanical power from the prime mover. In fact, it is the load current that makes the prime mover supply power.

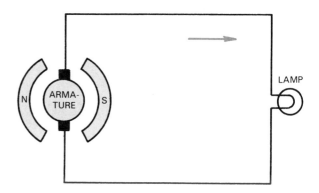

Fig. 8-12. Load current and armature current are the same current.

MECHANICAL-TO-ELECTRICAL POWER

Refer to Fig. 8-13. What happens is this:

1. As current flows through the armature coil, the rotor becomes an electromagnet.
2. Current, however, is alternating current.
   a. The rotor's magnetic poles change twice during each revolution.
   b. Thus, the north rotor pole is always approaching the north stator pole.
   c. Likewise, the south rotor pole is always approaching the south stator pole.
   d. Like poles tend to repel each other. Thus, while the prime mover turns the rotor in one direction, magnetic forces are trying to push the rotor the opposite way.

132

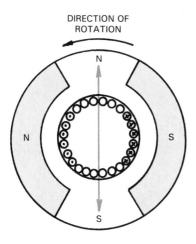

DIRECTION OF ROTATION

N

N

S

S

Fig. 8-13. At every instant, the north pole of the rotor is approaching the north pole of the main field, producing a counter torque.

We are led to expect this from Lenz's Law. *Lenz's Law says that induction always tends to oppose whatever causes it.* The prime mover supplies a torque. The rotor's magnetic field is therefore induced in a way that produces a counter torque.

COUNTER TORQUE

Now we see that the torque supplied by the prime mover has two jobs. One is overcoming rotational losses. The other is to overcome the counter torque.

Counter torque is the repulsion force between two magnetic fields. Counter torque is, therefore, proportional to the strength of those fields.

The symbol for the flux of the main field is ϕ. Rotor field strength, however, is given in terms of armature current, I_A. The equation is:

$$T_c = K_M \times \phi \times I_A$$
$\quad T_c$ is counter torque
$\quad \phi$ is flux
$\quad I_A$ is armature current in amperes

In this equation, the constant, K_M, depends on the design of the generator. The flux, ϕ, is

also constant in a permanent-magnet generator. The two factors, K_M and ϕ, can be combined into what is called the torque constant, K_T. This tells you the amount of counter torque produced for each ampere of load current.

The torque constant, K_T, can be given in newton metres per ampere, pound force-feet per ampere, ounce force-inches per ampere, or some similar torque expression.

Example 5

The permanent-magnet generator of Fig. 8-14 has a torque constant (K_T) of 0.0191 newton metres per ampere (2.7 ozf-in./amp). Terminal voltage is 4 volts. Speed is 2000 rpm. The load is a lamp with a resistance of 20 ohms. Find the counter torque and the power input to the generator in both SI metric and Conventional units and the power output to the lamp.

a. Compute load current.
$$I = \frac{E}{R} = \frac{4}{20} = 0.2 \text{ A}$$

b. Compute counter torque in SI metric units.
$$T_c = K_T \times I_A = 0.0191 \times 0.2$$
$$= 0.00382 \text{ N·m}$$

c. Compute input power in SI metric units.
$$P_M = \frac{\text{counter torque } (T_c) \times S}{9.55}$$
$$= \frac{.00382 \times 2000}{9.55} = 0.8 \text{ W}$$

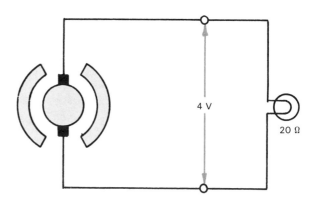

4 V

20 Ω

Fig. 8-14. Counter torque is proportional to field strength and armature current.

d. Compute counter torque in Conventional units.
$$T_c = K_T \times I_A = 2.7 \times 0.2$$
$$= 0.54 \text{ ozf-in.}$$

e. Convert ozf-in. to lbf-ft.
$$1 \text{ lbf-ft.} = 192 \text{ ozf-in.} \therefore \frac{0.54}{192}$$
$$= .0028 \text{ lbf-ft.}$$

f. Compute input power in Conventional units.
$$P_M = \frac{T_c \times S}{5252} = \frac{.0028 \times 2000}{5252}$$
$$= 0.00107 \text{ hp}$$

Note: 1 watt = 0.00134 hp. Therefore, 0.8 W = 0.00107 hp. The input power computed in Step c is the same, then, as that computed in Step f.

g. Compute power output.
$$P = E \times I = 4 \times 0.2 = 0.8 \text{ W}$$

COPPER LOSS

In *Example 5,* we assumed that the wire in the armature coil did not have any resistance. Eight-tenths of a watt (0.00107 hp) was supplied to the generator rotor. Then, 0.8 watts of electrical power was supplied to the load by the generator.

In real generators, current flowing through the armature coils causes an electrical power loss (I^2R). This COPPER LOSS becomes larger as the load current increases.

Fig. 8-15 shows what happens to the power delivered by a prime mover. Part of it is lost in windage and friction. These are called rotational losses. The rest is converted into electrical power. Then, part of the electrical power is lost in the winding resistance, called copper loss. In addition a small amount of power is lost in what is called STRAY POWER LOSS. Finally, electrical power is delivered to the load.

OTHER TYPES OF GENERATORS

The kind of permanent-magnet generators we have been studying are available only in small sizes. Larger single-phase alternators have a field coil on the stator. The armature coil is on the rotor. You must supply the field coil with direct current to produce the main magnetic field. This is shown in Fig. 8-16.

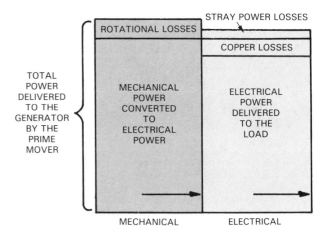

Fig. 8-15. Typically, about 85 to 90 percent of the mechanical power is delivered as electrical power.

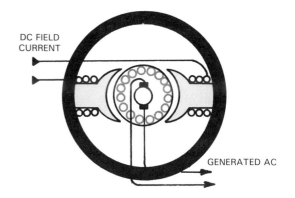

Fig. 8-16. Field coils wound on stator pole pieces produce the main field for larger dc generators.

Direct-current generators can have a permanent magnet field. Or, they can have a field coil, as shown in Fig. 8-17. For dc generators, you can supply the field with current generated by its own armature. The field coil can be connected with the armature in series. Or, it may be connected in parallel.

Sometimes two separate field coils are used. One is connected in series with the armature; the other in parallel.

Large three-phase alternators, like the one shown in Fig. 8-18, have coils on both the rotor and stator. However, the armature coil is on the stator. The field coil is on the rotor. Direct current is applied to the field coil to produce the main magnetic field. As the field sweeps around, its flux is cut by the armature windings. Voltage is thereby induced into the fixed armature windings.

MOTORS

The power input to a motor is electrical. A voltage is applied to a motor's terminals resulting in a current. The power output of a motor is mechanical. This power is transmitted by the rotor shaft as a torque. This torque tends to rotate a load, such as a fan or pump.

To drive a load at a particular speed, you need a certain amount of torque. If the motor's output is large enough, the load will turn. If the motor's output torque is too small, however, it will not drive the load. Torque requirement is one of the most important things to consider in selecting a motor.

As a motor drives a load, mechanical power is drawn from the motor. The motor, in turn, draws electrical power from the source at the same rate.

HOW MOTORS WORK

Motors convert electrical power to mechanical power by using magnetic forces. To see how this is done, we can examine a small dc motor, Fig. 8-19.

As shown in Fig. 8-20, the main field is supplied by a pair of permanent magnets. Notice the horizontal direction of this field is from north to south. The single loop represents the entire armature coil, which is wound on the

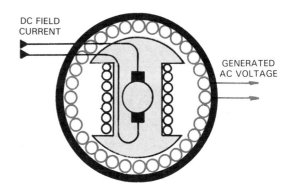

Fig. 8-18. On large alternators, the dc field coil is wound on the rotor.

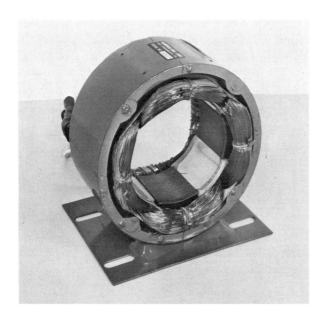

Fig. 8-17. Stator of a dc generator. Current to energize the field coil can be supplied by the generator itself.

Fig. 8-19. Permanent-magnet dc motor. Such motors may be used on portable appliances or motorized toys. (Ohio Electric Motors Div. of Magnetics International, Inc.)

rotor. Forces act on each coil loop. The total torque developed by the motor is the sum of the torques on each loop.

A voltage, V_A, is applied to the armature coil. This causes a current, called the armature current, I_A. The direct current causes the rotor to have a magnetic polarity. The polarity is determined by the right-hand rule for coils as shown in Fig. 8-21. Now the magnetic poles of the stator exert a force on the magnetic poles of the rotor. The stator south pole attracts the rotor north pole and repels the rotor south pole. The stator north pole attracts the stator south pole and repels the rotor north pole. This is illustrated in Fig. 8-22

LEFT-HAND RULE FOR MOTORS

Another way to determine the rotor's direction of rotation is the left-hand rule for motors. It is shown with cubes and arrows in Fig. 8-23.

You can also arrange the fingers and thumb of your left hand to show this. First hold the thumb, index finger and middle finger at right angles to each other. Then turn your hand so that your index finger points in the direction of field flux. Next, rotate your hand until the middle finger points in the direction of conventional current flow. Now your thumb is pointing in the direction of conductor motion.

Fig. 8-23 shows us an upward force on one side and a downward force on the other. This agrees with the torque direction we discovered in Fig. 8-22.

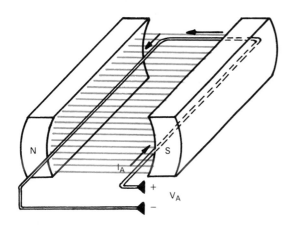

Fig. 8-20. The dc motor armature coil is represented by a single loop of wire. Often, construction of the dc motor and dc generator is identical.

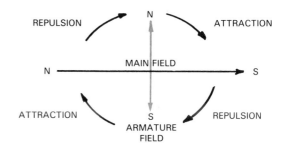

Fig. 8-22. The main field interacts with the rotor field according to the rule: Unlike poles attract; like poles repel.

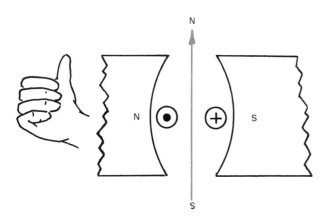

Fig. 8-21. Current through the armature coil produces a rotor field.

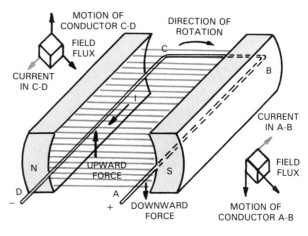

Fig. 8-23. There is a definite relationship between the direction of the main field, the direction of current flow and the direction of force.

THE COMMUTATOR

If real dc motors were like the one of Fig. 8-23, they would stop after a quarter turn. The rotor poles would line up with the stator poles, producing no rotational forces. Before that can happen, the rotor must reverse magnetic polarity. Something must cause the electric polarity of the armature to reverse. This is the job of the COMMUTATOR.

In dc machines (both motors and generators), electrical connections to the armature coils are made through the commutator. The commutator is mounted on the rotor shaft. It has two or more segments (pieces) which are insulated from each other. Fig. 8-24 represents a simplified motor with a single armature coil and a two-segment commutator. The motor terminals are connected to carbon brushes that make contact with the commutator. Notice that each brush always has the same polarity.

To see how the commutator affects the magnetic polarity of the rotor, study the three sketches of Fig. 8-25. End X of the armature coil is connected to commutator segment 1. End Y is connected to segment 2. View A, Fig. 8-25 shows the negative brush riding on segment 1. Current flows from Y to X. The rotor's north magnetic pole is at the top. This produces a clockwise torque and the rotor turns.

A quarter revolution later, in view B of Fig. 8-25, the rotor poles have tried to line up with the stator poles. However, the commutator has turned, too. The brushes are now bridging the insulated portion of the commutator. The armature coil is shorted. At this instant, there is no current flow through the armature coil. The rotor continues to turn, however, because of inertia.

In view C of Fig. 8-25, the brushes are once again touching commutator segments. This is a half revolution after view A.

Now the negative brush is riding on segment

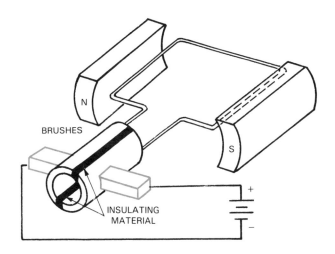

Fig. 8-24. The commutator conducts current to the armature coil through brushes.

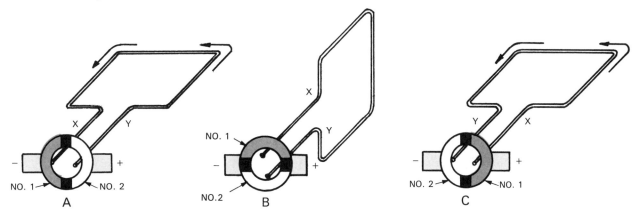

Fig. 8-25. The commutator reverses current flow through the armature coil as the rotor rotates. A — Negative brush rides on segment 1 of commutator. Current flows from Y to X. B — After quarter turn no current flows through armature coil. C — Negative brush rides on segment 2 of commutator. Current flows from X to Y.

2. Current flows from X to Y. The rotor's north magnetic pole is still at the top. The magnetic forces keep the rotor turning in the same direction.

TORQUE DEVELOPED BY A MOTOR

How much torque is produced by the magnetic forces? Torque depends on three things:
1. The strength of the stator field, ϕ.
2. The strength of the rotor field. (This is expressed in terms of the armature current, I_A.)
3. The design of the motor, including such things as the number of stator poles, the number of loops on the armature coil and the way the armature coil is wound and connected.

The equation is:
$$T = K_M \times \phi \times I_A$$

This equation may look familiar. The counter torque of a generator is computed in exactly the same way.

For permanent-magnet dc motors, you can combine the factors K_M and ϕ. Manufacturers may specify this as the TORQUE CONSTANT, K_T. They may give you the K_T of a motor in newton metres per ampere, pound force-feet per ampere, ounce force-feet per ampere or some similar torque expression.

Example 6

A permanent-magnet dc motor has a K_T of 0.05 newton metres/ampere (7.068 ozf-in./ampere). The resistance of the armature winding (R_A) is 2 ohms. Applied voltage (V_A) is 6 volts. Find the starting torque in both Conventional and SI units. (This is the torque developed by the motor at the instant voltage is applied.)
   a. Compute the armature current at the instant of start.
   $$I_A = \frac{V_A}{R_A} = \frac{6}{2} = 3 \text{ A}$$
   b. Compute the starting torque in SI metric units.
   $$T = K_T \times I_A = 0.05 \times 3 = 0.15 \text{ N·m}$$
   c. Compute the starting torque in Conventional units.
   $$T = K_T \times I_A = 7.068 \times 3$$
   $$= 21.204 \text{ ozf-in.}$$

ELECTRICAL-TO-MECHANICAL POWER

Magnetic forces produce a torque on the armature conductors. They, in turn, apply a torque to the rotor. If there is a load coupled to the motor, this torque drives the load. We are assuming the motor can put out enough torque. (If a load needs more torque than the motor can develop, the motor will stall.)

The motor draws electrical power from the source at the same time and at the same rate mechanical power is being used by the load. Motor current increases as the mechanical load gets heavier. It decreases as the load gets lighter. How does the electrical source know how much mechanical power is being used? It can tell by the amount of COUNTERELECTROMOTIVE FORCE (CEMF) being generated in the armature.

COUNTERELECTROMOTIVE FORCE (CEMF)

Electromotive force is another name for voltage. It usually describes the voltage induced in the armature coil of a generator. *Counterelectromotive force* is the voltage induced in the armature coil of a motor.

As the rotor turns, the armature conductors cut through the main magnetic field. Refer to Fig. 8-26. This is what induces a counter voltage into the armature windings.

CEMF is a real voltage, even though it can not be measured separately. It is induced in a way that opposes the voltage applied to the armature. It is like having a smaller voltage source connected series-opposing with the applied volt-

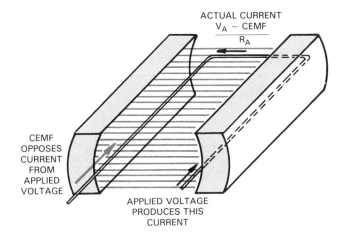

Fig. 8-26. Counterelectromotive force (CEMF) is generated in a motor's armature as the rotor rotates.

age. The current, I_A, that ends up flowing in the armature depends on the NET VOLTAGE. Net voltage equals the applied voltage *minus* CEMF. The equation is:

$$I_A = (\frac{V_A - CEMF}{R_A})$$

Where R_A is the resistance of the armature

CEMF is always less than applied voltage. The amount of CEMF induced depends on how much flux, ϕ, is cut and how fast. The equation is:

$$CEMF = K_G \times \phi \times S$$
CEMF is in volts
 ϕ is field strength
 S is speed in rpm

The factor, K_G, depends on the design of the motor. For permanent-magnet motors, manufacturers can provide a "generating constant," K_E. This is a combination of K_G and ϕ. The units of K_E may be volts per rpm or kilovolts (1000 volts) per rpm.

HOW MOTOR TORQUE IS RELATED TO CEMF

Assume you have a motor without voltage applied. The rotor is standing still. Then you turn on the switch. Let us freeze the instant you

apply power, as shown in Fig. 8-27. The rotor has not started to turn yet. Its speed is zero. Therefore, CEMF is zero. With no CEMF, the net voltage is as high as it can possibly be. That is, net voltage equals the applied voltage. This makes the armature current as high as it can be. It is the armature current, remember, that determines the torque ($T = K_T \times I_A$). Therefore, you have a strong starting torque.

Now, it is time to unfreeze the starting instant and move on. Refer to Fig. 8-28. Assume the motor is delivering enough starting torque to drive the load. The rotor will turn. As the rotor speeds up, more and more CEMF will be induced into the armature conductors. The NET

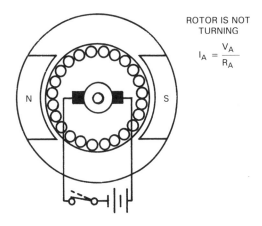

Fig. 8-27. Starting (or stall) current is very high because there is no CEMF.

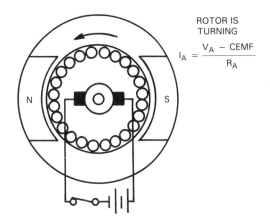

Fig. 8-28. Armature current is limited by CEMF when the motor is running.

VOLTAGE decreases. This makes the armature current drop. As a result, the torque output of the motor falls off.

At some point, the motor will stop building up speed. At that speed, a certain CEMF is induced. This CEMF results in the correct value of net voltage. It produces just the right amount of armature current. This current develops the torque needed to drive the load at that speed.

All of this happens automatically as a motor drives a load. If a load gets heavier, the motor needs more torque. This torque can only come from more armature current. So the motor slows down, inducing less CEMF. The result is a large net voltage and, consequently, more armature current.

Example 7

A dc permanent-magnet motor drives a load requiring 0.005 newton metres (0.7068 ozf-in.). Motor data is as follows: armature resistance, 2 ohms; applied voltage, 6 volts; torque constant, 0.05 newton metre/amp (7.068 ozf-in./amp); generating constant, 0.0052 volts/rpm. Find the motor's running speed.

a. Compute armature current.
$$T = K_T \times I_A \therefore I_A = \frac{T}{K_T} = \frac{.005}{.05}$$
$$= 0.1 \text{ A}$$
(also, $I_A = \frac{0.7068}{7.068} = 0.1$ A.)

b. Compute net voltage.
$$V_{NET} = I_A \times R_A = 0.1 \times 2 = 0.2 \text{ V}$$

c. Compute CEMF.
$$V_{NET} = V_A - CEMF \therefore CEMF$$
$$= V_A - V_{NET} = 6 - 0.2$$
$$= 5.8 \text{ V}$$

d. Compute motor speed.
$$CEMF = K_E \times S \therefore S = \frac{CEMF}{K_E}$$
$$= \frac{5.8}{.0052} = 1115 \text{ rpm}$$

POWER OUTPUT OF A MOTOR

A motor delivers torque to a load through its shaft. This torque could result in no power, a little power or a lot of power. *Mechanical power, remember, depends on torque and speed. If the motor delivers too little torque, the load will not turn.* In that case, power output is zero. If the load is turning, however, power output is proportional to torque times speed.

The equation, in SI metric, is:

$$P = \frac{T \times S}{9.55}$$
P is power in watts
T is torque in newton metres
S is speed in rpm

Or in Conventional units:

$$P = \frac{T \times S}{5252}$$
P is in horsepower
T is in pound force-feet
S is in rpm

TORQUE MEASUREMENT

Suppose you have a motor running without any load coupled to it. The rotor may be turning very fast but the motor is not delivering any mechanical power. Some of the electrical power input is lost in the armature windings. The part that is converted to mechanical power is used to overcome rotational losses. The torque output, therefore, is zero. Zero torque means zero power output.

Fig. 8-29 shows a device used for measuring torque. It is called a PRONY BRAKE. It puts a load on the motor. If you know the value of this load, you know the torque output of the motor. Then you measure the motor speed. Power output is computed from torque and speed.

The prony brake drum is coupled to a motor shaft. To load the motor, a canvas belt is tightened against the drum. This applies a force to the pivot arm.

The force is measured by a scale at a specific distance from the shaft's center. If the scale reads ounces of force and the distance is in inches, torque is in ounce-force inches. If the scale reads newtons and the distance is in metres, torque is in newton metres.

Another way to load a motor is with a dynamometer, like the one in Fig. 8-30. A dynamometer is a dc generator. When you load the generator, you load the motor coupled to it. The housing of the dynamometer is free to rotate slightly. The counter torque developed in the generator is applied to the stator. As the housing tries to turn, it is held back by a spring scale. The scale can be made to read torque in any desired units.

SPEED MEASUREMENT

Speed measuring devices are called TACHOMETERS. All read directly in revolutions per minute.

Mechanical type tachometers work like a car

speedometer. A rubber cone tip is held against the motor shaft. The dial indicates the speed of the spinning shaft.

The generator type tachometer has a small permanent-magnet generator driven by the motor shaft. Voltage output is proportional to speed. The indicator, therefore, is a voltmeter. Its scale, however, is marked in revolutions per minute.

An adjustable strobe light tachometer is shown in Fig. 8-31. The strobe light flashes on and off at the rate set on its dial. To measure speed, direct the light at the rotating motor shaft and adjust the dial. The shaft will appear to slow down, then stop. This is called the STROBOSCOPIC EFFECT. The dial setting when the shaft appears to be standing still is the speed in rpm.

Example 8

A dc permanent-magnet motor has a torque constant, K_T, of 0.0764 newton metres (0.0564 lbf-ft.) per ampere. Its armature re-

Fig. 8-29. A prony brake is a mechanical device for measuring the torque produced by a motor.
(Hampden Engineering Corp.)

Fig. 8-30. This dynamometer, used for measuring torque, is an electromechanical device.

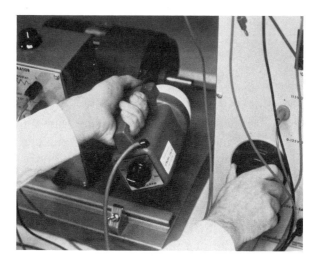

Fig. 8-31. Strobe light tachometer is used to measure rotational speed.

sistance, R_A, is 2 ohms. It is driving a load that requires 0.2 newton metres (0.1476 lbf-ft.) at its speed of 4700 rpm. Applied voltage, V_A, is 45 volts. Armature current, I_A, is 2.6 amperes. The rotational losses in the motor total 5 watts (0.0067 hp). Find the "power in," the electrical "copper" loss, the "power out" in SI metric and Conventional units, and the efficiency of the motor.

a. Compute the power in.
$$P_{IN} = E \times I = 45 \times 2.6 = 117 \text{ W}$$

b. Compute the copper losses, P_{CL}.
$$P_{CL} = I_A{}^2 \times R_A = 6.76 \times 2 \cong 13.5 \text{ W}$$

c. Compute the power out in SI metric units.
$$P_{OUT} = \frac{T \times S}{9.55} = \frac{0.2 \times 4700}{9.55}$$
$$\cong 98.5 \text{ W}$$

d. Compute the power out in Conventional units.
$$P_{OUT} = \frac{T \times S}{5252} = \frac{0.1476 \times 4700}{5252}$$
$$= 0.132 \text{ hp}$$

Note: The Conventional units of mechanical power (horsepower) must always be converted to SI metric units (watts) when computing the efficiency of a motor.

$$\text{Watts} = \text{hp} \times 746 = 0.132 \times 746$$
$$\cong 98.5 \text{ W}$$

e. Compute efficiency.
$$\% \text{ Eff.} = \frac{P_{OUT}}{P_{IN}} \times 100 = \frac{98.5}{117} \times 100$$
$$= 84.2\%$$

POWER LOSSES AND POWER DELIVERED

In *Example 8,* we could have computed the CEMF as follows:

$$V_{NET} = I_A \times R_A = 2.6 \times 2 = 5.2 \text{ V}$$
$$CEMF = V_A - V_{NET} = 45 - 5.2$$
$$= 39.8 \text{ V}$$

The CEMF power (CEMF $\times I_A$) is the power that gets converted to mechanical power.

$$CEMF \times I_A = 39.8 \times 2.6 = 103.5 \text{ W}$$

The total power in, therefore, equals the converted power plus the copper losses.

$$P_{IN} = (CEMF \times I_A) + I^2 R_A$$
$$= 103.5 + 13.5 = 117 \text{ W}$$

We have found that 103.5 watts of electrical power has been converted to mechanical power. However, only 98.5 watts of power is delivered to the load. The rest (5 watts) is the rotational losses and the stray power losses.

Fig. 8-32 shows that some of the incoming power is used inside the motor. The rest is delivered to the load.

OTHER TYPES OF MOTORS

Permanent-magnet motors are usually less than two horsepower (1492 watts). Larger dc motors have stators wound with field coils.

Alternating current motors always have wound stators. They can be either single-phase or three-phase. Most types of ac motors are INDUCTION MOTORS. Rotor current is the result of voltage induced in the rotor windings. Each type operates a little differently from the others. However, the same basic principles apply to all of them.

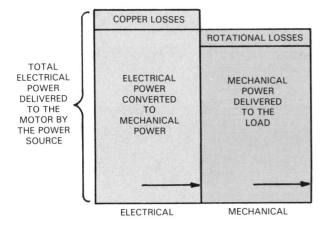

TOTAL ELECTRICAL POWER DELIVERED TO THE MOTOR BY THE POWER SOURCE

COPPER LOSSES

ROTATIONAL LOSSES

ELECTRICAL POWER CONVERTED TO MECHANICAL POWER

MECHANICAL POWER DELIVERED TO THE LOAD

ELECTRICAL   MECHANICAL

Fig. 8-32. Typically, 80 to 90 percent of the electrical power supplied to a motor is delivered as mechanical power to the load.

SUMMARY

Both generators and motors operate from induced voltage and magnetic forces.

Both generators and motors have stators which do not move and rotors which rotate.

The input of a generator, as well as the output of a motor, is torque (turning force). Torque times speed equals mechanical power.

The basic unit of torque in the SI metric system is the newton metre (N·m). In the Conventional system it is pound force-foot (lbf-ft.).

The basic unit of mechanical power in the SI metric system is the watt (W). In the Conventional system it is horsepower (hp).

Voltage is generated either by moving conductors that are linked to a magnetic field or by moving a field that is magnetically linked to conductors.

The amount of voltage generated depends on magnetic field strength and speed of rotation.

Both ac and dc generators generate an ac voltage. In dc generators, however, the ac is converted to dc by the commutator.

The frequency of the voltage generated in an alternator depends on speed of rotation and the number of magnetic poles.

Current in the armature conductors of a generator creates a magnetic field. This field works against the main field to produce a counter torque.

The prime mover supplies power to overcome the counter torque. This is the mechanical energy that is converted to electrical energy.

In a motor, the motion of the armature through the main magnetic field induces a counter-electromotive force (voltage) in the armature conductors.

The power source supplies power to overcome the counterelectromotive force. This is the electrical energy that is converted to mechanical energy.

The principal causes of lost power in motors and generators is the resistance of the wire and the mechanical losses of windage and friction.

TEST YOUR KNOWLEDGE

1. Give four examples of prime movers.
2. Explain what is meant by "generator action" and "motor action."
3. A generator converts (mechanical, electrical) energy to (mechanical, electrical) energy, while a motor converts (mechanical, electrical) energy to (mechanical, electrical) energy.
4. Both the _____ and the _____ of the voltage generated by an ac alternator is affected by its speed.
5. What is the purpose of the commutator in a dc motor? In a dc generator?
6. What is torque? How is it related to mechanical power?

PROBLEMS

1. Given: A prime mover is exerting a torque of 4 newton metres on the shaft of a generator, driving it at 1781 rpm. Find the mechanical power output of the prime mover in both watts and horsepower.

2. Given: A dc generator has a voltage constant, K_E, of 0.04 volts per rpm. Find the voltage when it is driven at 1500 rpm.

3. Given: A 4-pole ac alternator is driven at 600 rpm. Find the frequency of the generated voltage.

4. Given: A dc motor has a torque constant, K_T, of 0.2 newton metres per ampere. Find the torque output when current is 3 amperes.

5. Given: The motor of *Problem 5* is driving a load at 1800 rpm. Find the mechanical power delivered by the motor.

6. Given: A dc motor has a torque constant, K_T, of 0.4 newton metres per ampere, and an armature resistance of 2 ohms. Find the starting torque at the instant 120 volts is applied in both SI metric units and Conventional units.

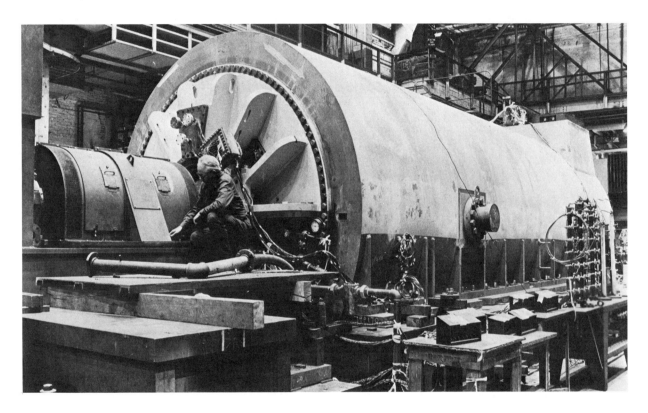

Huge turbine-generator is designed to be operated by a high-pressure steam turbine.   (Westinghouse Electric Corp.)

9 SINGLE-PHASE AC MOTORS

After studying this chapter, you will be able to:

☐ List the factors that affect the torque produced by an ac motor.

☐ Explain the principle of operation and characteristics of a series ac motor.

☐ Describe the construction and operation of a repulsion-start induction motor.

☐ List five types of single-phase induction motors.

Most of the electricity we use is in the form of single-phase alternating current. You can find single-phase ac in homes, schools, stores, offices and manufacturing plants. That is why there are more single-phase ac motors than any other kind. The average home has dozens of motor-driven devices. It might be fun to count the motors in your home.

THE MAGNETIC FIELD OF A MOTOR

Like all motors, single-phase ac motors have two main parts:
1. A rotor, which rotates on a shaft.
2. A stator which does not move.

Both the stator and rotor have coils through which current flows. Fig. 9-1 shows one type of rotor. It has a commutator like the dc motor in Chapter 8. The other type, Fig. 9-2, is known as a "squirrel-cage" rotor. Its coil is made up of aluminum bars short circuited at both ends.

AC MOTOR OPERATION

Electricity produces magnetism. Magnetism produces mechanical forces. These forces produce the torque that turns the rotor of a motor.

The principle of motor operation, just described, is simple. Beyond this, however, are a number of problems. For example, there is the problem of producing current flow in the rotor. Some single-phase ac motors use conduction. Others use induction.

Then there is the problem of phase. The strength of each field follows a sine wave. Ideal-

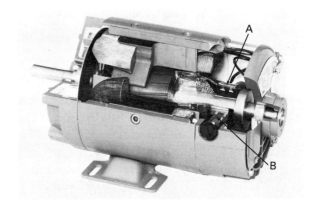

Fig. 9-1. Cutaway of alternating current motor with commutator type rotor. A—Commutator. B—One of brushes. (Gould Inc., Electric Motor Div.)

Fig. 9-2. This squirrel-cage rotor is from a small single-phase electric motor. (Hampden Engineering Corp.)

ly, they should reach peak strength at the same instant.

Finally, the angle between the fields is important. If the rotor field is at right angles to the stator field, maximum torque is produced. Any other angle produces less torque. If the poles of both fields are along the same line, for example, there is no torque produced at all.

SERIES AC MOTOR

The series motor has a commutator type rotor. The rotor coil is connected in series with the stator coil. Power is applied to this series combination, as shown in Fig. 9-3.

CONSTRUCTION OF A SERIES MOTOR

Fig. 9-4 is a diagram of the stator of a series motor. Notice the magnetic pole pieces, called shoes. The coils wound on these shoes are called the field coils. Their job is to produce the main magnetic field. When current passes through the field coils, one shoe becomes a north pole while the other becomes a south pole. The magnetic stator circuit is shown in Fig. 9-5. This is 60 Hz alternating current, however. The poles change magnetic polarity 120 times a second.

The rotor of a series motor is like the one shown in Fig. 9-1. Its core, like the stator core,

is made of steel laminations. Wires of the armature coils are placed into slots. Each end of each coil is connected to a segment of a commutator. Fixed brushes make contact with the commutator segments. As the rotor is turned, power is applied to one coil after the other. This is illustrated by Fig. 9-6.

So far, we have solved two of the problems. First, current is conducted to the rotor. Second, because the coils are in series, the same current produces both fields. This makes the rotor field in phase with the stator field at all times. Now we have to make sure the fields are lined up in the right direction.

DIRECTION OF MAGNETIC FIELDS

As was shown in Fig. 9-5, the stator's magnetic field is horizontal. The direction of the rotor

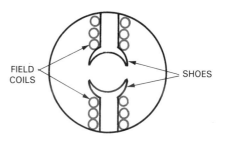

Fig. 9-4. Diagram of a stator for a series ac motor. Current in windings makes one shoe a north pole and the other shoe a south pole.

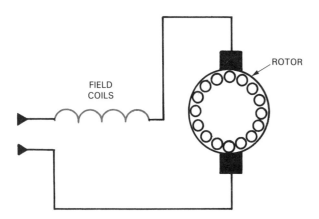

Fig. 9-3. Circuit diagram of a series ac motor. Stator and rotor coils are connected in series.

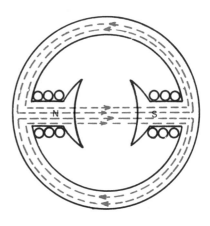

Fig. 9-5. Magnetic circuit of stator shows path of main field.

field depends on two things:
1. The position of the brushes.
2. The armature connection.

For best results, the brushes are located where they will produce a vertical rotor field. *It is important to remember that the rotor field does not rotate.* The commutator makes sure of this by changing the pattern of current through the armature coils. As a result, the rotor field is always vertical.

These two fields attract and repel each other. This produces a torque on the rotor. The torque may be clockwise or counterclockwise. Fig. 9-7 shows one possibility. The instantaneous polari-

ty causes the stator field to point from left to right. The rotor field points vertically downward. A clockwise torque results. What is more, as current reverses direction each cycle, the torque is still clockwise. This reversal is shown by Fig. 9-8.

We can reverse the motor's direction by interchanging the brush connections. Now the instantaneous polarity of the armature coil will be the opposite of before. As Fig. 9-9 and Fig. 9-10 show, torque reverses, too. Now there is a counterclockwise torque.

TORQUE IN A SERIES MOTOR

We now know how the series motor produces torque. Now let us see how much torque. The standard "torque equation" for all kinds of ac motors is:

$$T = K_T \times \phi \times I_R \times \cos \Theta$$
K_T is a design constant
  ϕ is stator field strength (flux)
  I_R is rotor current
  Θ is the displacement angle

When the rotor field is at right angles with the stator field, the displacement angle, Θ, is zero. On the other hand, the largest possible displacement angle is 90 degrees. That occurs when the rotor field is lined up parallel with the stator

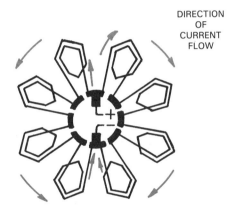

Fig. 9-6. As the rotor rotates, one coil after the other is connected to the power source.

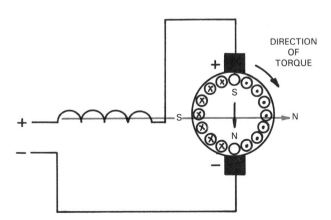

Fig. 9-7. The north pole of the rotor's field is attracted by the south pole and repelled by the north pole of the stator's field.

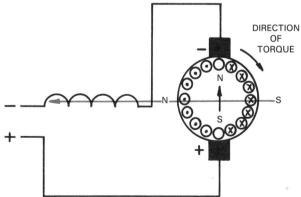

Fig. 9-8. Since both fields reverse polarity together, the torque direction does not change.

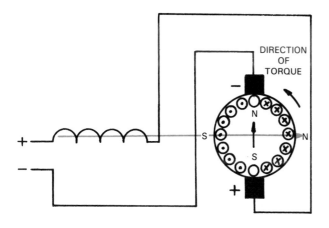

Fig. 9-9. The motor can be reversed by switching the armature connections.

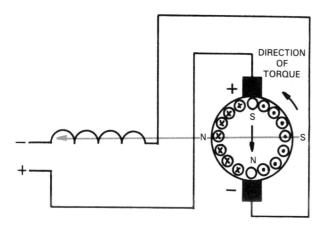

Fig. 9-10. Torque is in the same direction during both halves of every ac cycle.

field. In a series motor, Θ is always 0 degrees. Therefore, cos Θ is always 1. The stator field strength, ϕ, is proportional to field current. Field current, remember, is also rotor current, I_R. By using another constant, K_S, we can rewrite the torque equation for series motors as follows:

$$T = K_S \times I_R^2$$

This equation tells us that the torque developed by a series motor is proportional to the current squared.

Example

A series ac motor develops 0.4 newton metres (0.3 lbf-ft.) when current is 4 am-peres. Find the torque developed when current is 2 amperes.

a. Compute K_S.
$$T = K_S \times I_R^2$$
$$K_S = \frac{T}{I_R^2} = \frac{0.4}{16} = 0.25$$

b. Compute torque when $I_A = 2$ amperes.
$$T = K_S \times I_R^2 = 0.25 \times 4$$
$$= 0.1 \text{ N·m. } (0.75 \text{ lbf-ft.})$$

STARTING AND ACCELERATING TORQUE

A series motor develops the most torque at the instant you turn it on. That is when current is the greatest. Winding resistance is the only thing to limit current when the rotor is not turning. Part of this starting torque is used by motor and load inertia. The rest of the torque speeds up the motor.

The difference between the torque actually being used by the load and that being delivered by the motor is called ACCELERATING TORQUE. Accelerating torque causes an increase in speed. For example, assume 4 N·m (3 lbf-ft.) of torque is needed to overcome inertia. A given motor, however, has a starting torque of 5 N·m (3.7 lbf-ft.). Accelerating torque at the instant of start is, therefore, 1 N·m (0.7 lbf-ft.).

COUNTERELECTROMOTIVE FORCE

The large starting current lasts only a short time. As soon as the rotor starts to turn, another factor appears. Armature conductors cut through the magnetic field of the stator. This induces a counterelectromotive force (CEMF) into the armature coil. At every instant, CEMF opposes the armature voltage. The equivalent circuit is shown in Fig. 9-11. To the power source, counterelectromotive force looks like another source of power that has been connected series-opposing.

Counterelectromotive force is a motion voltage. It depends on two things:

1. Speed at which the armature conductors are cutting the stator field.
2. Strength of the field.

The equation is:

$$CEMF = K_E \times \phi \times S$$

K_E is a design constant
ϕ is stator field strength
S is the rotor speed

EFFECT OF CEMF

There is no CEMF when the rotor is not turning. But CEMF is created as soon as the rotor begins to turn. The faster the rotor turns the greater the CEMF. This reduces current, which makes the motor develop less torque. There is no further increase in speed when the motor torque matches the load torque—in other words, when there is no accelerating torque.

CEMF depends also, on the strength of the stator field. In turn, CEMF determines what that strength will be. Fortunately, the motor automatically works all of this out by itself. The load establishes how much torque the motor must put out. The motor then decides the speed that permits the right value of current to produce that torque. It then draws the right amount of current from the source.

OPERATING CHARACTERISTICS OF SERIES MOTORS

A series motor can run at very high or very low rpm. It is the only type of single-phase motor that can do this. The speed may be determined automatically by the load, or manually by varying current.

SPEED CONTROL FROM LOAD

When lightly loaded, a series motor rotates rapidly. The load requires only a little torque. Therefore, current must be low. That means the stator field must be weak. The rotor has to turn rapidly to generate the CEMF that limits the

current. In fact, with no load, a series motor can "run away." The rotor spins so rapidly that the conductors are thrown from the rotor core.

Some motors built for school laboratories have extra losses built in. These losses act like a mechanical load. The rotor will turn at high speed, but will not run away. Fig. 9-12 shows such a motor. The name, UNIVERSAL, means that it runs on either ac or dc. Most series motors, however, are designed to run on ac only.

Assume a heavy load is placed on the motor.

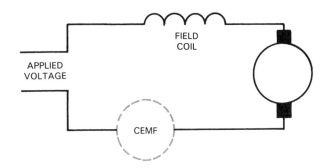

Fig. 9-11. CEMF is a real voltage. It is generated in the armature windings and opposes the applied voltage.

Fig. 9-12. A "universal" ac series motor used in school laboratories. It turns at high speed on either alternating or direct current.

The rotor slows, causing less CEMF to be generated. This increases the current and the torque. The final speed will be such that the right amount of torque is being delivered. Fig. 9-13 shows how the speed of a series motor changes with its torque output.

STALL TORQUE

Normally, a series motor does not mind an overload. That is, if the overload is not severe or if it does not last too long. But, if the load requires more torque than the motor can deliver, the motor STALLS. Stalling means coming to a complete stop. The rotor is standing still, as it was at the instant of start. Therefore, its stall torque is the same as its starting torque. Current rises rapidly when the motor stalls. Circuit breakers usually interrupt the circuit before the motor windings can burn up.

APPLICATIONS OF SERIES MOTORS

There are three ways that speed-load characteristic of series motors are put to good use:

1. In diesel electric locomotives. Fig. 9-14 shows a modern locomotive. A diesel engine drives an ac alternator. This ac power runs series motors geared to drive axles. At start, the load is heavy and the speed is slow. As the cars begin to roll, less torque is required.

The motors speed up. When the train comes to a steep grade, the motors run slower and produce extra torque.

2. In tools. Fig. 9-15 shows a hand-held drill driven by a series motor. Brief overloads are common when drilling. The motor simply slows until its torque has increased enough to overcome the extra load.

3. In air-moving equipment. A household vacuum cleaner, Fig. 9-16, works well with a series motor. Because series motors can operate at such high speeds—19,000 rpm is common—a small fan blade can move as much air as a large blade turning more slowly.

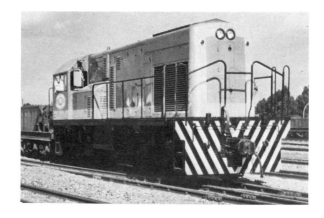

Fig. 9-14. Diesel locomotives generate electrical power to operate series motors which, in turn, drive the wheels. (Caterpillar Tractor Co.)

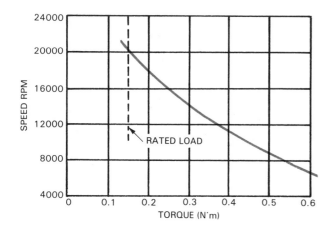

Fig. 9-13. The speed of a series motor goes down as its torque output increases.

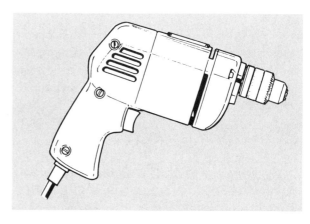

Fig. 9-15. Many electric drills make use of the series motor's ability to produce a large amount of torque.

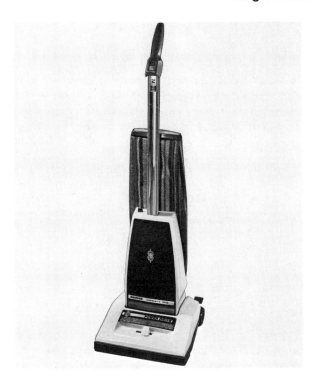

Fig. 9-16. Vacuum cleaners take advantage of the series motor's high speed. (The Hoover Co.)

Fig. 9-17. Rotor of repulsion-start induction motor has a commutator.

Fig. 9-18. Stator of repulsion-start induction motor. Same type stator is also used for squirrel-cage induction motors.

REPULSION-START INDUCTION MOTORS

The repulsion-start induction motor was the first kind of single-phase motor that used induction for producing the rotor field. It is mostly used today in the larger sizes, up to 15 horsepower (12 kilowatts).

CONSTRUCTION

The rotor is identical to that of the series motor. It has armature windings and a commutator. The brushes, however, are mounted on a movable yoke. The position of the brushes determines the starting torque as well as the direction of rotation. A typical rotor of a repulsion-start induction motor is shown in Fig. 9-17.

The stator of a repulsion-start motor is shown in Fig. 9-18. There are no salient poles. Field windings are inserted into slots. They are, however, wound so as to produce magnetic poles when current flows.

INDUCED ROTOR VOLTAGE

Fig. 9-19 represents the field and rotor coils. Current in the field coils induces a voltage into the rotor coils by transformer action. The short-circuited brushes establish the pattern of current flow in the rotor. When they are in direct line with the stator poles, current in half the rotor coils opposes current in the other half. Net rotor current is zero.

In Fig. 9-20, we have rotated the brushes 90 degrees in space. Now we have the largest possible rotor current. That is not all we need, however. Note the direction of the rotor field. It is in a direct line with the stator field. No torque is produced.

Note: In Figs. 9-19, 9-20 and 9-21, brushes are shown inside the commutator so that electrical path is easier to see. In practice, brushes always ride on the outside of the commutator.

To review, when the brushes are 90 degrees from the center of the stator field, there are high rotor currents but no torque. The rotor field and stator field are in line. This is called the HARD NEUTRAL position of the brushes.

As the brushes are rotated, less current is induced while the rotor field shifts. The current goes to zero when the brushes are lined up with the stator field. Here, there is no rotor field because there is no rotor current. Therefore, there is no torque either. This point is called the SOFT NEUTRAL position.

PRODUCTION OF STARTING TORQUE

We cannot have both maximum current and maximum angle at the same time. The brushes have to be in a position that will produce a starting torque. The best position is shown in Fig. 9-21. We have reduced the rotor current, but we have shifted the field. Now the north rotor pole is repelled by the north stator pole. A 15-degree shift from hard neutral gives us the maximum starting torque. Shifting the brushes one way produces a clockwise rotation. The opposite way produces a counterclockwise rotation. This is how repulsion start motors are reversed.

MOTION VOLTAGE

Now, after the rotor starts turning, a curious thing happens. Flux from the armature windings cuts through the alternating field of the stator. This induces a motion voltage. At about 75 percent of rated speed, a ring short-circuits all of the commutator segments. At this point, the pattern of current flow through the armature

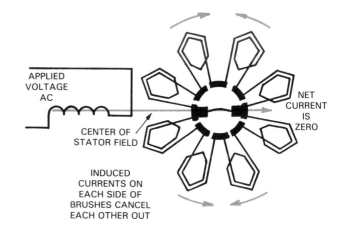

Fig. 9-19. There is no rotor field when brushes are in a direct line with the stator field. Therefore, there is no torque at this "soft neutral" position.

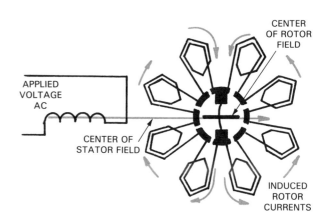

Fig. 9-20. With brushes at right angles to stator field, there is current but no torque. This is the "hard neutral" position.

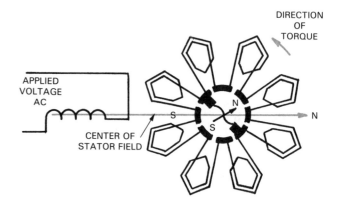

Fig. 9-21. Maximum starting torque is produced when brushes are located 15 degrees from "hard neutral" position.

does not depend on brush position. In fact, the brushes are usually lifted off the commutator to reduce wear.

PRODUCTION OF TORQUE

The torque is in the same direction the rotor is already turning. That might seem strange until we remember Lenz's Law which is explained in Chapter 4. Induction, you see, always opposes whatever causes it. The only way it can stop this induction is by making the rotor turn as fast as the field is alternating.

For example, suppose that a 60-hertz alternating current is applied to the motor at the instant shown in view A, Fig. 9-22. Conductor C is under a north stator pole. View B of Fig. 9-22 shows what happens 1/120th of a second later. The opposite pole has become north. If the rotor has turned exactly one-half revolution, conductor C is still under the north pole. One-half revolution at 1/120th of a second works out to be 3600 revolutions per minute:

$$1 \text{ revolution per } 1/60 \text{ sec.} \begin{cases} = 60 \text{ revolutions per second} \\ = 60 \times 60 \text{ revolutions per minute} \\ = 3600 \text{ rpm} \end{cases}$$

The rotor can never reach that speed, however. If it did, all induction would stop. When induction stops, so does torque. Some torque, of course, is needed to overcome rotational losses, even for an unloaded motor.

DIRECTION OF FIELDS

We saw, in Fig. 9-21, that the rotor field was about 15 mechanical degrees from the stator field at the start. In an induction motor, however, the position of the field depends on rotor speed. At start, the frequency of the rotor current was 60 hertz, the same as stator current. This frequency went down, however, as speed increased. Frequency would drop to 0 hertz when the rotor reached 3600 rpm. Assume for a second that it could. That would mean that the armature windings would have no inductive reactance ($X_L = 6.28 \times f \times L$ with $f = 0$). There would be no lag in the rotor field.

At this point, the rotor field would be at right angles to the stator field. This is shown in Fig. 9-23. Real motors come close to this right-angle relationship.

AMOUNT OF TORQUE PRODUCED

Now, we shall see how the torque equation ($T = K_T \times \phi \times I_R \times \cos \Theta$) applies to induction motors. The field strength, ϕ, hardly changes at all between "no-load" and "full load." That leaves rotor current, I_R, and the displacement angle, Θ. If you look at the trigonometric tables, you will find very little change

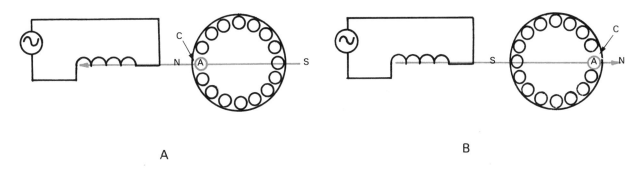

A   B

Fig. 9-22. A—The conductor, C, rotating at 3600 rpm would have no voltage induced in it. B—The instant one half-cycle behind A. There is no change in the flux linking C.

in the value of the cosine of angles between 0 and 10 degrees. It is practically constant over the operating range of the motor.

As an induction motor is loaded, the rotor slows a little. This increases the value of the induced voltage. The result is an increase in rotor current. The rotor field is strengthened and extra torque is produced. Torque, therefore, is directly proportional to rotor current over the operating range of the motor.

NUMBER OF STATOR POLES

The induction motor we have been discussing has two poles, one north and one south. The rated speed is less than 3600 rpm. The figure, 3600 rpm, was computed from a frequency of 60 hertz. It is known as synchronous speed. It is the speed at which the rotor would be synchronized with the line frequency.

Induction motors may have a different number of poles, however. Four-pole motors are very common in the fractional horsepower sizes. Fig. 9-24 shows the resulting stator field pattern. Note that the rotor needs only a half revolution in going from north to south and back again to north. Its synchronous speed, therefore, is half that of the two-pole motor. Synchronous speed may be computed from the following equation:

$$S_s \text{ (rpm)} = \frac{\text{frequency} \times 120}{\text{number of poles}}$$

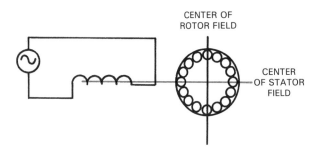

CENTER OF ROTOR FIELD

CENTER OF STATOR FIELD

Fig. 9-23. This right-angle relationship between the two fields is best for producing torque.

Typical four-pole induction motors are rated at 1725 rpm.

SQUIRREL-CAGE SINGLE-PHASE MOTORS

The idea of inducing voltage into the rotor, instead of applying voltage to an armature coil, was a good one. The next logical step was to eliminate the commutator and brushes. That led to the development of the squirrel-cage rotor.

CONSTRUCTION OF A SQUIRREL-CAGE ROTOR

The squirrel cage is the simplest type of rotor construction. Probably 90 percent of the electric motors in use today are squirrel-cage induction motors. They are used in most major household appliances.

Fig. 9-25 shows a typical squirrel-cage rotor before and after the core has been attached. The "before" picture does look somewhat like the cages used to exercise squirrels and other small pets. The "cage" is made up of aluminum or brass bars. They run through the core, just below the surface. The bars are short-circuited at each end by the rings.

TYPES OF SQUIRREL-CAGE INDUCTION MOTORS

We have seen that an induction motor will develop torque once the rotor starts turning. The problem is in getting the rotor to turn in the first place. Fig. 9-26 shows what happens when we put a squirrel-cage rotor into a stator having a single winding. It acts just like a transformer.

In a transformer, remember, the primary current magnetized the core and induces a secondary voltage. Current flow in the secondary then created a magnetomotive force which opposed the primary magnetomotive force. The secondary field was said to be 180 degrees out of phase with the primary field. The only real difference is that the stator's exciting current must magne-

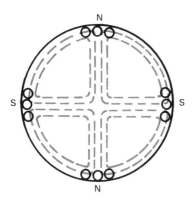

Fig. 9-24. Magnetic circuit of four-pole stator. Colored lines mark the four magnetic flux paths or "fields."

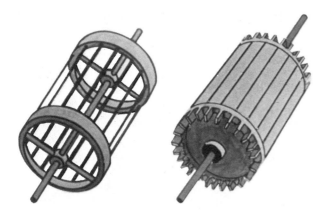

Fig. 9-25. Rotor of a squirrel-cage induction motor. Left. Current-carrying portion. Rings short-circuit the core. Right. Complete rotor with core in place.

STATOR CONSTRUCTION

Fig. 9-27 shows a stator typical of a squirrel-cage induction motor. It has two windings:
1. The main winding. This one is always connected to the power source.
2. The phase winding or sometimes, the start winding. This winding may be disconnected from the power source after the rotor starts turning. It is placed halfway between the coils of the main winding. The stator is shown schematically in Fig. 9-28.

The reason for the second stator winding is to produce a revolving magnetic field. This revolving field, then, is cut by the nonmoving squirrel-

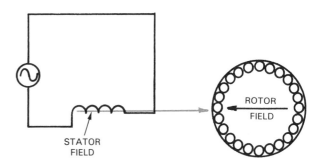

Fig. 9-26. Single coil on stator of single-phase induction motor produces no torque.

tize both cores and the air gap between them.

Fig. 9-26 shows that the rotor field is 180 degrees out of phase in both space and time. The fields are parallel. There is no starting torque.

The five kinds of squirrel-cage induction motors have different schemes for starting. Their names indicate how the starting is done:

1. Split-phase (resistance-start).
2. Capacitor-start.
3. Permanent-capacitor.
4. Two-capacitor.
5. Shaded-pole.

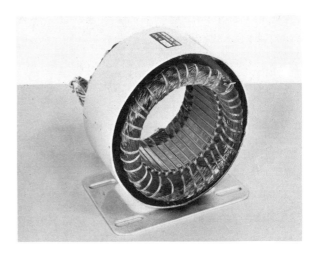

Fig. 9-27. Stator of single-phase induction motor has two windings separate from each other.

cage bars. This produces the same kind of rotor field as when the bars were moving.

The trick in producing a revolving stator field is getting PHASE DISPLACEMENT. The current through the main winding must be out of phase with the current through the start winding.

SPLIT-PHASE (RESISTANCE-START) MOTOR

The start winding of a split-phase motor uses smaller wire. This gives it a higher resistance than the main winding. Sometimes there is also a series resistance added. The same single-phase voltage is applied to both. This is illustrated in Fig. 9-29. Start winding current lags the voltage by a smaller angle than main winding current.

The phasor diagram for stator currents in a typical split-phase motor is shown in view A, Fig. 9-30. In the start winding, current lags voltage by 20 degrees. In the main winding, current lags voltage by 50 degrees. That puts the start winding current 30 degrees ahead of the main winding current. This is also shown by the wave forms of Fig. 9-30, view B.

HOW THE STATOR FIELD REVOLVES

The fields produced by the two windings combine into a single stator field. As the two fields

increase, decrease and change direction, this stator field revolves.

Each of the sketches of Fig. 9-31 represent an instant as marked in Fig. 9-30, view B. One ac cycle produces one revolution of the field. Black vectors show the direction and strength of the individual fields. The blue vector shows the strength and direction of the combined field. You will notice that the field does not revolve smoothly. Both speed and strength change as it goes around. Still, it does get cut by the rotor bars. Enough rotor voltage is induced to get the rotor going.

It is interesting to note that, if the main winding current could lag the start winding current by 90 degrees, the stator field would revolve at a uniform speed and with uniform strength.

PRODUCTION OF STARTING TORQUE

Here is what occurs at the instant you turn on the switch:

1. A magnetic field is revolving at synchronous speed, one revolution per cycle of ac.
2. This induces a 60 hertz voltage into the rotor.

The result is a rotor field that also revolves at synchronous speed. The amount of starting

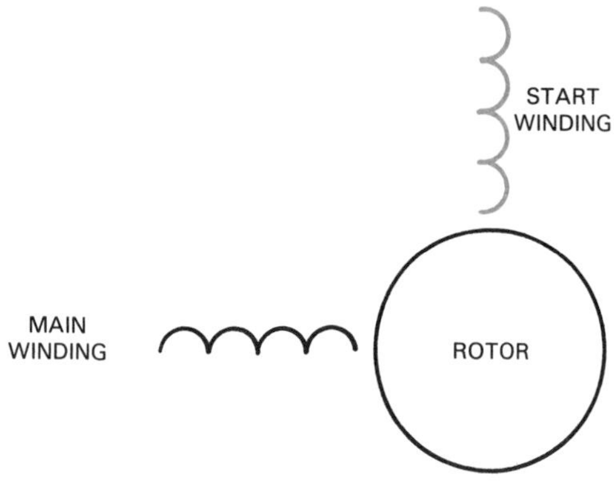

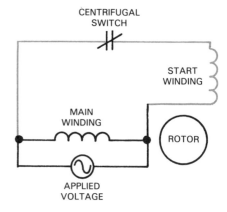

Fig. 9-28. Schematic representation of single-phase motor. Stator coils are wound so that the angle between their magnetic fields is 90 degrees.

Fig. 9-29. Circuit diagram of split-phase motor. Start winding is in circuit during start out; a centrifugal switch disconnects it when motor is running.

Single-Phase AC Motors

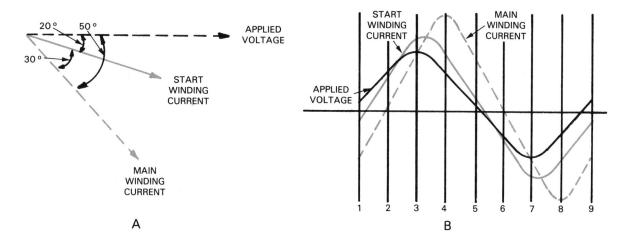

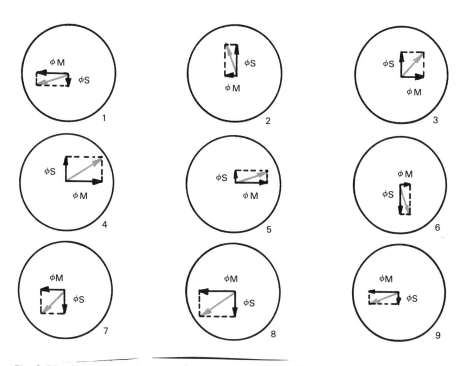

Fig. 9-30. Applied voltage and currents of single-phase motor stator. A—Phasor representation. B—Wave forms.

Fig. 9-31. Stator field at instants from sine wave in Fig. 9-30, view B. Field varies in strength (represented by vector length) as it revolves.

torque produced depends on two things: the rotor current and the displacement angle.

Now, consider the two extremes. In the first, the rotor has no inductance; it is entirely resistive. The induced rotor field, if it existed, would be at right angles to the stator field. The displacement angle is 0 degrees. That would be

good, because the cosine of 0 degrees is 1. However, current would be zero because there would be no induction.

In the other extreme, the rotor is totally inductive. Rotor current, as a result, lags rotor voltage by 90 degrees. Current lag, in turn, makes the rotor field lag the stator field by an additional 90 degrees. Now the rotor field parallels the stator field. The displacement angle is 90 degrees. Since the cosine of 90 degrees is zero, there is no torque.

Actually, rotors have both resistance and inductance. Rotor current lags rotor voltage about 85 degrees. That puts the rotor field's north pole about 15 degrees behind the south stator pole, Fig. 9-32. The rotor is, thereby, made to revolve in the same direction as the stator field. Now, the question is: "Which way does the stator field revolve?"

DIRECTION OF ROTATION

The field will always move from a start winding pole to the nearest main winding pole having the same instantaneous polarity. It does so because start winding current is in the lead. You can reverse directions, by interchanging connections at the start winding terminals. However, these are normally wired in on most appliances since there is no need to reverse motor direction.

CENTRIFUGAL SWITCH

By the time a split-phase motor is running at 75 percent of full speed, the main winding alone develops more torque than the starting winding and the run winding together. At this point, it is a good idea to disconnect the start winding.

The start winding is disconnected with a centrifugal switch, view A, Fig. 9-33. Its operating principle is illustrated in view B, Fig. 9-33. The movable part mounts on the rotor shift. Rotor speed forces weights outward by centrifugal force. Their movement opens a snap-action switch mounted on the motor's end bell. The electrical connection of this switch is shown schematically in Fig. 9-29.

SPLIT-PHASE MOTOR APPLICATIONS

Being simple to manufacture, split-phase motors are relatively low cost. Starting torque is

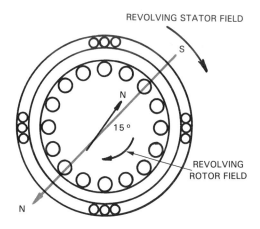

Fig. 9-32. At instant of start, the induced rotor field lags the revolving stator field by 15 degrees in space.

low, however, limiting use to easily started loads. The high starting current required is a second limiting factor. Split-phase motors larger than 250 watts (1/3 horsepower) are not used where the motor must be started often. They put too much drain on the power line.

CAPACITOR-START MOTORS

In a split-phase motor, current in the main winding lags 30 degrees behind start winding current. Capacitor-start motors, on the other hand, have a full 90 degrees between currents.

PURPOSE OF THE CAPACITOR

A capacitor is placed in series with the start winding as shown in Fig. 9-34. A typical motor is also shown. The capacitor makes start winding current lead the applied voltage. A capacitance is selected to cause a 40 degree lead. Main winding current, however, still lags voltage by 50 degrees. The phasor diagram is shown in Fig. 9-35. Main winding current lags start winding current by 90 degrees.

The capacitor is cut out of the circuit, along with the start winding, at 75 percent of rated speed. This happens only a few seconds after the motor starts. The capacitor, then, is rated for short duty. A large-value capacitor, 200 micro-farads, is very small — less than 4 cm in diameter. Often it is enclosed within the motor housing.

EFFECTS OF THE CAPACITOR

The capacitor affects motor operation only during the start period. The 90 degrees between

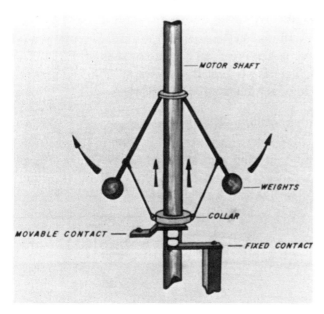

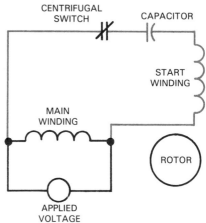

Fig. 9-34. Capacitor in series with start winding causes the start-winding current to lead applied voltage. Top. Typical capacitor-start motor. (Leeson Electric Corp.) Bottom. Electrical schematic.

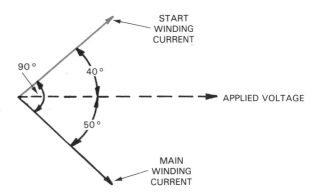

Fig. 9-33. Split-phase motors use rotation speed to disconnect brushes from start windings. Top. Simple illustration of how centrifugal switch works. Bottom. Appearance of centrifugal switch on small electric motor.

Fig. 9-35. With start-winding current leading and main-winding current lagging applied voltage, angle between them approaches 90 degrees.

the two stator fields makes the combined field revolve at a constant speed. This improves starting characteristics three ways:
1. Lower current.
2. Higher torque.
3. Improved power factor.

Fig. 9-36 is a tabulation of data from the same motor. The only difference is that a resistor was used in one instance and a capacitor in the other.

CAPACITOR-START MOTOR APPLICATIONS

The capacitor-start motor is the most widely used single-phase motor. It runs pumps, compressors, refrigeration units, conveyors and washing machines. It is also used in a wide range of machine tools. Capacitor-start motors are available in sizes up to 6 kilowatts (7.5 horsepower).

PERMANENT-CAPACITOR MOTOR

The capacitor of capacitor-start motor improves the starting power factor. But it has no effect on the power factor once the motor is up to speed. Now, suppose we make two changes. Let us make the second winding, called the PHASE WINDING, identical to the main winding. That way, the capacitor can stay in the circuit at all times.

We can still use a capacitor to produce the phase shift between current in the main and phase windings. The capacitor will have to be larger, however. Also, it must be rated for continuous duty since it remains in the circuit.

The permanent-capacitor motor is represented by Fig. 9-37. The capacitor improves starting and running power factors. At the same time, it provides several other advantages.

ADVANTAGES OF A PERMANENT-CAPACITOR MOTOR

Permanent-capacitor motors have many advantages. You can see one of them in Fig. 9-37. There is no centrifugal switch. This eliminates one possible source of trouble.

A second advantage is reversibility. You cannot reverse most types of squirrel-cage induction motors while they are running. You have to turn them off and wait until the centrifugal switch

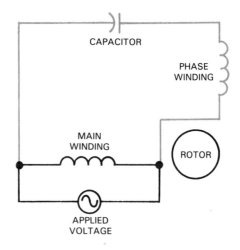

Fig. 9-37. Schematic of permanent-capacitor motor.

CAPACITOR-START MOTOR

TYPE	STARTING TORQUE AS A % OF FULL LOAD TORQUE	STARTING CURRENT	POWER FACTOR
Resistance-start	64	25.3 A	0.53
Capacitance-start	316	22.6 A	0.93

After centrifugal switch trips, operation is identical to split-phase motor.

Fig. 9-36. Resistors or capacitors have different effect on operation of same motor.

returns to its normally closed position. Only then can you interchange leads and operate the motor in the opposite direction.

A permanent-capacitor motor, however, has a continuously revolving stator field. To reverse the rotor's direction, all you need do is reverse the direction of the revolving field. This is accomplished by switching the capacitor from one winding to the other. This connection is shown in Fig. 9-38.

A third advantage is its quietness. Since the windings are the same, their fields are identical. Since the revolving field is fairly uniform, you

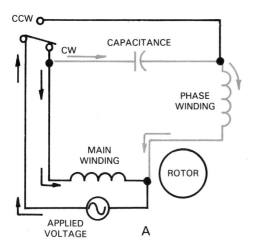

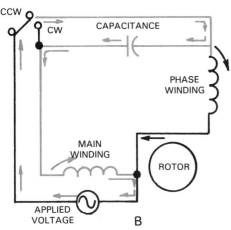

Fig. 9-38. Circuit used for reversing the direction of permanent-capacitor motors. A—Capacitor in line with phase winding. B—Switch shifted, capacitor is in line with main winding. Arrows represent current flow for an instant during one cycle.

do not get the pulsating hum that you do with either split-phase or capacitor-start motors.

Finally, a permanent-capacitor motor is more sensitive to voltage variations than other squirrel-cage induction motors. Speed can be reduced by as much as 50 percent by reducing the applied voltage.

DISADVANTAGES OF PERMANENT-CAPACITOR MOTOR

The permanent-capacitor motor has some disadvantages:

1. The last-mentioned advantage, speed control, may cause problems. If line voltage drops, the motor will slow down. Also, there is a greater change in speed from no-load to full-load than with other squirrel-cage induction motors.
2. Since the capacitor is large, it may have to be mounted outside the motor housing.
3. The permanent-capacitor motor has a lower starting torque than a capacitor-start motor. There is a reason for this. Starting capacitors range in value from 85 to 300 microfarads. This is far from the ideal value for running where ideal values range from 3 to 16 microfarads. A compromise between the two values is not ideal for either starting or running.

PERMANENT-CAPACITOR MOTOR APPLICATIONS

Permanent-capacitor motors are used mainly for low-starting-torque loads. The quiet operation and adjustable speed characteristics make them suitable for shaft-mounted fans and blowers. They range in size from 50 to 560 watts (1/20 to 3/4 horsepower).

TWO-CAPACITOR MOTORS

The capacitor-start motor has high starting torque. The permanent-capacitor motor operates quietly and permits control of speed. You can get both advantages by using two capacitors. Capacitance of one value is used for starting. Another value is used for running. The two-capacitor motor is shown schematically in Fig. 9-39. The start capacitor is of the same high-value electrolytic type, used in a capacitor-start motor. During start, it is in parallel with an oil-type running capacitor. The total starting capacitance is the sum of the two. At 75 percent of rated speed, a centrifugal switch cuts the starting capacitor out of the circuit. The running capacitor remains to improve the operating characteristics.

REVERSING THE TWO-CAPACITOR MOTOR

The two-capacitor motor cannot be reversed without turning it off. The centrifugal switch must have a chance to close. Then the leads of the phase-winding circuit can be interchanged.

USE OF AUTOTRANSFORMER

Some two-capacitor motors operate with only one capacitor and a tapped autotransformer. This is shown in Fig. 9-40. By transforming voltage and current, transformers change the equivalent impedance. That makes the capacitor look larger in proportion to the turns ratio squared.

For example, assume the "start" tap produces a turns ratio of 8. The capacitor has a value of 4 microfarads. The autotransformer makes this act like a 256 microfarad capacitor ($4 \times 8^2 = 256$) during start. This time, the centrifugal switch is used to change the tap. The "running" tap makes the turns ratio 1.5. Now the 4 microfarad capacitor looks like a 9 microfarad capacitor ($1.5^2 \times 4$). Unfortunately, voltage is stepped-up by the turns ratio. Eight times line voltage is applied across the capacitor during start. The capacitor must be rated for 1000 volts.

APPLICATIONS

Two-capacitor motors are noted for quiet operation. Starting torque is high and running

torque is good. They are used to drive the compressor of small air-conditioning units.

SHADED-POLE MOTORS

The shaded-pole motor is a type of single-phase squirrel-cage induction motor. It is in a different class than the four types just described:
1. It is low-cost.
2. It is made only in small sizes from 1 1/2 to 200 watts (1/500 to 1/4 horsepower).

CONSTRUCTION OF SHADED-POLE MOTOR

The stator pole pieces are made a special way. As shown in view A, Fig. 9-41, each pole has two sections created by a slot in the pole's end. The magnetic flux in both sections comes from current in the stator winding. In addition, the smaller section has a copper ring, called a shading coil, view B, Fig. 9-41. Usually the ring is a single wrap of heavy copper wire but there can be more than one turn. Induced current in the shading coil also affects the flux in the smaller section.

PLOTTING WAVE FORMS AND PHASOR DIAGRAMS

Fig. 9-42 shows the wave forms and phasor diagram of the stator field. The black line is the field from the main winding. This is also the field in the larger section of the stator pole. The dotted line is the field induced by the shading coil. The blue line is the field in the shaded section. It is the combination of the other two.

Note that field in the shaded section lags the main field. When these two combine, the effect is a revolving field. The phase angle, however, is nowhere near the ideal value of 90 degrees. Still, it does produce a starting torque. Induced rotor voltage is quite low, so starting current is low.

ADVANTAGES AND DISADVANTAGES

Shaded pole motors are inefficient and have a low power factor. These disadvantages are insig-

Single-Phase AC Motors

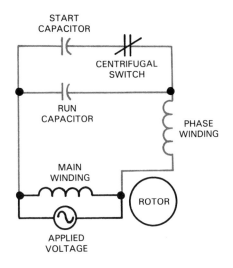

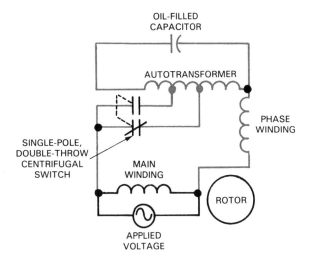

Fig. 9-39. Schematic of two-capacitor motor. Centrifugal switch removes capacitor from circuit after motor has started.

Fig. 9-40. Schematic of autotransformer-type, two-capacitor motor.

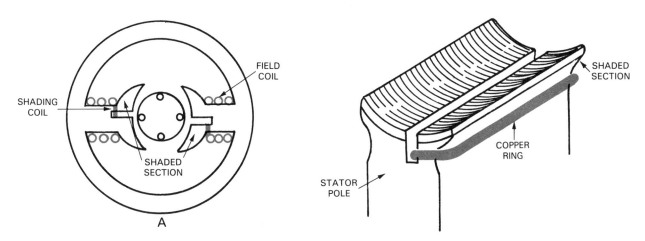

Fig. 9-41. Construction of shaded-pole single-phase ac motor. A—Cutaway of end view of rotor and stator. B—Shaded winding is usually a single coil of heavy copper wire.

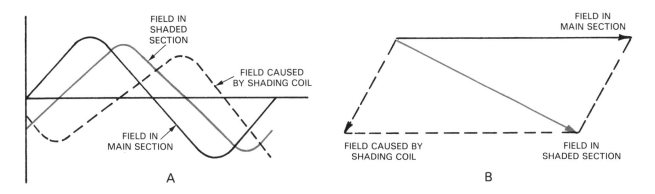

Fig. 9-42. Plotting sine wave and phase angle of shaded-pole motor. A—Wave forms. B—Phasor diagram of magnetic fields in the stator of a shaded-pole motor.

nificant since they consume very little power anyhow. Torque is very low. Their low cost, rugged construction and ease of maintenance make them useful for fans and blowers. They can be used wherever there are light loads such as on phonograph turntables, movie projectors and vending machines.

SUMMARY

In an ac motor, magnetic forces that are created by electricity produce torque.

Torque depends not only on the magnitude of the magnetomotive forces but on their direction and phase relationships as well.

A series ac motor has rotor coils that are connected in series with the stator field coils through a commutator and brushes.

A series ac motor has a high starting torque and the ability to handle overloads.

Although the rotor of a repulsion-start induction motor has a commutator, its current is caused by induction.

The most popular type single-phase motor has a squirrel-cage rotor, whose current is caused by induction.

The stator of a squirrel-cage motor must have two windings in order to produce starting torque.

Current in the two windings must be out of phase (ideally by 90 degrees) in order to produce a revolving stator field. The revolving field induces a voltage into the squirrel-cage rotor bars.

After the rotor is turning, voltage is induced into the squirrel-cage bars by motion. Therefore, one of the stator windings may be disconnected.

A capacitor-start motor has a capacitor in series with the second winding. Both the winding and capacitor are cut out at 75 percent speed.

A permanent-capacitor motor also has a capacitor in series with the seconding. However, both remain in the circuit at all times.

A two-capacitor motor uses one value of capacitance for starting and another value for running. The second winding is not cut out.

Shaded-pole motors produce a revolving magnetic field by inducing a lagging current into a shading coil which surrounds a portion of each pole piece.

TEST YOUR KNOWLEDGE

1. Why is it a good idea to rigidly couple a series motor to its load?
2. Name one advantage and one disadvantage of a series ac motor compared to an induction motor.
3. How is a repulsion-start induction motor similar to an ac series motor? How is it different?
4. The purpose of the capacitor in series with a start winding is to produce a 90 degrees phase displacement between current in the _____ _____ and current in the _____ _____.
5. Name one advantage and one disadvantage of a permanent-capacitor motor compared to a capacitor-start motor.
6. Describe the starting sequence of a two-capacitor motor.

PROBLEMS

1. Given: The armature winding of a series ac motor has a resistance of 4 ohms. The series winding has a resistance of 6 ohms. Applied voltage is 120 volts. If a heavy load is placed on this motor and it stalls, find the armature current.
2. Given: A six-pole induction motor has 120 volts, 60 hertz applied. Find the synchronous speed.
3. Given: A capacitor-start motor is tested at an applied voltage of 38 volts, producing the following values at the instant of start: Start winding current: 2.8 amperes; start winding active power: 86 watts; main winding current: 4.96 amperes; main winding active power: 112 watts. Find the phase angle between main winding current and start winding current.
4. Given: A single-phase induction motor has

Motors come in all sizes. Large motors, such as the one pictured, are not uncommon in industry.
(Westinghouse Industrial Systems Div.)

an applied voltage of 230 volts. Total current is 2.5 amperes. Active power input is 460 watts. Find the power factor of the motor.

5. Given: The motor of *Problem 4* is delivering 2.4 newton metres of torque at 1725 rpm. Find the total output power in watts and the power losses.

6. Given: The motor of *Problems 4 and 5.* Find the efficiency.

10

THREE-PHASE ALTERNATORS

After studying this chapter, you will be able to:

☐ Explain the significance of the saturation curve of an alternator.
☐ Describe the damper windings of an alternator rotor and explain their purpose.
☐ Define synchronous impedance and explain how it is determined.
☐ Compute the voltage regulation of an alternator with different types of loads.

Household power is single-phase. In fact, most of the lights and other electrical equipment we use every day operate on single-phase ac. Heavy users of power, however, operate with three-phase current. You will normally find three-phase in commercial buildings, manufacturing plants and retail stores.

Huge alternators at power generating plants produce three-phase power. It has several advantages:
1. It is cheaper to transmit than single phase.
2. Three-phase motors are simpler, less expensive and more powerful.

ALTERNATORS

There are two general types of alternators:
1. The REVOLVING-ARMATURE type. It has a stationary magnetic field and rotating alternating current windings.
2. The REVOLVING-FIELD type. Armature is stationary and field windings turn.

The revolving-field type has many advantages. It is usually found in all three-phase alternators of any size. This type includes:
1. The SALIENT-POLE alternator. The ro-

tating member has projections to hold the field coil windings.
2. The DISTRIBUTED-FIELD or turbo-construction alternator with field coils buried in the face of the rotor.

An instant of electrical generation is illustrated in Fig. 10-1. A revolving field alternator is shown. The field coil on the rotor must be supplied with direct current.

The ends of the field coil are connected to two slip rings. Carbon brushes are always in contact with the slip rings. Direct current is conducted through the brushes and rings to the field coil. This current produces the needed magnetic field.

GENERATION OF A SINGLE-PHASE VOLTAGE

A prime mover (mechanical force) drives the rotor. As the rotor turns, its magnetic field

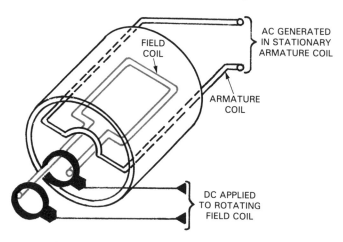

FIELD COIL

AC GENERATED IN STATIONARY ARMATURE COIL

ARMATURE COIL

DC APPLIED TO ROTATING FIELD COIL

Fig. 10-1. A basic revolving field alternator. Changing flux linkage induces voltage into armature coil. At field coil position shown here, maximum voltage is being induced.

sweeps around the air gap. This field has a constant strength.

There is, however, a change in linkage flux. When the field is directly opposite the coil, the maximum change in flux linkage occurs. Maximum voltage is induced into the coil at that instant. This instant is shown by Fig. 10-1.

The other extreme is shown in Fig. 10-2. There is no flux linking the conductors at that instant. Induced voltage is zero. As the rotor turns, the field sweeps around. The induced voltage rises, falls, and changes direction in a sine wave pattern. This is shown by Fig. 10-3.

The moving magnetic field exerts a force on the electrons in the armature coil. Electrons begin to pile up, first at one terminal, then at the other.

This action does not produce a current. It merely produces a voltage—the ability to make current flow. There will be current only when a load is connected to the alternator terminals.

GENERATION OF THREE-PHASE VOLTAGE

Fig. 10-4 shows three armature coils on the stator. The physical location of the coils is important. Coil B is 120 mechanical degrees from coil A. Likewise, coil C is 120 mechanical degrees from coil B. The rotating magnetic field induces a voltage into each coil separately. This is shown by Fig. 10-5.

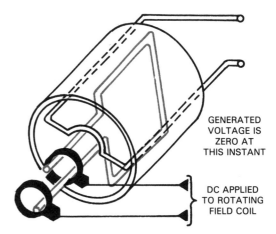

GENERATED VOLTAGE IS ZERO AT THIS INSTANT

DC APPLIED TO ROTATING FIELD COIL

Fig. 10-2. Slip rings and brushes are used to conduct dc to field coil on rotor. At the instant shown here, no voltage is being induced.

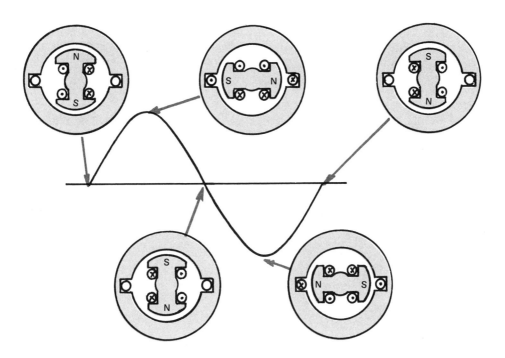

Fig. 10-3. One complete cycle of two-pole rotating field produces one cycle of ac voltage.

Notice the time relationship of these three voltages. Coil B voltage lags that of coil A by 120 degrees. Coil C lags coil B by another 120 degrees. The phasor diagram of these voltages is shown in Fig. 10-6.

DELTA CONNECTION

A single-phase voltage is generated in each coil. The coils, however, may be interconnected. This allows us to transmit all three voltages over fewer wires.

Fig. 10-7 shows the schematic of a delta connection. The instantaneous polarity is important. The positive end of each coil is connected to the negative end of the one next to it.

Now, look at the voltage between lines L1 and L2. On one hand, it is the voltage generated in coil A. On the other, it is the phasor sum of the voltages generated in coils B and C. Either way, it is the same value. See Fig. 10-8.

Voltage between L2 and L3 is generated in coil B. In the same way, coil C voltage appears between lines L1 and L3.

The delta connection has two disadvantages.
1. You can obtain only one voltage value, the coil voltages.

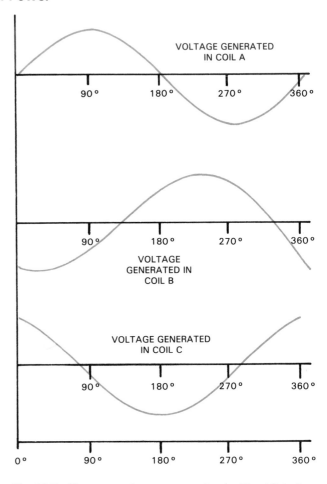

Fig. 10-5. Sine wave of generator action for Fig. 10-4. A separate ac voltage is induced into each coil. Each of these voltages is displaced 120 electrical degrees from the other two.

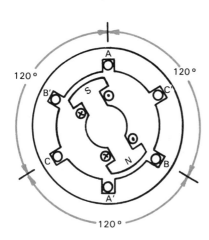

Fig. 10-4. Three-phase alternators have three armature coils spaced 120 mechanical degrees apart on the stator.

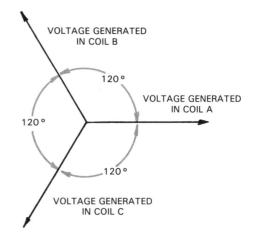

Fig. 10-6. Phasor representation of the voltages induced in the three stator coils of Fig. 10-4.

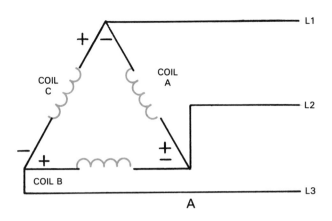

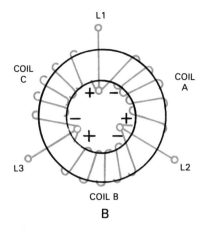

Fig. 10-7. A delta connection has coils connected minus-to-plus. A—Schematic. B—Simplified drawing of delta connection.

2. An unbalanced load unbalances all three voltages.

Assume one phase is more heavily loaded. This also affects current in the other two coils.

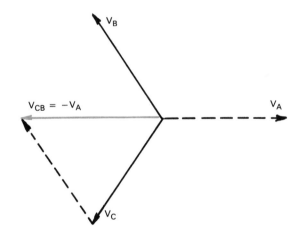

Fig. 10-8. The voltage, V_{CB}, (the phasor sum of V_C plus V_B) is equal to $-V_A$.

A larger current increases their internal voltage drop. The terminal voltage is thereby lowered. For this reason, you will seldom find delta-connected alternators.

WYE-CONNECTION

Three-phase alternators are usually wye-connected. Ends having the same polarity are connected to the same point. This is shown in Fig. 10-9. Now, the coil voltage is between one line and the neutral point. For example, assume that 120 volts is generated in each coil. There is 120 volts between L1 and N, L2 and N, and L3 and N.

Between any two lines you have the phasor sum of two coil voltages. This is 1.73 times the voltage generated in one coil. For example, assume each coil generates 120 volts. Line voltage is 208 volts. The phasor diagram is shown in Fig. 10-10.

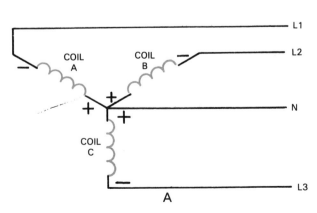

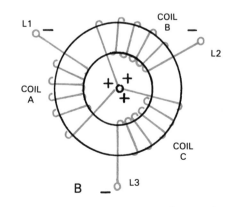

Fig. 10-9. A wye connection has terminals of like polarity wired together. A—Schematic of wye connection. B—Simplified drawing of wye-connected coils on a stator.

In a wye connection, remember, the coils are connected plus to plus. That means the vector of one coil must be subtracted from the others to find the phasor sum. To subtract a vector, simple reverse its direction on the vector diagram and add (extend the vector line in the new direction).

FREQUENCY OF GENERATED VOLTAGE

The alternator of Fig. 10-11 is a two-pole machine. The rotor field has one pair of poles. On the stator, each phase consists of one coil producing one pair of poles. *Each time the field completes one revolution, one cycle of ac is produced. Sixty revolutions per second (3600 rpm) produces voltage with a 60 hertz frequency.*

However, alternators may have two or more pairs of poles. The one in Fig. 10-12 has four poles. It produces two cycles each revolution. Thirty revolutions per second (1800 rpm) produces a voltage at a frequency of 60 hertz. The frequency of the generated voltage, therefore, depends on two things:

1. The number of pairs of poles.
2. Speed.

The equation is:

$$f = \frac{N \times S}{60}$$

f is the frequency in hertz
N is the number of pairs of poles
S is the speed in revolutions per minute

The table in Fig. 10-13 gives the speed of various alternators producing 60 Hz alternating current.

VALUE OF GENERATED VOLTAGE

Induced voltage depends on the rate of change in flux linkages. The voltage generated in a given alternator, then, depends on the strength of the rotor field and its speed.

The equation is:

$$E_g = K_E \times \phi \times S$$
E_g is voltage generated in volts

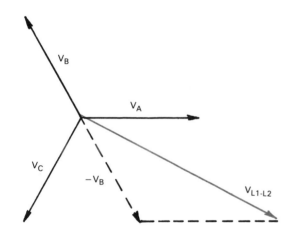

Fig. 10-10. For a wye connection, voltage from line-to-line is the phasor sum of voltages across two coils.

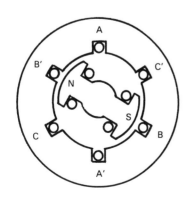

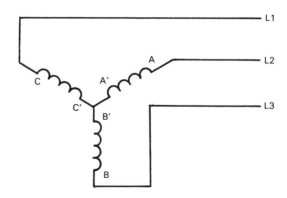

Fig. 10-11. A two-pole alternator has one pair of N-S poles on the rotor and one pair in each phase on the stator.

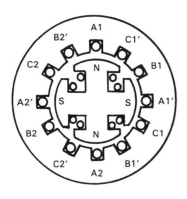 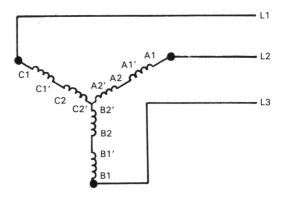

Fig. 10-12. A four-pole alternator has two pairs of poles on the rotor and two pairs in each phase on the stator.

NO. POLES	ROTOR PAIRS	SPEED RPM	PAIRS FOR EACH PHASE ON STATOR
2	1	3600	1
4	2	1800	2
6	3	1200	3
8	4	900	4
10	5	720	5
12	6	600	6
14	7	514	7
16	8	450	8
18	9	400	9
20	10	360	10
22	11	327	11
24	12	300	12
26	13	277	13
28	14	257	14
30	15	240	15

Fig. 10-13. The more pairs of poles in an alternator, the lower the rotational speed needed to produce 60 Hz.

ϕ is field strength in teslas

S is speed of rotor in revolutions per minutes.

The field strength, ϕ, is directly proportional to field current. We can, therefore, use a different constant, K, and rewrite the equation:

$E_g = K \times I_f \times S$ where I_f is field current in amperes.

We really do not have any choice of speed. The speed depends on the frequency wanted. The only way to change the output voltage is to change field current.

SATURATION CURVE

Field current may be varied in either of two ways:
1. The field supply may come from a variable voltage source, as shown in Fig. 10-14. In this case, the field voltage may be changed.
2. Another method is shown in Fig. 10-15. A field rheostat is connected in series with the field coil.

Magnetic field strength is directly proportional to current through the field coil. As field current increases, more and more of the rotor's magnetic domains become lined up. At some current, the domains become hard to arrange. The core has become SATURATED. Further increases in current produce progressively smaller increases in field strength.

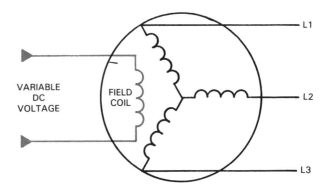

Fig. 10-14. Output voltage of an alternator depends on strength of rotor field. One way to change that strength is by changing field voltage.

We cannot measure field strength directly. What we can do is measure generated voltage. First we make sure the alternator is turning at a constant speed. With no field current, only a little voltage is generated. This is due to residual magnetism in the rotor core. Then we slowly increase field current. Terminal voltage increases. The curve is shown in Fig. 10-16.

The point where the curve begins to flatten out is called the knee of the curve. Alternators are designed to operate around that point.

By plotting the saturation curve, you learn two things:
1. The operating voltage of the alternator.
2. The value of field current needed to produce that voltage.

ALTERNATOR ROTOR CONSTRUCTION

Rotors have the same number of poles as each phase of the stator. A salient-pole rotor is shown in Fig. 10-17. The core of this rotor is called a SPIDER. It is used for slow-speed alternators like the ones driven by hydroturbines.

Steam turbines are more common prime movers, however. They run at high speed. Rotors of high-speed alternators, Fig. 10-18, are cylindrical. This design reduces windage (wind friction) losses. The field winding of a cylin-

drical rotor is distributed in slots. However, they produce magnetic poles just like the salient pole rotors. We will use the spider type rotor to represent both kinds.

ALTERNATOR LOADS

Current flows through the armature coils when a load is connected. This current sets up a revolving stator field. The stator field reacts with the rotor field to produce a counter torque.

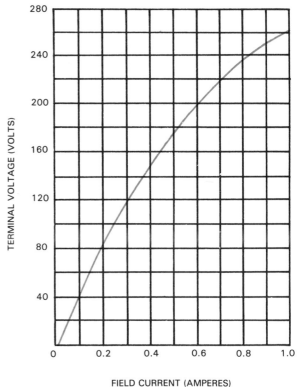

Fig. 10-16. Typical saturation curve of an alternator rated at 120 volt-amperes.

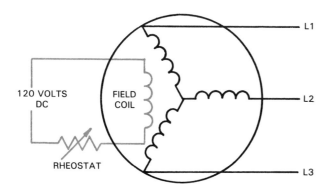

Fig. 10-15. With a constant field voltage source, a rheostat can be used to vary field current to change the alternator's voltage.

Fig. 10-17. Salient-pole rotor of the type used in slow-speed alternators.

The more current drawn from the alternator, the harder it is to turn the rotor. The prime mover, therefore, has to supply more mechanical power. It is assumed that the electrical load is always balanced. This is usually the case. In a wye-connected alternator, line current is equal to coil current.

POWER USED BY THREE-PHASE LOADS

Assume you have a resistive load connected to an alternator. This is shown in Fig. 10-19. An

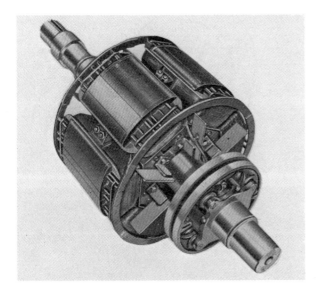

Fig. 10-18. Cylindrical rotor of the type used in high-speed alternators.

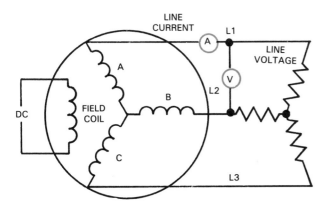

Fig. 10-19. Total power delivered by a three-phase alternator is 1.73 times the product of line voltage and line current.

ammeter measures current in line L1. A voltmeter measures the voltage between L1 and L2. Assume that we want to know the total power supplied by the alternator. It is three times the power supplied by one phase.

The same current flows through coil A and line L1. Therefore, we can use the ammeter reading as the phase current. Line voltage, E_{LINE}, is 1.73 times phase voltage E_{ph}. Therefore, E_{ph} equals line voltage divided by 1.73. The power supplied by one phase, P_{ph}, is:

$$P_{ph} = I_{ph} \times E_{ph} = I_{LINE} \times \frac{E_{LINE}}{1.73}$$

To find total power, multiply by three:

$$P_{TOT} = 3 \times I_{LINE} \times \frac{E_{LINE}}{1.73}$$

$$= 1.73 \times I_{LINE} \times E_{LINE}$$

This relationship holds true for all balanced three-phase resistive loads, whether delta or wye connected. Total power is 1.73 times line current times line voltage.

Example 1

The alternator of Fig. 10-20 has a balanced resistive load. Line voltage is 208 volts. Line current is 3.5 amperes. Find the total power drawn by the load.

$$P_{TOT} = 1.73 \times I_{LINE} \times E_{LINE}$$
$$= 1.73 \times 3.5 \times 208 = 1260 \text{ W}$$

If the load has inductance (or capacitance), the above equation gives you the apparent power. To find the active power, multiply apparent power by the power factor.

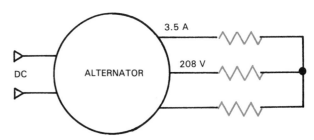

Fig. 10-20. Alternator loads are always assumed to be balanced. This means that all three phases are loaded equally.

Example 2

The alternator of Fig. 10-21 has a three-phase inductive load with a 0.8 lagging power factor. Line voltage is 208 volts. Line current is 3.5 amperes. Find the active power drawn by the load.

$$P_{TOT} = 1.73 \times I_{LINE} \times E_{LINE} \times \text{Power Factor}$$
$$= 1.73 \times 3.5 \times 208 \times 0.8 = 1008 \text{ W}$$

ARMATURE MAGNETIC FIELD

As the rotor field sweeps around, voltage is induced into the armature conductors on the stator. When a load is connected, current flows through these conductors.

Current in each conductor sets up a magnetic field. The field around each conductor expands, collapses, and reverses directions.

However, the combined effect from all the fields of all the armature conductors is a single field. This armature field revolves inside the alternator. As shown in Fig. 10-22, the armature field is at right angles to the rotor field. What is more, it revolves in SYNCHRONISM with the rotor field. (That is, both travel at the same speed.) This type of alternator is, therefore, called a SYNCHRONOUS ALTERNATOR.

DAMPER WINDINGS

The speed of an alternator must not change. If it did, frequency would change. A prime mover, however, may pulsate. To maintain a constant speed, alternators use DAMPER WINDINGS. These are copper bars just below the surface of the spider poles. As shown in Fig. 10-23, the damper windings are short-circuited at each end.

As long as the rotor is turning at a constant speed, the damper windings do not cut through any magnetic lines of force. The rotor field, ar-

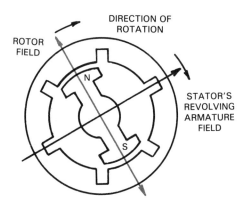

Fig. 10-22. Load current through the three armature coils produces a revolving magnetic field. This stator field revolves in step with the rotor field.

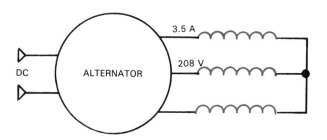

Fig. 10-21. When the power factor of the load is other than unity, the alternator must supply reactive power as well as active power.

Fig. 10-23. Besides the field coil, the alternator rotor also contains damper windings, which are bars, short-circuited at both ends.

mature field, and damper windings all move together.

Suppose, however, that the rotor changes speed. Now there is a change in the armature flux linking the windings. This induces a voltage into the bars. Current immediately begins to flow, setting up another field. The direction of this field is determined by Lenz's Law: *induction tends to oppose whatever causes it.*

If the rotor tries to speed up, the damper winding flux adds to the counter torque. This keeps it from speeding up. If the rotor tries to slow down, the damper winding flux helps overcome some of the counter torque. This keeps the rotor from slowing down.

HOW VOLTAGE VARIES WITH LOAD

If there is no load connected to an alternator, its terminal voltage is exactly the same as the generated voltage. When a load is connected,

however, load current flows through the armature coils. This brings into play the internal impedance of the alternator.

Internal impedance is made up of three parts:
1. Resistance of the armature conductors.
2. Inductive reactance of the armature coils.
3. Armature reaction.

These three factors affect terminal voltage. How they do it depends on the power factor of the load.

RESISTANCE

Resistance of the armature conductors produces an "IR" voltage drop. This voltage drop is in phase with the load current. To see the effect of armature resistance, refer to Fig. 10-24. The three phasor diagrams shown are for the same alternator. Phase voltage (which is 120 volts) and coil resistance, (at 24 ohms) are the same in each case.

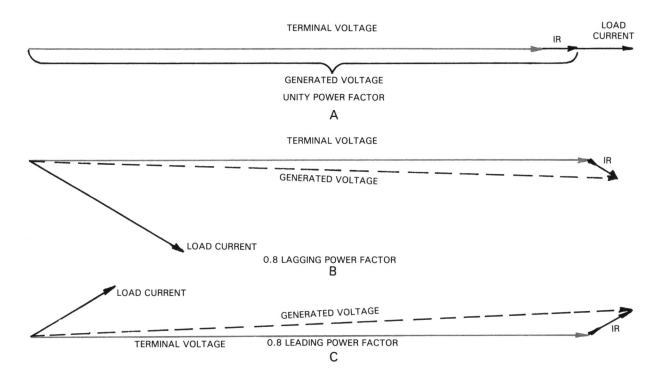

Fig. 10-24. Effect of armature resistance on generated voltage. A—For a resistive load. B—For an inductive load. C—For a capacitive load.

The vector representing this "IR" drop is drawn parallel to the load current vector in every case. This shows that it is in phase with load current. The dotted line merely shows the extra volts that must be generated to overcome the voltage drop due to armature resistance. Besides this, however, there is also inductive reactance and armature reaction.

INDUCTIVE REACTANCE

Inductive reactance is different from resistance. It comes from the self-inductance of the armature coil. Inductance, remember, is the ability to generate a back voltage. This voltage, shown as line IX in Fig. 10-25, view A, leads the current that causes it by 90 degrees.

As ac load current flows through the armature conductors, it creates inductive reactance. When load current is in phase with terminal voltage, IX voltage opposes the generated voltage. This is shown also in Fig. 10-25, view A.

When load current lags terminal voltage, Fig. 10-25, view B, inductive reactance has even greater effect. Additional voltage must be generated to overcome this IX back voltage.

Fig. 10-25, view C, shows what happens when load current leads terminal voltage. The situation is reversed. Now the inductive reactance voltage, IX, actually aids generated voltage. Less voltage must be generated than was needed when there was only resistance to consider as in Fig. 10-24, view C.

ARMATURE REACTION

The third reason that terminal voltage is different from generated voltage is armature reaction which is caused by the revolving stator field. This field, remember, results from load current in the armature coils.

The revolving stator field combines with the rotor field. This has the effect of changing the amount of voltage generated in the armature coils.

The armature reaction voltage does not exist as a separate voltage. It can be represented, however, as a back voltage, E_{ar}, leading load current by 90 degrees. That makes it in-phase with inductive reactance.

Fig. 10-26 shows all three factors. With a unity power factor load, 128.4 volts must be generated in each coil to produce a terminal phase voltage of 120 volts.

With a 0.8 *lagging* power factor load, 135.2 volts must be generated to produce the same terminal voltage. However, with a 0.8 *leading* power factor load, only 118.2 volts need be generated. This is less than the terminal voltage. The curves of Fig. 10-27 show how terminal voltage changes with load current for different types of loads.

VOLTAGE REGULATION

Voltage regulation is usually expressed in a percentage. It tells us how much change to expect in terminal voltage between no-load and full-load. The equation is:

% Voltage Regulation =

$$\frac{\text{NO-LOAD VOLTS} - \text{FULL-LOAD VOLTS}}{\text{FULL-LOAD VOLTS}}$$

The volts may be either line or phase volts. The following examples use the same data as in Fig. 10-26:

With unity power factor:
$$\% \text{ Voltage Regulation} = \frac{128.4 - 120}{120} \times 100$$
$$= 7\%$$

With 0.8 lagging power factor:
$$\% \text{ Voltage Regulation} = \frac{132.5 - 120}{120} \times 100$$
$$= 10.4\%$$

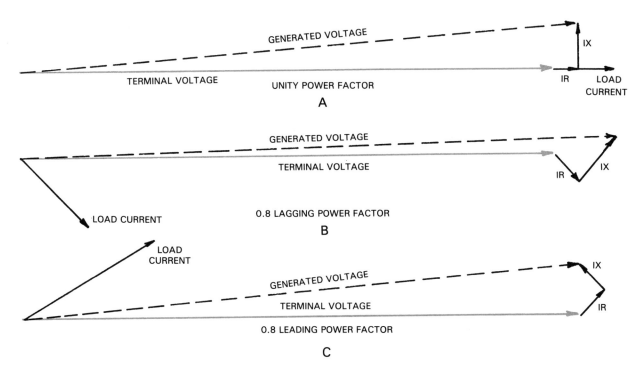

Fig. 10-25. Combined effect of armature resistance and armature reactance on generated voltage. A—For a resistive load. B—For an inductive load. C—For a capacitive load.

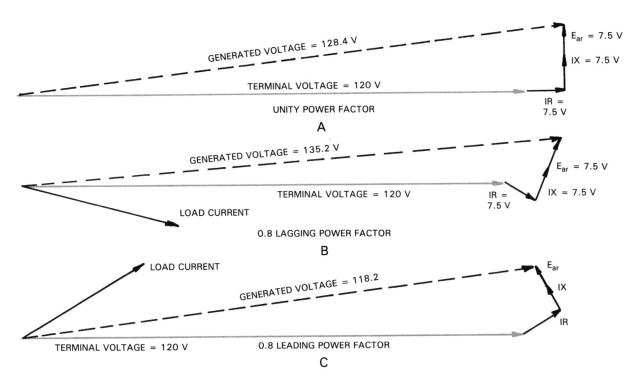

Fig. 10-26. Combined effect of armature resistance, armature reactance, and armature reaction on generated voltage. A—For a resistive load. B—For an inductive load. C—For a capacitive load.

With 0.8 leading power factor:

$$\text{\% Voltage Regulation} = \frac{118.2\text{-}120}{120} \times 100$$
$$= -1.5\%$$

From these examples, you can see why you must include the load power factor when talking about the voltage regulation of an alternator.

ALTERNATOR TESTING

There are two ways to determine the voltage regulation of an alternator:

1. If the alternator is small enough, you can simply measure voltage at no-load and full-load.
2. But often this cannot be done. Certain tests must be made. Data from these tests can be used in computing voltage regulation.

The object of testing is to determine the alternator's SYNCHRONOUS IMPEDANCE. *Synchronous impedance is the combined effect of armature resistance, inductive reactance and armature reaction.*

The tests that we make are called the:

1. Short-circuit test.
2. Open-circuit test.
3. Effective resistance test.

SHORT-CIRCUIT TEST

The short-circuit test determines a value of field current to use in the open-circuit test. The circuit for this test is shown in Fig. 10-27. First, the terminals of the alternator are short-circuited. The alternator is then run at rated speed. The dc field excitation is slowly increased until some load current flows.

The exact value is not important. It must be recorded, however, for use later.

Under these conditions, a certain voltage is being generated. Since the terminals are

shorted, there is no terminal voltage. This means that all of the generated voltage is being used to overcome the synchronous impedance.

OPEN-CIRCUIT TEST

The open-circuit test finds the value of generated voltage that we could not measure in the short-circuit test.

Again, we run the alternator at rated speed. Then we supply the field with the same dc current that we did in the short-circuit test. Now we know that the same voltage is being generated. We simply measure this no-load voltage. *The open-circuit terminal voltage IS the generated voltage.*

COMPUTING SYNCHRONOUS IMPEDANCE

We can use the voltage measured in the open-circuit test to compute synchronous impedance. The equation is:

$$Z_{ph} = \frac{E_{g(PH)}}{I_{ph}}$$

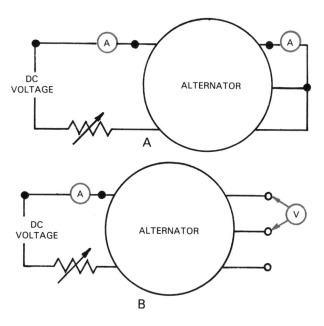

Fig. 10-27. Circuits for alternator testing. A—Short-circuit test. B—Open-circuit test.

Z_{ph} is the synchronous impedance per phase in ohms

$E_{g(PH)}$ is the generated voltage per phase in volts

I_{ph} is the short-circuit current in amperes

Example 3

A short-circuit test on a certain wye-connected alternator is performed with a current of 0.3 amperes. The line (terminal) voltage measured in the open-circuit test is 29 volts. Find the synchronous impedance.

a. Convert line voltage to phase voltage.

$$E_{ph} = \frac{E_{LINE}}{1.73} = \frac{29}{1.73} = 16.8 \text{ V}$$

b. Compute synchronous impedance per phase.

$$Z_{ph} = \frac{E_{g(PH)}}{I_{ph}} = \frac{16.8}{0.3} = 56 \ \Omega$$

EFFECTIVE RESISTANCE TEST

While it is good to know the total synchronous impedance, that alone will not help us predict voltage regulation. We must also know how it breaks down into resistance and reactance.

Most alternators are wye-connected internally. The resistance between any two terminals is the resistance of two coils. The setup for a resistance test is shown in Fig. 10-28. First, a dc voltage is applied. The dc resistance is computed from voltmeter and ammeter readings $(R = \frac{E}{I})$. This number is divided by two to get the resistance of one coil. Finally, to determine the effective ac resistance, multiply the dc resistance by 1.5.

Example 4

A resistance test performed on the alternator of *Example 3* shows an applied voltage of 20 volts, dc. Current is 0.6 amperes, dc. Find the effective ac resistance for each phase of the alternator. To solve:

a. Compute the dc resistance between terminals.

$$R = \frac{E}{I} = \frac{20}{0.6} = 33.3 \ \Omega$$

b. Compute the dc resistance per phase.

$$R_{ph} = \frac{R_t}{2} = \frac{33.3}{2} = 16.7 \ \Omega$$

c. Compute the effective ac resistance per phase.

$$R_{EFF} = 16.7 \times 1.5 = 25 \ \Omega$$

COMPUTING REACTANCE

We can think of the synchronous impedance as being the hypotenuse of a right triangle. This is shown in Fig. 10-29. The adjacent side is resistance. The opposite side is reactance. Using the values from *Example 3 and 4,* the solution follows:

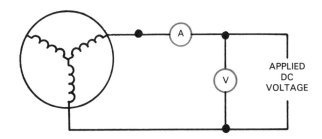

Fig. 10-28. Circuit for determining the effective resistance of the armature windings.

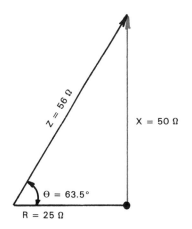

Fig. 10-29. Phasor diagram used for determining the reactance of the armature windings.

θ is the angle whose cosine equals resistance divided by impedance.

$$\cos θ = \frac{25}{56} = 0.446$$

$$θ = 63.5°$$

Reactance, X_s, $= Z × \sin 63.5°$

$$X_s = 56 × 0.895 = 50 \, Ω$$

LOSSES AND EFFICIENCY OF ALTERNATORS

Some of the mechanical power that drives an alternator never gets converted to electrical power. It is used to overcome friction and wind resistance. These rotational losses do not change with load.

There are two types of electrical losses:
1. Those due to the resistance of the field coil.
2. Those due to the resistance of the armature coils.

Field loss is constant. The armature loss varies with load.

The only other power loss in an armature is in the core. This loss is due to hysteresis and eddy currents. Core loss does not vary with load.

FIELD LOSS

Direct current is applied to the field. This makes field loss the easiest to determine. Apply a dc voltage to the field and measure the current. Then compute the power loss from the equation, $P = EI$.

Example 5A

A variable dc voltage is applied to the field coil. Voltage is increased until rated field current of 0.58 amperes flows. Voltage at this point is 46 volts. Find the field loss, P_{FL}.

$$P_{FL} = E × I = 46 × 0.58 = 26.7 \, W$$

ROTATIONAL LOSSES

To determine rotational losses, it helps to have a motor whose losses are already known. If none is available, you must also find this value. For example, assume you are using a dc motor as a prime mover. Connect meters to the motor as shown in Fig. 10-30. Start the motor and adjust it to the alternator's rated speed. Then multiply the motor's voltage by its current. This is the power lost in the motor alone.

Then couple the motor to the alternator. Do not excite the alternator's field. Again, run the motor at rated speed. Now the power input to the motor is the total of the motor's losses and the alternator's rotational losses.

Example 5B

A dc motor running uncoupled has voltage of 125 volts, and a current of 0.62 amperes. When coupled to an unexcited alternator, readings are 125 volts and 0.95 amperes. Find the alternator's rotational losses.

a. Compute the motor's losses.

$$P_M = E × I = 125 × 0.62 = 77.5 \, W$$

b. Compute the total (motor plus alternator rotational) losses.

$$P_T = E × I = 125 × 0.95 = 118.75 \, W$$

c. Compute the alternator rotational losses.

$$P_R = P_T - P_M = 118.75 - 77.5$$
$$= 41.25 \, W$$

CORE LOSSES

We can not determine the core losses directly. However, we can determine the combined rotational and core losses. Then, by subtracting

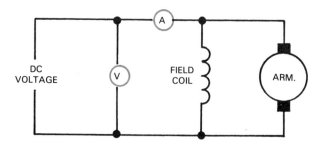

Fig. 10-30. Circuit used for determining the rotational losses of a test motor.

rotational losses, we are left with the value of the core losses.

The alternator is once again driven at rated speed without load. This time, however, voltage is applied to the field. Now the rotor, stator, and air gap are magnetized. Power input to the motor now includes motor losses, alternator rotational losses and core losses.

Example 5C

With the alternator field excited, motor values are 115 volts and 1.13 amperes. Find core losses.

a. Compute total power (motor, plus rotational, plus core losses).
$$P_{TT} = E \times I = 125 \times 1.13$$
$$= 141.25 \text{ W}$$

b. Compute the core losses.
$$P_C = P_{TT} - P_M - P_R$$
$$= 141.25 - 77.5 - 41.25$$
$$= 22.5 \text{ W}$$

ARMATURE COPPER LOSSES

The power loss in the armature windings of each coil is three times the current squared multiplied by the resistance ($I^2 \times R$). The current is load current. When there is no load, the efficiency is 0 percent. All of the input power is used to overcome losses. As load current increases, the alternator delivers more electrical power. At the same time, armature copper loss increases.

Armature resistance can be determined from voltage and current measurements. The ac resistance is always somewhat larger than the dc resistance. It varies from 1.2 times to 1.8 times, depending on:

1. Size and shape of core slots.
2. Size and insulation of the conductors.

There are many other factors which will affect armature resistance. We have used 1.5 as a convenient figure.

Example 5D

An alternator rated at 208 volts is supplying a load current of 0.5 amperes. The effective alternating current resistance is 24 ohms for each phase. Find the copper loss assuming unity power factor load.

a. From the information given above compute the full-load copper loss in each coil.
$$P_{PH} = I^2R = 0.5^2 \times 24 = 6 \text{ W}$$

b. With the full-load copper losses known, compute the total armature copper loss at full load.
$$P_A = 3 \times P_{PH} = 3 \times 6 = 18 \text{ W}$$

EFFICIENCY

Efficiency of an alternator is its ratio of output to input. This ratio is expressed as a percentage. The equation is:

$$\% \text{ Efficiency} = \frac{\text{Output power}}{\text{Input power}} \times 100$$

For output power, we can use the alternator's rating. Input power is the sum of all the losses plus the output power. The equation can be written as follows:

$$\% \text{ Efficiency} = \frac{\text{Output power} \times 100}{\text{Output Power} + \text{Losses}}$$

All of these factors must be in the same units. You must multiply the volt-amperes rating by the load power factor to convert to watts. The same procedure would be used to convert KVA to KW.

Example 5E

Using the losses as computed in *Examples 5A through 5D,* find the efficiency of this alternator.

a. Compute total losses.
$$P_T = P_{FL} + P_R + P_C + P_A$$
$$= 26.7 + 41.25 + 22.5 + 18$$
$$= 108.45$$

b. Compute the total output power.

$$P_o = 1.73 \times E \times I \times P.F.$$
$$= 1.73 \times 208 \times 0.5 \times 1 = 180 \text{ W}$$

c. Compute efficiency.

$$\% \text{ Eff} = \frac{P_o}{P_o + P_T} \times 100$$
$$= \frac{180}{180 + 108.45} \times 100$$
$$= 62.4\%$$

Efficiency is poor at low loads. There the losses are a large share of the total power. Efficiency increases steadily until the I^2R losses get too high. Maximum efficiency occurs when the variable losses equal the fixed loss.

Example 5F

Using the alternator of *Examples 5A through 5E*, find the load current at the point of maximum efficiency and the efficiency at that point.

a. Compute the total fixed losses.
$$P_{FL} + P_R + P_C = 26.7 + 41.25 + 22.5$$
$$= 90.45 \text{ W}$$

b. Compute the load current at maximum efficiency.
$$I^2R = 90.45$$
$$I = \sqrt{\frac{90.45}{72}} = 1.121 \text{ A}$$

c. Compute the total losses.
$$P_T = 90.45 + 90.45 = 180.9 \text{ W}$$

d. Compute the output power at 1.121 amperes.
$$P_o = 1.73 \times 208 \times 1.121 = 403.4 \text{ W}$$

e. Compute efficiency at unity power factor load.
$$\% \text{ Eff} = \frac{P_o}{P_o + P_T} = \frac{403.4}{403.4 \times 180.9}$$
$$= \frac{403.4}{584.3} = .69 \times 100$$
$$= 69\%$$

f. Compute efficiency at 0.8 lagging power factor load.
$$\% \text{ Eff} = \frac{P_o \times P.F.}{(P_o \times P.F.) + P_T}$$
$$= \frac{322.72}{322.72 + 180.9}$$
$$= \frac{322.72}{503.62} = .64 \times 100 = 64\%$$

PARALLELING ALTERNATORS

Most power lines are fed by several alternators. From time to time, each alternator must be shut down for maintenance. It must then be brought back on the line. That takes more than just starting it up. It must be "paralleled" with alternators already generating. The output of an alternator being paralleled must match the "bus" voltage in four ways:

1. Voltage.
2. Frequency.
3. Phase sequence.
4. Phase relationship.

Paralleling is something like meshing two gears. Refer to Fig. 10-31. The teeth of both gears must have the same size and shape. This is like voltage. The voltage must have the same shape wave and RMS value.

The gears must be turning at the same speed. This is like frequency. They must also be turning in the right direction. This is like phase sequence. Finally they must be properly lined up. Gears will not mesh if a tooth of one gear is opposite the tooth of another. This is like phase relationship.

PARALLELING PROCEDURE

A convenient aid to paralleling is a bank of three lamps, as shown in the circuit of Fig. 10-33. One side of each lamp is connected to each of the three lines of the bus; the other side to the incoming alternator. In parallel with the lamps is a three-pole, single-throw "paralleling" switch. The switch connects the incoming alternator to the bus.

FREQUENCY AND PHASE

When the switch is open (before paralleling), the voltage across each lamp at every instant is the difference between the instantaneous values of the bus and alternator voltages. When the voltages are 180 degrees out of phase, as shown in Fig. 10-32, lamp voltage is twice line voltage;

the lamps glow brightly. When the two voltages are in phase, there is no drop across the lamps; the lamps are out.

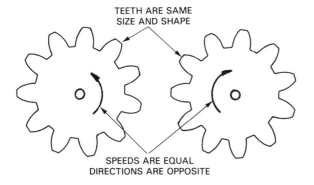

Fig. 10-31. Paralleling alternators is like meshing two rotating gears.

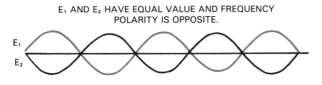

Fig. 10-32. Wave forms of one phase of two alternators whose voltages have the right magnitude, frequency, and polarity for paralleling.

The frequency of generated voltage depends on the speed of the alternator. If you had to depend on a speed measurement alone, you would find it hard to set the prime mover perfectly. With the lamps, however, it is easy to match the incoming alternator's frequency with that of the bus.

As you adjust the prime mover's speed and approach the correct frequency, the alternator's voltage will gradually drift in and out of phase, as shown in Fig. 10-34. You will see the lamps slowly become bright, then dark, then light, then dark. As we will explain, paralleling is done when the lamps are dark — when the two voltages are at the same frequency and are in phase.

VOLTAGE

The terminal voltage of an incoming alternator must equal that of the bus. Voltage, remember, is proportional to both speed and excitation. When paralleling, you won't be able to adjust voltage by changing speed; you have to use speed to set frequency. You therefore adjust voltage by increasing or decreasing the dc excitation current.

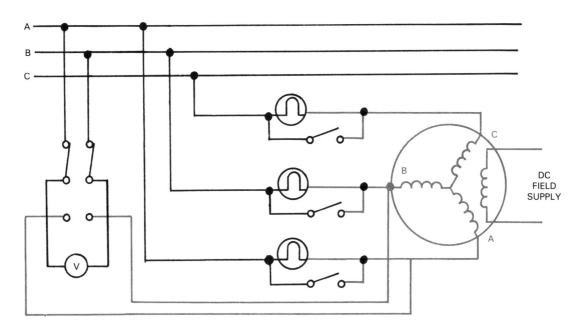

Fig. 10-33. Circuit is used for paralleling alternator with power company's bus.

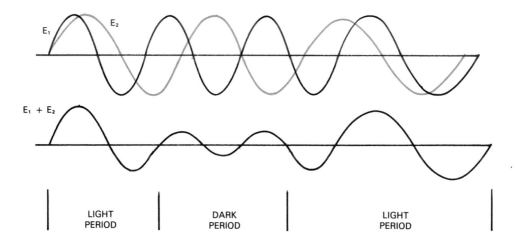

Fig. 10-34. Wave forms of one phase of two alternators having slightly different frequencies.

It's a good idea to measure both alternator and bus voltage with the same voltmeter, switching it as shown in Fig. 10-33. That way, you won't have to worry about differences in accuracies between two meters.

PHASE ROTATION

The phase rotation of the incoming alternator must be the same as the phase rotation of the bus. This is easy to detect when you use the three-lamp method of paralleling. The lights blink ON and OFF together when the incoming alternator's phase rotation is correct. When it is opposite, they blink rapidly in sequence. This will be the first thing you will notice when you excite the alternator.

To reverse the phase rotation of the alternator, you must first turn OFF the prime mover. Then simply interchange any two leads of the alternator.

When you have the phase rotation correct, next adjust the prime mover to give you close to the correct frequency. Then check the voltage of the bus and adjust the alternator's excitation until it produces the same voltage. Finally, when the two voltages are in phase (the middle of a "lamp dark" period), close the paralleling switch to connect the incoming alternator to the power line bus.

THE SYNCHROSCOPE

The "three-lamp" method is especially useful for comparing phase rotations. However, it is not a precise way of identifying the exact point when the frequency and phase angle of the alternator match those of the bus. For that purpose, technicians use a SYNCHROSCOPE, Fig. 10-35.

The synchroscope is a phase angle meter whose needle is free to rotate. Its rotations of the needle indicates the frequency and phase relations between the voltage of the incoming alternator and the voltage of the running alternator. When the frequency of the incoming alternator is larger than that of the bus, the needle rotates clockwise. When its frequency is lower than that of the bus, the needle rotates counterclockwise. When the needle stops, the two voltages have the same frequency.

Further, the needle position indicates the phase angle between the two voltages. There is only one line on a synchroscope dial—in the twelve o'clock position. When the needle points to this line, the voltages are in-phase. The paralleling steps are the same as with the lamps: (1) make incoming frequency close to bus frequency (the needle rotates slowly), (2) match incoming voltage with bus voltage, then (3) when the needle is straight up, close the paralleling switch.

FLOATING ON THE LINE

Once the alternator has been paralleled, its voltage and speed are determined by the bus. At the instant you throw the switch, the alternator is connected to the bus, but is not delivering power. It is said to be "floating" on the line. If you try to reduce the speed of the prime mover, the alternator continues to run at rated speed. How is that possible? There is not enough mechanical power to drive the alternator that fast. Therefore, the alternator draws electrical power from the bus. It actually runs as a motor.

If you try to increase voltage by increasing direct current excitation, terminal voltage will remain the same. You actually increase the generated voltage.

Why is it that the terminal voltage does not change? It may not rise above bus voltage. The extra generated voltage produces a reactive current flow into the bus. This current does not deliver any active power.

Fig. 10-35. Synchroscope is used for paralleling alternators.

ACCEPTING A SHARE OF THE LOAD

The alternator can be made to deliver power current. The correct procedure is to adjust the prime mover. Try to increase its speed. The speed will not change but it will deliver mechanical power to the alternator. The alternator will now produce in-phase load current (active power). Not counting the losses, the alternator delivers as much electrical power as it receives mechanical power.

Large electrical power systems are controlled by computers. The computer senses when more power is needed. Then it decides which alternator in the system should provide it. The computer can automatically adjust the controls on the prime mover.

Stored in the computer are the operating characteristics of every alternator in the system. Power will be drawn from the one that can supply it most efficiently.

SUMMARY

An alternator can generate a single ac voltage or several voltages having a specific phase relationship.

Voltage is induced in the armature of an alternator because of the relative motion between conductors and a magnetic field.

Armature windings are arranged so they produce a sinusoidal wave form.

The frequency of generated ac depends on the number of poles and the rotation speed of the alternator.

Alternators require direct current excitation of the field coils.

Small alternators have their field coils on the stator. Large alternators have field coils on the rotor and armature coils on the stator.

A three-phase delta connection produces a three-wire system. The voltage across any two of the three wires is equal to the voltage across one phase.

A three-phase wye connection produces a four-wire system. The voltage between line and neutral equals phase voltage. The voltage across any two lines equals 1.73 times phase voltage.

The terminal voltage of an alternator will decrease as a unity power factor load or a lagging power factor load increases.

The terminal voltage of an alternator will increase as a leading power factor load increases.

TEST YOUR KNOWLEDGE

1. The phase angle between the phases of three-phase power is _____ degrees.
2. Increasing the speed of an alternator would:
   a. Cause a decrease in frequency.
   b. Cause increase in frequency.
   c. Have no effect on frequency.
3. Why do alternators have damper windings? Tell how they work.
4. The combined effect of armature resistance, inductive reactance and armature reaction is known as _____ _____.
5. Following are the power losses associated with alternators. Tell whether each is electrical or mechanical and whether it is fixed or variable.
   a. Friction.
   b. Field copper loss.
   c. Core loss.
   d. Armature copper loss.
   e. Windage.
6. Indicate which four of the following characteristics of two alternators must be the same in order to parallel them:
   a. Current.
   b. Frequency.
   c. Phase relationship.
   d. Phase sequence.
   e. Voltage.

PROBLEMS

1. Given: The phase voltage of an alternator is 240 volts. Find the line voltage.
2. Given: A 10-pole alternator is being driven at 1440 rpm. Find the frequency of the generated voltage.
3. Given: A balanced resistive load is connected to a three-phase alternator. Line voltage is 416 volts. Line current is 2.5 amperes. Find the total power.
4. Given: A short-circuited test is performed with a current of 0.25 amperes. Line voltage in the open circuit test is 65 volts. Find the synchronous impedance per phase.
5. Given: A resistance test is performed on the alternator of *Problem 4*. With 15 volts applied, current is 0.6 amperes. Find the effective resistance per phase.
6. Given: From the values computed in *Problems 4* and *5* of synchronous impedance and effective resistance, find the reactance.

11 THREE-PHASE MOTORS

After studying this chapter, you will be able to:

☐ Explain how a revolving magnetic field is produced in the stator of a three-phase motor and how it produces torque.
☐ Define the terms, synchronous speed and slip.
☐ Compute starting torque and starting current from locked rotor tests.
☐ Describe the construction and purpose of wound-rotor motors.
☐ Explain the procedure for synchronizing a three-phase synchronous motor.

Some three-phase motors may be as large as 1500 horsepower (over 1000 kilowatts). The smallest are 1 horsepower (746 watts).

There are two main types of three-phase motors:
1. Induction.
2. Synchronous

Induction motors may have either a squirrel-cage rotor or a wound rotor. The squirrel-cage motor is, by far, the most popular.

STATOR CONSTRUCTION

The stators of all types of three-phase motors are the same. A stator is made from sheets of steel called LAMINATIONS. This steel is highly refined to produce good magnetic properties with minimal losses. Laminations are insulated from each other by an OXIDE that forms on the surface during heat treating. A typical stator lamination is shown in Fig. 11-1.

Each lamination has notches around the inner edge. When laminations are stacked together,

these notches line up to form slots. The conductors of the stator coils are placed in these slots.

Fig. 11-2 shows the stator of the three-phase motor. Coils are arranged around the stator in groups. Each group produces a north and south pole when current flows.

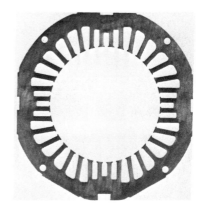

Fig. 11-1. Single lamination from stator core of three-phase motor. Slots along inner edge will hold stator windings.

Fig. 11-2. Stator of three-phase motor. Coils are insulated from laminations.

In a two-pole motor there are three such coil groups—one group per phase. We will label these windings A, B and C, as in Fig. 11-3. Although the windings themselves overlap, the center of their magnetic fields are 120 mechanical degrees apart.

A four-pole motor has twice as many coil groups. Their centers, therefore, are placed 60 mechanical degrees apart. Mostly, we will be describing a two-pole motor, because it is easier to show. The same principles, however, apply to all three-phase motors.

REVOLVING STATOR FIELD

All three-phase motors must have a revolving stator field. The stator, itself, does not move. It is the combined magnetic field from the stator coils that revolves. Fig. 11-4 demonstrates how it is done. One full cycle of three-phase current is shown. The location of the field is indicated for four instants of time. For this example, we have chosen 10 amperes for peak current.

Assume current flowing out from the motor terminal is positive. Current flowing into a motor terminal is negative.

At *Instant 1,* current from phase B terminal is 10 amperes positive. At the same instant, 5 amperes is flowing into phase A terminal and phase C terminal. The instantaneous current in phase B is the sum of the currents flowing into the neutral point.

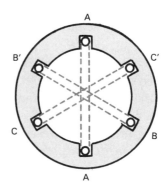

Fig. 11-3. Three coil groups are located 120 degrees apart, producing two poles for each phase.

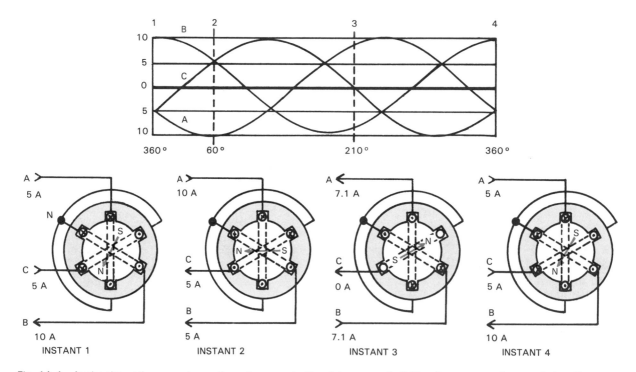

Fig. 11-4. As the alternating current goes through one cycle, the stator magnetic field makes one complete revolution. Compare the positions in the stator drawings with the sine wave above.

The stator field at *Instant 1* is downward to the left as shown by the blue arrow. Most of this field is from phase B coil. However, the fields from both phase A and phase C (half strength) add to it.

At *Instant 2,* the situation is as follows: Phase B current has fallen to 5 amperes positive. Phase C current has reversed direction and is now also 5 amperes positive. Phase A current is 10 amperes negative. Current flows into phase terminal A, splits, and flows from terminals B and C.

The field at *Instant 2* has revolved 60 degrees clockwise. Most of the field is from phase A coil. The rest is from phases B and C.

Instant 3 is taken at 210 degrees. At that instant, Phase C, current is zero. Phase A current is equal and opposite to phase B current. Current flows into terminal B and out terminal A. Now the combined A and B field is horizontal to the left.

Finally, *Instant 4* is at 360 degrees. The currents are back where they started. The stator field, too, is back where it started. In a two-pole motor the stator field makes one revolution per cycle of alternating current.

SQUIRREL-CAGE MOTORS

Squirrel-cage motors, Fig. 11-5, are less expensive to build and maintain than other types. There are no commutator, slip rings or brushes. There are no electrical connections to the rotor. The link is strictly magnetic.

ROTOR CONSTRUCTION

Like the stator, the rotor is made from steel laminations. In fact, the discs punched out of the center of the stator laminations can be used. The discs are heat treated and pressed onto the rotor shaft. As shown in Fig. 11-6, evenly spaced holes are cut around the outside of the disc. When discs are laminated, the holes line up to form slots. Into these slots are pressed bars of copper or aluminum.

The squirrel-cage winding with only a single lamination of the core is shown in Fig. 11-7. The bars are shorted at each end. When voltage is induced in the bars, current flows because it has a closed path. This current will lag the induced voltage. The amount of lag will depend on the rotor's resistance and inductive reactance.

SYNCHRONOUS SPEED

The instant you turn on a three-phase motor, current flows in the stator coils. The stator magnetic field begins to revolve. The field's rate

Fig. 11-5. Typical squirrel-cage rotor is from a three-phase motor. (Westinghouse Electric Co.)

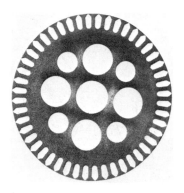

Fig. 11-6. One lamination from the rotor core. Holes along outer edge will hold copper or aluminum bars.

of travel, called SYNCHRONOUS SPEED, depends on two things:
1. The number of poles.
2. The frequency of the applied voltage.

In a two-pole motor, the stator field revolves once each cycle. In a four-pole motor, it goes through only half a revolution each cycle. Fig. 11-8 shows the pattern of the stator magnetic fields of two-pole and four-pole motors.

The table in Fig. 11-9 gives the synchronous speed with 60 hertz voltage applied according to the number of poles.

ROTOR MAGNETIC FIELD

As the stator's field sweeps around the air gap, it induces a voltage into the rotor bars.

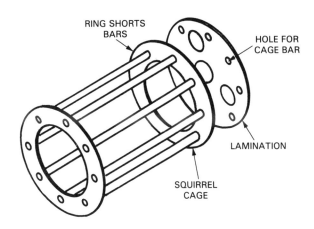

Fig. 11-7. Cage carries current and holds laminations in squirrel-cage rotor.

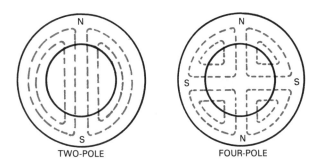

Fig. 11-8. Pattern of stator magnetic flux in two and four-pole motors. Flux pattern is in color.

This induced voltage is proportional to the rate of change in flux linkage. At the instant of start, the flux linkage is changing at the maximum rate. That is because the rotor has not started turning yet.

Current starts flowing in the rotor bars, one after the other. This produces a revolving rotor field. If the rotor bars had no inductive reactance, this revolving rotor field would be at right angles to the stator field. This is shown by view A, Fig. 11-10. However, the rotor does have inductive reactance which makes the rotor field lag the stator field by more than 90 degrees, as shown in view B, Fig. 11-10.

FREQUENCY OF INDUCED VOLTAGE

Frequency varies with rotor speed. When the rotor is not turning, the induced frequency is the same as that of the applied voltage. Frequency, like voltage, depends on the rate at which the stator field is being cut by the rotor bars. When the rotor is turning, the field is cut at a slower speed. *The important thing is the difference between rotor speed and synchronous* speed. This difference is known as SLIP SPEED.

SLIP SPEED

Slip speed is the relative speed between rotor speed and synchronous speed (speed of the stator's magnetic field). It is like two runners, Joe and Jim, on a circular track. Joe, in the outside lane, goes around the track twice in one

No. of Poles	Synchronous Speed (RPM)	No. of Poles	Synchronous Speed (RPM)
2	3600	14	514
4	1800	16	450
6	1200	18	400
8	900	20	360
10	720	22	372
12	600	24	300

Fig. 11-9. Rotor speed depends on number of poles.

minute. At first, Jim is standing still. Joe's speed is 2 rpm faster than Jim's. You could say that the slip speed is 2 rpm. This is shown in view A, Fig. 11-11.

Now, suppose Jim starts running too. Jim is slower, going only halfway around in one minute (0.5 rpm). This is shown in view B, Fig. 11-11. Now Joe's speed relative to Jim's (that is, the slip speed), is 1.5 rpm. The effect is the same as it would be if Jim was not moving and Joe's speed was 1.5 rpm. Joe is like the stator field, which revolves at synchronous speed. Jim is like the rotor. The rotor always turns slower than synchronous speed. The stator field has to keep on passing the rotor bars to induce voltage into them.

Example 1

A three-phase induction motor has a syn-

chronous speed of 1800 rpm. Find the slip speed when the rotor is turning 1750 rpm.

Slip speed = synchronous speed − rotor speed
= 1800 − 1750 = 50 rpm

PERCENT SLIP

Slip speed is not used in computations. Instead, we use the ratio of slip speed to synchronous speed. This is simply called SLIP (symbol: s). In talking about slip, however, the term, PERCENT SLIP, is often used. This is slip multiplied by 100.

Example 2

Using the motor of *Example 1* (synchronous speed, 1800 rpm and rated speed, 1750 rpm), find the slip and percent slip.

a. Compute slip.

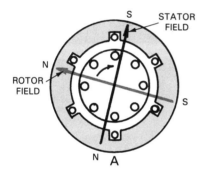
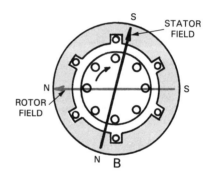

Fig. 11-10. Rotor field rotates in a fixed relationship to the stator field. A—Idealized angle is 90 degrees. B—Actual angle is greater than 90 degrees.

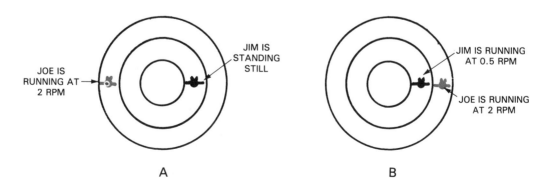

Fig. 11-11. Slip speed and synchronous speed are like runners going different speeds. A—Joe's speed, relative to Jim's, is 2 revolutions per minute. B—Joe's speed, relative to Jim's, is 1.5 revolutions per minute.

$$s = \frac{\text{synchronous speed} - \text{rotor speed}}{\text{synchronous speed}}$$

$$= \frac{1800 - 1750}{1800}$$

$$s = \frac{50}{1800} = 0.028$$

   b. Compute percent slip.

     $0.028 \times 100 = 2.8\%$

A motor with 2.8 percent slip is a low-slip motor. High-slip motors have up to 15 percent slip.

It should be noted that when the rotor is not turning, slip is unity (100 percent). This occurs at the instant of start. It can also happen if the motor stalls.

EFFECTS OF SLIP

The speed of an induction motor is always less than its synchronous speed. There is always slip. The amount of slip affects rotor voltage, frequency, inductive reactance, impedance and current. The amount of torque produced, then, is directly related to slip.

HOW SLIP AFFECTS VOLTAGE

At start (or whenever the rotor is not turning) slip is unity ($s = 1$). There is no way to measure the voltage induced in the rotor bars. We do know, however, that rotor voltage is maximum at that time. As the rotor speeds up, slip goes down. The induced voltage goes down, too, in direct proportion to slip.

The equation is:

$E_i = E_{max} \times s$

   E_i is the induced voltage at any rotor speed.

   E_{max} is the induced voltage when the rotor is not turning.

   s is slip, expressed as a decimal.

For example, if slip is 0.5, the voltage induced in the rotor is one-half that induced when the rotor is standing still.

HOW SLIP AFFECTS FREQUENCY

The frequency of the induced voltage goes down in direct proportion to slip. The equation is:

$f_i = f \times s$

f_i is frequency at any rotor speed

f is frequency of the applied voltage

s is slip, expressed as a decimal

HOW SLIP AFFECTS INDUCTIVE REACTANCE

The squirrel-cage winding has both resistance and inductance. Both are built in; they do not change. Inductive reactance, however, depends on frequency ($X_L = 6.28 \times f \times L$). As frequency goes down, so does inductive reactance.

When the rotor is not turning, frequency of the rotor current is maximum. That makes the inductive reactance of rotor bars maximum, too. When the rotor is turning, inductive reactance is directly proportional to slip. The equation is:

$X_L = X_{L_{max}} \times s$

   X_L is inductive reactance at any speed

   $X_{L_{max}}$ is inductive reactance when the rotor is not turning

   s is slip, expressed as a decimal

HOW SLIP AFFECTS ROTOR IMPEDANCE

Rotor impedance is the phasor sum of resistance and inductive reactance. Resistance is constant. Inductive reactance, on the other hand, changes with slip.

At the instant of start, inductive reactance is high. The impedance is mostly inductive. The rotor has a low, lagging power factor.

As the rotor picks up speed, its inductive reactance goes down. At some point it will equal the resistance. At operating speed, inductive reactance is low. Rotor power factor becomes close to unity. Impedance goes down as the rotor speeds up, but it cannot get any smaller than the resistance. The equation is:

$$Z_R = \sqrt{R_R{}^2 + X_{LR}{}^2}$$

Z_R is rotor impedance
R_R is rotor resistance
X_{LR} is rotor inductive reactance

HOW SLIP AFFECTS ROTOR CURRENT

From Ohm's Law, we know that, at any rotor speed:

$$I_R = \frac{E_i}{Z_R}$$

I_R is rotor current
E_i is induced rotor voltage
Z_R is rotor impedance

The faster the rotor turns, the less the induced rotor voltage becomes. This smaller voltage should reduce rotor current. However, rotor impedance is going down, too, but at a different rate. As it turns out, the induced voltage affects the value of rotor current more than impedance.

At start, rotor current is quite high. This is because voltage is so high.

When running at high speed without load, there is almost no rotor current. This is because of the small amount of voltage induced.

Over the operating range of a motor (typically from 0.5 to 8 percent slip) there is a big change in rotor current. This is due almost entirely to the change in induced voltage.

Example 2

A squirrel-cage induction motor has a full-load speed of 1750 rpm (2.8 percent slip). The rotor's resistance is 0.4 ohms. The rotor's inductive reactance at 60 Hz is 2 ohms.
We will not go through the math this time. The table in Fig. 11-12 shows the frequency, inductive reactance, impedance, rotor

current, and power factor for unity slip (instant of start), 1 percent slip (no-load) and 2.8 percent slip (full-load).

SQUIRREL-CAGE MOTOR TORQUE

What all this leads up to is torque. A motor's job, after all, is to deliver torque. There are specific names for the torque developed under certain conditions.

Following are the definitions of the most common torques. It is assumed that rated voltage is applied to the motor at rated frequency.

LOCKED-ROTOR TORQUE

Locked-rotor torque is also called BREAK-AWAY torque and STARTING torque. This is the torque developed when the rotor is not turning.

FULL-LOAD TORQUE

The power of a motor is usually given either in horsepower (Conventional units) or watts (SI metric units). The torque needed to produce rated power at full-load speed is called FULL-LOAD TORQUE. It can be computed from the equation:

$$T_{FL} = K \times \frac{P_D}{S_{FL}}$$

In Conventional units:

T_{FL} is in lbf-ft.
K is 5252
P_D is in horsepower
S_{FL} is full-load speed in rpm

Condition	% Slip	Freq. Hz.	X_i Ohms	Z_R Ohms	I_R Amps	Rotor P.F.
Start	100	60	2	2.04	49	0.2
No-load	1	0.6	0.02	0.401	2.494	0.999
Full-load	2.8	1.7	0.056	0.404	6.93	0.990

Fig. 11-12. Motor performance at different instants of time.

In SI metric units:

T_{FL} is in N·m
K is 9.55
P_D is in watts
S_{FL} is full-load speed in rpm

ACCELERATING TORQUE

Accelerating torque never lasts very long. *It is the difference between the torque delivered by the rotor and the torque needed by the load.* Accelerating torque speeds up the motor. When the rotor and load reach the correct speed, there is no more accelerating torque.

BREAKDOWN TORQUE

As the load on a motor increases, the rotor delivers more torque. At some value of torque, however, the motor stalls. This is the BREAK-DOWN torque. *It is the maximum torque the motor can develop without an abrupt drop in speed.*

FACTORS AFFECTING TORQUE

The standard torque equation is:

$$T = K \times \phi \times I_R \times PF_R$$

T is torque
K is a design constant
ϕ is stator field strength
I_R is rotor current
PF_R is rotor power factor

In this equation, two of the factors are fixed and two are variable. The constant, K, is fixed by the design of the motor. Field strength, ϕ, depends on the applied voltage and does not change. Both rotor current and rotor power factor depend on slip.

CONDITIONS AT START

Starting current can be five or six times full-load rated current. Starting torque, however, is usually less than twice full-load torque. The

reason you do not get more torque from all the current is the low power factor of the rotor.

Fig. 11-13 shows the location of the two revolving fields. Note that the rotor field is lagging the stator field by 165 degrees. The best angle for producing torque would be 90 degrees. The additional 75 degrees is due to the rotor's inductive reactance, which is highest at start.

LOCKED ROTOR TEST

Tests made with a locked rotor can determine both current and torque at start. Fig. 11-14 is a photo of the setup. The rotor is first clamped so

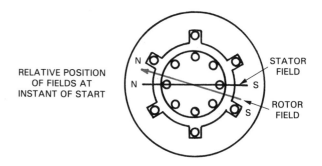

Fig. 11-13. At the instant of start, rotor field lags stator by 165 degrees.

Fig. 11-14. Starting conditions can be determined by locking the rotor so it cannot turn.

that it cannot move. Then reduced voltage is applied to the motor. Readings are taken of the torque produced and the current at this reduced voltage.

Current is directly proportional to voltage. First find the ratio between full voltage and reduced voltage. Then multiply this ratio times the measured current. The result is starting current.

Example 3

A motor is rated at 208 volts, 2.4 amperes. With the rotor locked and 104 volts applied, current is 6.0 amperes. Find the starting current and percentage of rated current.

a. Compute the voltage ratio.
$$\frac{208}{104} = 2$$
b. Compute starting current.
$$6.0 \times 2 = 12.0 \text{ A}$$
c. Compute percent rated current.
$$\frac{12}{2.4} \times 100 = 500\%$$

Therefore, starting current is five times rated current.

Torque is proportional to the voltage squared. Two of the factors in the torque equation depend on voltage. One is the field strength, ϕ. The other is the rotor current, I_R. To find starting torque:

1. Square the voltage ratio.
2. Multiply the voltage by the measured torque.

Example 4

A locked-rotor test is performed at 104 volts on a 208 volt motor. The torque reading is 0.52 newton metres (0.38 lbf-ft.). Find the starting torque. The full-load rated torque is 1.385 N·m (1.025 lbf-ft.). Find the percent starting torque as compared to rated torque.

a. Compute the voltage ratio squared.
$$\left(\frac{208}{104}\right)^2 = 2^2 = 4$$

b. Compute starting torque in SI metric units.
$$0.52 \times 4 = 2.08 \text{ N·m}$$
c. Compute starting torque in Conventional units.
$$0.38 \times 4 = 1.52 \text{ lbf-ft.}$$
d. Compute percentage of rated torque.
$$\frac{2.08}{1.384} \times 100 \text{ or } \frac{1.52}{1.025}$$
$$\times 100 = 150\%$$

NO-LOAD CONDITIONS

Assume now that the rotor is not coupled to a load. The motor is running fast. In fact, the rotor turns almost as fast as the revolving stator field. Fig. 11-15 shows that the rotor field is only a few degrees from the best torque-producing angle. It is almost at right angles to the stator field.

Very little torque is produced, however. Nor is there much rotor current. The closer rotor speed gets to synchronous speed, the less voltage induced in the rotor winding. Only enough torque is produced to overcome mechanical losses.

The power factor of the motor current is poor (less than 0.5) at no-load. The in-phase component, which supplies electrical and mechanical losses, is very small. By comparison, the 90 degree out-of-phase component, which magnetizes the cores and air gap, is large. Stator power

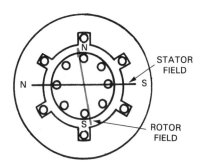

Fig. 11-15. When a motor is running with no load, rotor magnetic field is approximately 90 degrees from the stator field.

factor improves as the motor is loaded. Typically, it is 0.8 at full load.

FULL-LOAD CONDITION

As motor load increases, the rotor slows down. The slowdown induces a larger voltage into the rotor. A big change occurs in rotor current. Fig. 11-16 shows how torque changes with slip over the operating range of the motor. This change is due to increasing rotor current. Even though the angle between the rotor field and stator field increases as the motor is loaded, this does not affect torque much until the motor is overloaded.

OVERLOAD CONDITION

Normally, induction motors do not mind small overloads. The rotor simply slows down. Rotor current and torque output increase to handle the load. Additional current drawn by the stator could burn up the windings if the overload lasts too long.

Torque tends to increase as slip increases. At the same time, rotor power factor decreases. A typical motor will produce maximum torque

around 20 percent slip. If the load needs more torque, the motor will stall. At that point, stator current rises to its starting value. Most three-phase motors are equipped with overload devices to remove power when the winding temperature gets too high.

INDUCTION MOTOR EFFICIENCY

Efficiency is the ratio of output to input. Only part of the electric power going into a motor shows up as mechanical power delivered to the load. Some is lost in the resistance of the stator windings. A little is lost in the stator core. The rest is transmitted across the air gap to the rotor. Then, some of the rotor power is lost in rotor resistance. Finally, the power needed to overcome windage and friction losses reduces the mechanical output still further.

DETERMINING MOTOR EFFICIENCY

Copper losses are proportional to the current squared ($P = I^2R$). These are only the variable losses. Rotational and core losses do not change as the motor is loaded.

You can easily measure the electrical input and mechanical output of small motors. This may be hard to do with larger motors. In that case, you can determine the motor input and losses. Input minus losses, then, is the output. The efficiency equation is:

$$\% \text{ efficiency} = \frac{\text{input} - \text{losses}}{\text{input}}$$

You can find the values of the fixed and variable losses with the *locked-rotor test* and the *no-load test*. Use the same circuit for both tests, Fig. 11-17.

LOCKED-ROTOR TEST

First lock the rotor in position. This will eliminate any rotational losses. Then slowly increase voltage until rated current is drawn by the motor. This will be a low voltage value. At

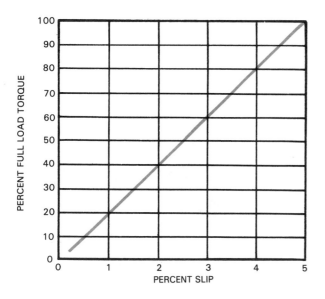

Fig. 11-16. As a motor is loaded, slip increases. Torque increases directly with slip.

this point, all of the power going into the motor is lost in the motor windings. This is the copper loss at full load.

From this data we can compute the equivalent resistance of the motor as seen from the power source. Then we can compute the copper loss at any load.

Example 5

A locked-rotor test is made on a three-phase induction motor rated at 3.16 amperes. Total power in is 150 watts. Find the equivalent resistance.

$$P = I^2 \times R \therefore R = \frac{P}{I^2} = \frac{150}{3.16^2} = \frac{150}{10}$$
$$= 15 \ \Omega$$

NO-LOAD TEST

To make a no-load test, operate a motor at full voltage with no load coupled to it. The power input is made up of two parts:

1. The rotational and core losses.
2. The copper loss at no-load current.

By subtracting copper loss from input power, we arrive at the fixed losses.

Example 6

Under a no-load test the motor of *Example*

5 has power input of 85 watts. Current is 2.2 amperes. Find the fixed losses.

a. Compute copper loss at 2.2 amperes.
$$P_c = I^2 \times R = 2.2^2 \times 15$$
$$= 4.84 \times 15 = 72.6 \ W$$
b. Compute fixed losses.
$$P_f = P_{IN} = P_c = 85 - 72.6 = 12.4 \ W$$

Now we have enough information to compute the efficiency at full load:

1. Total power in is 150 watts.
2. Fixed losses are 12.4 watts.
3. Resistance is 15 ohms.
4. In a wye-connected three-phase motor, the voltage across any two lines is 1.73 times the voltage in one line.

Example 7

The motor of *Examples 5 and 6* is rated at: 208 volts, 3.16 amperes, 746 watts (one horsepower), 0.8 power factor. Find the full-load efficiency.

a. Compute full-load power input.
$$P_{IN} = 1.73 \times E \times I \times PF$$
$$= 1.73 \times 208 \times 3.16 \times .8$$
$$P_{IN} = 909.7 \ W$$
b. Compute total losses.
$$P_L = P_c + P_f = [(3.16)^2 \times 15] + 12.4$$
$$= 150 + 12.4 = 162.4 \ W$$
c. Compute full-load efficiency.

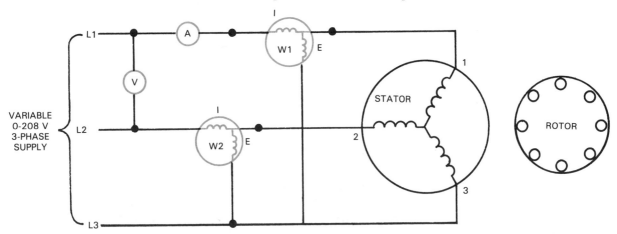

Fig. 11-17. Same circuit is used for locked-rotor and no-load motor tests. Hook up meters as shown. Refer to Fig. 7-33 and Fig. 7-34 for more information on wattmeter hookup.

$$\% \text{ eff.} = \frac{P_{IN} - P_L}{P_{IN}} \times 100$$

$$= \frac{909.7 - 162.4}{909.7} \times 100$$

$$= \frac{747.3}{909.7} \times 100$$

efficiency = 82%

ROTOR RESISTANCE

The resistance of the squirrel cage has an effect on the operation of the motor. Large induction motors are built with different resistances to produce different characteristics. The torque curves for several standard types are shown in Fig. 11-18. The letters, B, C and D, are assigned by the National Electrical Manufacturers Association (NEMA).

NEMA DESIGN B

This is the standard type. It has what is considered normal starting torque, low starting current, and low slip at full load.

NEMA DESIGN C

This type has greater rotor resistance. This improves the rotor power factor at the start, giving more starting torque. When loaded, however, this extra resistance causes greater slip.

NEMA DESIGN D

This type has even more resistance. The starting torque is also the maximum, or breakdown, torque.

WOUND-ROTOR MOTORS

The resistance of the rotor circuit is important. A high-resistance rotor develops a high starting torque at low starting current. A low-resistance rotor develops low slip and high efficiency at full load. You can get both of these advantages by using a wound rotor. The stator of a wound-rotor motor is the same as the stator of a squirrel-cage motor.

A typical wound rotor is shown in Fig. 11-19. The rotor is wound to have the same number of magnetic poles as the stator. Rotor windings are wye-connected and terminate at slip rings.

Wound-rotor motors are sometimes called slip-ring motors. Connections to the rotor windings are made through brushes.

Variable resistances are connected to the rotor brushes. That way, the resistance of the rotor circuit can be changed while the motor is running. A typical external resistance box is shown in Fig. 11-20. The connection diagram is shown in Fig. 11-21.

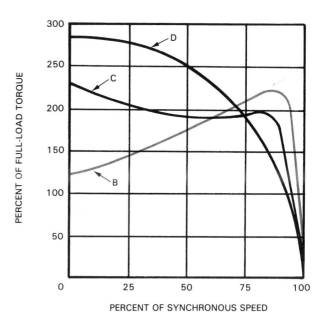

Fig. 11-18. Curves show how motor torque varies with rotor speed for three types of electric motor.

Fig. 11-19. Wound-rotor motor. This type is used in school machine laboratories.

A wound-rotor motor is normally started with full resistance in the circuit. Then resistance is gradually cut out, either manually or automatically. When all resistance is out, a wound-rotor motor has the characteristics of a squirrel-cage motor. Fig. 11-22 shows how torque changes with speed for different values of resistances.

SPEED CONTROL

A wound-rotor motor costs more than a squirrel-cage motor of the same size. Also, its slip rings and brushes must be maintained.

However, you can control its speed. For a given torque output, the motor will run at the speed determined by the circuit resistance of the rotor.

You cannot make the motor run any faster than a squirrel-cage motor with the same synchronous speed, but you can slow it down. By adding resistance, you increase motor slip. This is not an exact speed control, however. If load changes, so will speed.

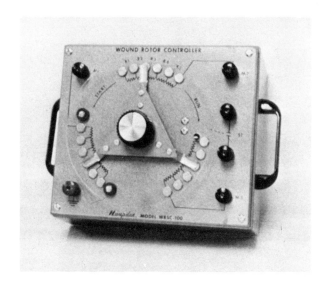

Fig. 11-20. Typical resistance controller. This unit is used with wound-rotor motors.

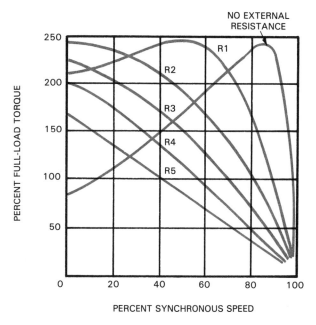

Fig. 11-22. Curves showing how torque changes with speed for different values of resistance added to the rotor circuit.

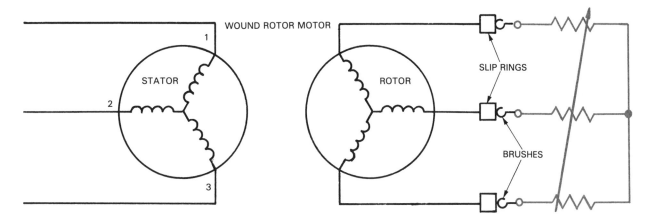

Fig. 11-21. Resistance is added to the rotor circuit of a wound-rotor motor to change its characteristics.

Wound-rotor motors are used on cranes, hoists and elevators, where frequent starting, stopping and reversing, as well as speed control, is required.

SYNCHRONOUS MOTORS

Induction motors (both squirrel-cage and wound-rotor) always run slower than synchronous speed. Synchronous motors run at exactly synchronous speed. The difference is in the way the rotor field is produced. Direct current is fed directly to the rotor windings through brushes and slip rings.

The rotor of a synchronous motor is shown in Fig. 11-23. The core, called the "spider," contains salient poles. The spider is wound to produce north-south poles when voltage is applied. For each stator magnetic pole there is a rotor pole. When the motor is running, the rotor field "locks" on the stator field. The rotor, thus, is made to turn at the same (synchronous) speed.

The synchronous motor is not self starting. Even if it had rotor current at the instant of start, the stator field would be sweeping around so fast that the rotor could not lock in. For this reason, the spider has a squirrel-cage winding in addition to the dc rotor winding. Bars are embedded in the surface and shorted at each end. Direct current is applied to the rotor winding only after the motor is running at full speed.

STARTING A SYNCHRONOUS MOTOR

During the start period, a synchronous motor acts like a squirrel-cage induction motor. The revolving stator field induces voltage into the cage (starting) winding. It also induces a voltage into the dc "field" winding. During this time, a resistor is connected to the dc field winding.

At the instant of start, the induced rotor field lags the stator field about 175 degrees. This is shown by Fig. 11-24. As the rotor speeds up, its field shifts until it is almost at right angles to the stator field. What is more, the rotor is turn-

ing almost as fast as its field. At this point, shown by Fig. 11-25, the motor is ready for SYNCHRONIZING.

SYNCHRONIZING A SYNCHRONOUS MOTOR

First, the internal resistor is disconnected from the field coils. Then direct current is applied to the fields. Both operations can be handled by a single relay. When actuated, the relay opens the resistor circuit. An instant later, it closes the direct current supply circuit.

Fig. 11-23. Stator of a synchronous motor. Core is wound to set up electromagnetic field when voltage is applied.

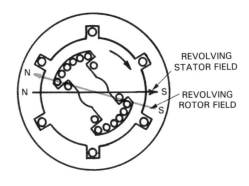

REVOLVING STATOR FIELD

REVOLVING ROTOR FIELD

Fig. 11-24. Relationship of stator and rotor fields of a synchronous motor at instant of start. Rotor field lags behind by nearly 180 degrees.

The relay can be actuated manually or automatically. Automatic operation is usually based on rotor speed. Direct current flowing in the field coils is called EXCITATION CURRENT. The field coils are said to be EXCITED.

Refer to Fig. 11-26. With the field coils excited, one pole is north and the opposite coil is south. Now we have a strong rotor magnetic field revolving almost as fast as the stator field. As the south stator magnetic pole gets near the north rotor pole, there is a strong attraction. The rotor, therefore, begins to rotate at synchronous speed. It is pulled around by the stator field.

EXCITATION SUPPLY

The dc excitation current is separate from the three-phase ac supply. On large motors, a small dc generator may be used. Sometimes the generator is coupled to the rotor shaft.

Another method is the "brushless" system. A small "revolving armature" alternator is a part of the motor. A "solid state rectifier" is mounted on the rotor shaft. Direct current can, thereby, be fed to the rotor without brushes. The rotor of a brushless synchronous motor is shown in Fig. 11-27.

OPERATING CHARACTERISTICS

The rotor turns at synchronous speed. When there is no load, the center of the rotor field is lined up with the center of the stator field.

Actually there is a slight angle between the

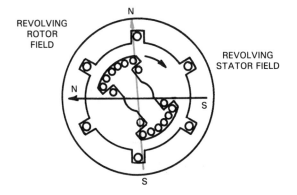

Fig. 11-25. As synchronous motor's rotor field approaches 90 degrees from the stator field, motor can be synchronized.

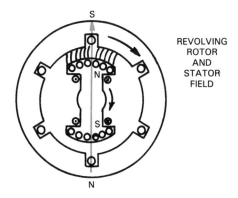

Fig. 11-26. When a synchronous motor is synchronized, the rotor field revolves at same speed as stator field.

Fig. 11-27. Direct current field current can be supplied from an alternator and rectifier on the rotor.

fields. This is known as the torque angle. As the motor is loaded, this angle increases, producing more torque. The rotor continues to turn at the same speed as before, however. This is shown in Fig. 11-28.

Normally, the torque angle is around 30 degrees at full load. At about one-and-a-half to two times full load, the angle becomes too great. Then the motor pulls out of synchronism. *The pull-out torque of a synchronous motor is defined as the maximum torque developed by a motor, for one minute, before it pulls out of step due to overload.*

When a synchronous motor pulls out of step, it does not run smoothly. The stator field keeps bumping into the rotor's dc field as it revolves. The motor must be shut down immediately. It will not resynchronize itself. The load must be removed and the direct current disconnected before restarting.

COMPARING POWER FACTORS

Induction motors always run with a lagging power factor. This comes from drawing their magnetizing current from the ac power source.

Synchronous motors, on the other hand, may run with a lagging, unity or leading power factor. It depends on how much of the magnetizing current is drawn from the dc excitation supply.

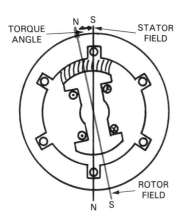

Fig. 11-28. As a synchronous motor is loaded, the torque angle increases.

Fig. 11-29 represents the power input of an induction motor. Part of the input power is active power. Part is reactive power. The reactive power produces the magnetic fields in the stator core, the rotor core and the air gap.

In a synchronous motor, the magnetizing power may come from either the dc or ac supply. If only part of the magnetizing is done by the alternating current, the motor is said to be UNDER EXCITED. In normal excitation, current in the dc field coil supplies just the right amount of magnetization. A synchronous motor is said to be OVER EXCITED when the dc field supplies more magnetization than necessary. This results in a leading current that can supply reactive power to induction motors, as shown in Fig. 11-30.

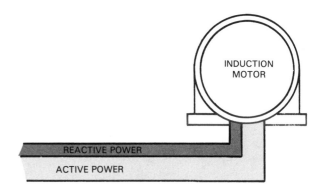

Fig. 11-29. For an induction motor, total power input consists of active power and reactive power.

UNDER EXCITED FIELD

Synchronous motors are seldom run with an under excited field coil. It would produce a lagging power factor, like an induction motor. What is more, the full-load torque and pull-out torque would be low.

NORMAL EXCITED FIELD

A synchronous motor with a normal excited field runs at unity power factor. The incoming alternating current is in phase with the applied voltage. Except for the losses, all of the ac

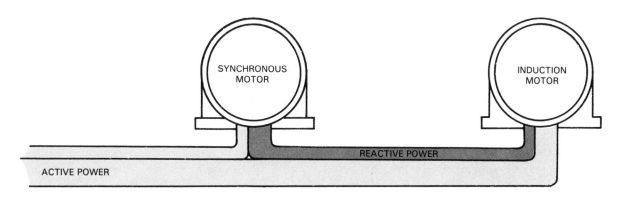

Fig. 11-30. When a synchronous motor is on the same line as an induction motor, input power is all active power. Reactive power is supplied by the synchronous motor.

power is delivered to the load as mechanical power.

OVER EXCITED FIELD

The revolving magnetic field in the air gap induces a back-voltage into the stator coil. This happens in a synchronous motor as well as in an induction motor. When the rotor field is made extra strong, additional back-voltage is induced. This results in a 90 degree, out-of-phase current. When the field coil is over excited, it feeds magnetizing current back into the ac power lines. It runs with a leading power factor. To the power source, it looks like a capacitor.

SYNCHRONOUS MOTOR V CURVES

The three curves shown in Fig. 11-31 are known as "V curves." They show how the motor current reacts when you change the dc field excitation current.

The top curve shows the relationship at full load. When the excitation current is low, the rotor field is under excited. The motor current is high. Much of it is magnetizing current.

As the excitation current is raised, motor current goes down. It reaches a low point, then starts increasing again.

When motor current is lowest, it is in phase with the voltage. The motor is operating at unity power factor. On the right side of the curve, current is leading the voltage.

The middle curve was plotted from data taken at half-load. The lower curve shows the no-load conditions.

IMPROVING POWER FACTOR

Synchronous motors can be used for the same applications as Design B squirrel-cage motors. Factories often choose synchronous motors to improve power factor. They can be operated at leading power factor to supply magnetizing current to induction motors in the same factory. Refer, again, to Fig. 11-30. It shows how the magnetizing current shuttles between motors. The overall power factor of the plant is improved.

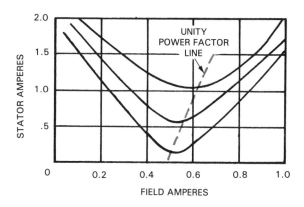

Fig. 11-31. Curves showing how stator current goes from lagging to leading as field current increases.

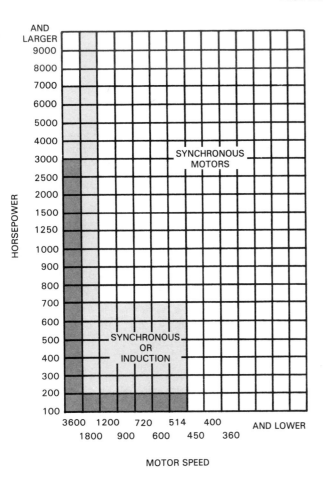

Fig. 11-32. Selection chart for three-phase motors. Synchronous motors are preferred for large horsepower and low speed.

APPLICATION

Generally, synchronous motors are used for high horsepower or low speed applications. The graph in Fig. 11-32 is a rough guide used by some engineers to help choose between squirrel-cage and synchronous motors.

SUMMARY

There are three types of three-phase motors. All have the same type of wound stator: 1. Squirrel-cage induction motors. 2. Wound-rotor induction motors. 3. Synchronous motors.

Three-phase motor stators are wound in coil groups. The minimum number of coil groups is three, producing one pair of magnetic poles per phase.

When three-phase voltage is applied to a three-phase stator, a revolving stator magnetic field is produced.

The speed of this revolving field is synchronized with line frequency and is called synchronous speed.

The revolving field induces a voltage into the rotor windings to produce a starting torque.

Squirrel-cage and wound-rotor motors always run slower than synchronous speed so that voltage can continue to be induced into the rotor windings.

The difference between synchronous speed and rotor speed is called slip speed, which is usually expressed as a percent of synchronous speed.

A synchronous motor runs at exactly sychronous speed because its rotor field is locked into the revolving magnetic field.

The torque output of a squirrel-cage motor varies with rotor current and rotor power factor. Both depend on slip.

The greater the rotor resistance, the more torque is produced at start for each ampere of current. High resistance, however, produces high slip at full load.

Resistance is added to the rotor of wound-rotor motors during start then removed as the motor is loaded.

Synchronous motor rotors contain squirrel-cage bars which start it as an induction motor. At full induction motor speed, dc is applied to rotor windings.

TEST YOUR KNOWLEDGE

1. Speed at which the stator field is revolving is known as _____ speed.
2. A four-pole induction motor would run (faster, slower) than a four-pole synchronous motor.
3. What happens to induced rotor voltage as the rotor of an induction motor slows due to load? Why?
4. By adding _____ _____ to

the wound-rotor motor circuit, you can vary the running speed of the motor.

5. The synchronous motor is self starting. True or false?

6. What happens when a synchronous motor is "synchronized"?

7. Name two disadvantages of a low power factor.

8. Synchronous motors can be operated at a (leading, lagging) power factor to supply magnetizing current to induction motors in the same factory.

PROBLEMS

1. Given: A two-pole three-phase induction motor has a full-load rated speed of 3450 rpm. Find the percentage of slip at full load.

2. Given: A locked rotor test on a three-phase squirrel-cage induction motor provides the following data: voltage, 560 volts; current, 2.4 amperes; power in, 160 watts. Find the total equivalent resistance of the stator.

3. Given: A no-load test on the motor of *Problem 2* provides the following data: voltage, 208 volts; current, 2.2 amperes; total power in, 144 watts. Find the total fixed losses.

4. Given: The motor of *Problems 2 and 3* having a current at a rated load of 2.5 amperes assuming a 0.8 power factor. Find the efficiency at rated load.

5. Given: A wound-rotor motor with no resistance in the circuit has a starting torque of 4.32 newton metres and a starting current of 5.4 amperes. When resistance is added the torque is 2.4 newton metres and current is 2 amperes. Find the percentage of improvement in newton metres per ampere.

6. Given: A synchronous motor with a dc rotor field excitation current of 0.6 amperes which makes it run at unity power factor. The excitation current is reversed to 0.8 amperes. The voltage is 208 volts, stator current is 3 amperes, input power is 500 watts. Find the power factor.

12 DC MOTORS AND GENERATORS

After studying this chapter, you will be able to:

☐ Describe the characteristics of series, shunt and compound dc motors.

☐ Explain the effect of a loss of load on a dc series motor and the loss of field on a dc shunt motor.

☐ List the factors that may prevent voltage buildup in a self-excited dc generator.

☐ Describe the characteristics of over-compounded, under-compounded, and flat-compounded dc generators.

The first alternating current power plant in the United States was built at Niagara Falls, New York. In August, 1895, power was transmitted to the first customer, Pittsburgh Reduction Company which is now known as Aluminum Company of America (ALCOA). The alternating current was used to run an induction

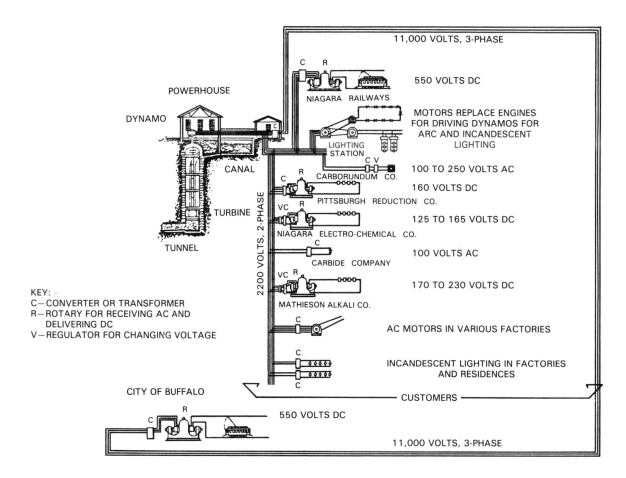

Fig. 12-1. Diagram shows some of the circuits of the first Niagara Falls generating plant in 1897. (Copyright © 1972 by the Institute of Electrical and Electronics Engineers, Inc. Reprinted, by permission, from IEEE SPECTRUM, vol. 9, no. 6, June 1972.)

Fig. 12-2. Off-road vehicles and trucks are often powered by direct current motors. Left. In-plant vehicle designed for hauling personnel and materials operates from batteries at 36 or 48 volts. (OMC Lincoln) Right. Trucks have electric drive wheels powered by diesel-driven generator. (WABCO/creative services)

motor. The motor, in turn, drove a direct current generator. Several other companies did the same thing. Fig. 12-1 shows some of the circuits of the Niagara Falls power plant in 1897.

DIRECT CURRENT MOTORS

Direct current motors have many applications where they can do the job better than ac motors. For example, they perform better in traction equipment, such as golf carts and other small off-road vehicles and in large quarry and mining trucks and locomotives. Smaller equipment will operate from batteries. Large equipment, like trucks and locomotives, may use diesel engines to operate a dc generator. Current is then fed to electric drive motors. See Fig. 12-2.

Direct current motors are also used in manufacturing and processing situations where easy motor speed control is needed. Either generators or rectifiers supply the current. Sometimes, the rectifier and speed controller are put into the same MOTOR DRIVE package.

DC MOTOR CONSTRUCTION

Like alternating current motors, dc motors have a stator and rotor. Rotors are more often called ARMATURES. The armature core is made up of notched laminations. With laminations stacked, the notches form slots. Armature winding conductors are pressed into the slots. A typical armature is shown in Fig. 12-3.

COMMUTATOR

The end of each coil is connected to a segment of the COMMUTATOR. Fig. 12-4 is a simpli-

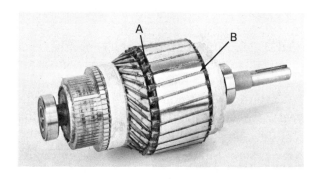

Fig. 12-3. Typical armature for a dc motor. A—Conductor winding. B—Slot. (Hampden Engineering Corp.)

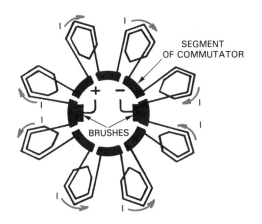

Fig. 12-4. Brushes contact the commutator mounted on the armature shaft. Arrow shows direction of current. (To avoid hiding electrical connections, the brushes are shown inside the commutator, a position they never take.)

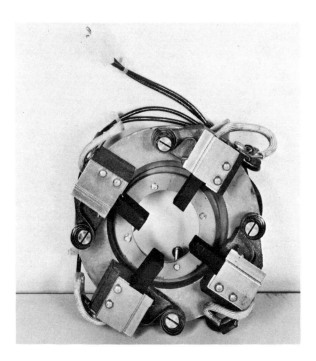

Fig. 12-5. Brush-holder assembly for dc motor. Hold down springs are flat coils of steel seen alongside each brush. (Hampden Engineering Corp.)

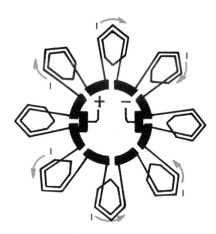

Fig. 12-6. As commutator rotates, each coil is short-circuited for a moment. (Note: Brushes are always on the outside of the commutator, in practice. They are placed inside on some drawings so the electrical short-circuit is more easily seen.)

fied sketch of the connections. Notice that each segment has one end of two coils connected to it.

Carbon brushes complete the electrical contact between the stator and the commutator. Springs, shown in Fig. 12-5, hold the brushes firmly against the commutator. As the armature rotates, the brushes rub across one segment after another, making electrical contact. Fig. 12-6 shows that each brush briefly short-circuits each coil as the commutator segments pass under it. Next, the current is reversed in that coil. As we shall see, reversing the current makes sure that torque will always be produced in the same direction.

STATOR CONSTRUCTION

The stator of a dc motor consists of a yoke (frame or housing) which has field pole pieces bolted to it. In small motors, the pole pieces are permanent magnets. In large motors they are electromagnets.

Most often, the pole pieces are made of laminations of high-grade magnetic steel. The coils

wound on these pole pieces are called FIELD COILS.

DC MOTOR FIELD

The term FIELD may be used in two ways. It may refer to:

1. The main magnetic field.
2. The coils that cause it.

There are two types of field coils. One is wound with many turns of fine wire. Called the SHUNT FIELD, it connects in parallel (shunt) with the armature. The other type is wound with larger wire and fewer turns. Being wired in series with the armature, it is called the SERIES FIELD.

A motor with only a series field is called a SERIES MOTOR. One with only a shunt field is called a SHUNT MOTOR. One with both types of field coils is called a COMPOUND MOTOR. Fig. 12-7 shows a typical stator of a compound dc motor.

MOTOR TORQUE

With voltage in the field coils, the pole pieces become magnetized. One becomes a north pole while the other becomes a south pole. Fig. 12-8 shows the path of the field flux.

Voltage is also applied to the armature coils through the commutator and brushes. This sets up a magnetic field around each conductor. The result, shown in Fig. 12-9, is a torque.

The total torque developed by the magnetic forces of the two fields is proportional to two things:
1. Field strength, ϕ.
2. Armature current, I_A.

The equation is:

$$T = K \times \phi \times I_A$$
T is torque

ϕ is field strength
I_A is armature current
K is a design constant

COUNTERELECTROMOTIVE FORCE (CEMF)

Assume that there is enough torque to make the armature rotate. Armature conductors cut through the main magnetic field. There is a counterelectromotive force (CEMF) induced in the armature coils. To the power source, the CEMF looks like another power source connected series-opposing. This is shown by Fig. 12-10.

There is no way to measure CEMF directly. You can determine its value, however, from the effect it has on armature current. You see, armature current results from NET VOLTAGE. Net voltage equals applied voltage minus CEMF. Following Ohm's Law, armature current, I_A, is net voltage divided by armature resistance. The equation is:

Fig. 12-7. Typical stator of a dc compound motor. Both field coils are wound on the same pole pieces.

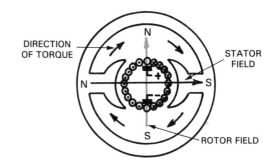

Fig. 12-9. Magnetic forces of attraction and repulsion in a dc motor result in turning force or torque.

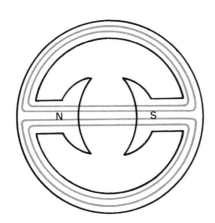

Fig. 12-8. There are two paths of magnetic flux in a two-pole dc motor.

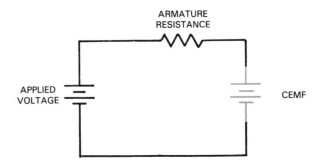

Fig. 12-10. Counterelectromotive force is like a voltage source connected series-opposing with the applied voltage.

$$I_A = \frac{V_{NET}}{R_A} = \frac{V_A - CEMF}{R_A}$$

INTERRELATIONSHIPS

Speed, CEMF, armature current and torque all depend on each other. Armature current (with the main field) produces the torque that makes the rotor turn. The turning rotor (with the main field) produces CEMF. The CEMF then reduces armature current. A motor will balance all of these factors at a certain speed. At this speed, exactly the right amount of CEMF is produced to allow the right amount of armature current. That armature current produces the torque needed to drive the load at that speed. To see how all this works out in practice, we can look at the three types of dc motors: series, shunt and compound.

THE SERIES MOTOR

Fig. 12-11 shows the connections of a series motor. The field coil is in series with the armature coil. Note that armature current, field current and load current are all the same current. This means that the load you put on a series motor greatly affects the torque it develops.

ARMATURE CURRENT AT START

At the instant you turn it on, the rotor is not moving. At that instant, then, there is no CEMF

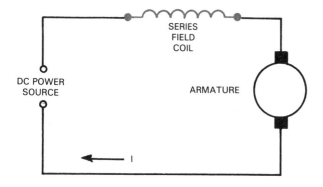

Fig. 12-11. In a "series" dc motor the field coil is connected in series with the armature coil.

to limit current. Armature current is equal to the applied voltage divided by the total resistance of the armature coil and field coil.

Example 1

The armature of a dc series motor has a resistance of 5 ohms. The series field coil also has a resistance of 5 ohms. Find the current through the armature and field coils at the instant 120 volts is applied.

a. Compute total resistance.
$$R = R_A + R_F = 5 + 5 = 10 \, \Omega$$

b. Compute current at the instant of start.
$$I = \frac{V}{R} = \frac{120}{10} = 12 \, A$$

STARTING TORQUE

Torque is proportional to both armature current, I_A, and field strength, ϕ ($T = K \times I_A \times \phi$). Field strength also depends on armature current, however. We can, therefore, rewrite the torque equation for series motors as follows:
$$T = K_T \times I_A{}^2 \text{ with } K_T \text{ being the torque constant.}$$

Starting current produces a particularly large starting torque in series motors. That is why they are used for traction equipment which demands a great deal of start-up torque.

Example 2

A series dc motor's constant, K_T, is 0.1356 in SI metric units (0.1 in Conventional units). Find the torque produced by 12 amperes.

a. Compute torque in SI metric units.
$$T = K_T \times I_A{}^2 = 0.1356 \times 144$$
$$= 19.5 \, N \cdot m$$

b. Compute torque in Conventional units.
$$T = K_T \times I_A{}^2 = 0.1 \times 12^2$$
$$= 0.1 \times 144 = 14.4 \, lbf\text{-}ft.$$

SERIES MOTOR SPEED

Starting torque makes the armature turn. As it does, CEMF reduces armature current. Lower current produces less torque. The motor will still speed up as long as it develops more torque than the load needs. The additional torque is known as accelerating torque.

Fig. 12-12 shows how the speed of a series motor varies with the load on it. When lightly loaded, it runs fast. If the load increases, the motor simply slows down and produces more torque.

The table in Fig. 12-13 shows some values for a typical series motor. The motor is rated at 120 volts, 3.5 amperes, developing 1.6 newton metres (1.18 lbf-ft.) at 1800 rpm. This motor needs 0.015 newton metres (0.011 lbf-ft.) to overcome windage and friction.

OVERLOAD CONDITION

The bottom line of the table in Fig. 12-13 shows what happens if the motor is overloaded. This overload is about 62 percent. The motor

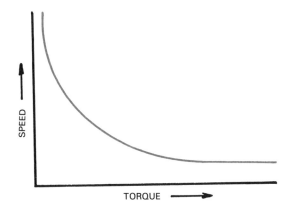

Fig. 12-12. Curve shows how speed of a dc series motor decreases as its load increases.

TORQUE DEVELOPED		ARMATURE CURRENT	CEMF	SPEED
N·m	lbf-ft	amperes	volts	rpm
0.015	0.011	0.34	116.6	25,494
0.8	0.59	2.5	95.0	2825
1.6	1.18	3.5	85.0	1800
2.6	1.9	4.5	75.0	1115

Fig. 12-13. Table of values for typical series motor. Bottom line is for an overloaded motor.

slows down causing less CEMF. Lower CEMF allows enough current to produce the torque. It takes less than 30 percent over-current for the 62 percent extra torque.

PARTIAL LOAD CONDITION

The second line from the top in the table in Fig. 12-13 shows only half-rated torque. The rotor is turning faster. Even though the field current is lower (and the magnetic field weaker), more CEMF is generated. This reduces the current to that needed to drive the half load.

RUNAWAY CONDITION

The top line of the table in Fig. 12-13 shows what happens when no load at all is coupled to the motor. As it speeds up, more CEMF is generated. This reduces current through the field and armature coils. Actually, having lower armature current causes the motor to produce less torque.

An unloaded motor needs so little torque, that there is still plenty of accelerating torque. Therefore, the motor continues to speed up.

Meanwhile, the field has become weaker because of lower field current. Had it not weakened, the faster speed would have generated enough CEMF to reduce armature current quickly. The problem is that the series motor is generating less CEMF for each rpm. Current decreases slowly while speed goes up rapidly. A series motor without load is said to "run away." In our example, final speed is 25,494 rpm.

Some motors may tear themselves to pieces before reaching final speed. Centrifugal forces on the spinning rotor can actually throw armature coils through the motor housing.

CEMF RELATED TO POWER

Under all conditions, incoming electrical power has two jobs to do. It must first overcome the electrical (I^2R) losses. The rest becomes mechanical power. The mechanical power, in watts, equals CEMF times current. It can be computed from applied voltage and current. The complete power equation is:

$$V_A \times I = (I^2R) + (I \times CEMF)$$

It is left as a student exercise to confirm this power balance from the data in the table.

THE SHUNT MOTOR

Fig. 12-14 shows the connections of a shunt motor. The field coil is in parallel (shunt) with the armature coil. Note that the voltage applied to the armature is also applied to the field coil.

Incoming current divides. Some of it flows through the field coil but most of it goes through the armature. Fig. 12-15 is an EQUIVALENT circuit. (This means it is a "model" of the real circuit.) The armature is shown both as a resistance and as a voltage source (CEMF). Changing armature current, caused by changing CEMF, will affect the total current. It will have very little effect, however, on field current. Therefore, the field strength, ϕ, is almost constant over the operating range of a shunt motor.

ARMATURE CURRENT AT START

At the instant of start, CEMF is zero. Net voltage, then, must be the same as applied voltage. Armature current is equal to applied voltage divided by armature resistance ($I_A = \dfrac{V_A}{R_A}$).

Example 3

The armature of a dc shunt motor has a resistance of 5 ohms. Find the armature current at the instant 120 volts is applied.

$$I_A = \frac{V_A}{R_A} = \frac{120}{5} = 24 \text{ A}$$

STARTING TORQUE

As with series motors, torque is proportional to both armature current, I_A and field strength,

ϕ ($T = K \times I_A \times \phi$). Field strength will remain fairly constant. Therefore, we can rewrite the torque equation for shunt motors as follows:

$T = K_T \times I_A$ where K_T is a torque constant which may be given in newton metres per ampere or pounds of force-feet per ampere.

The shunt motor of *Example 3* had a very large starting current. In fact, it is double that of a similar series motor. We shall see how much torque it can develop.

Example 4

The dc shunt motor of *Example 3* has a torque constant of 0.5 newton metres per ampere (0.37 lbf-ft.). Find the starting torque when starting current is 24 amperes.

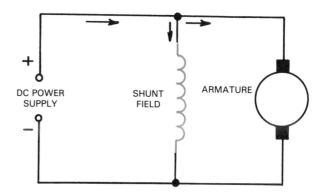

Fig. 12-14. The field coil of a shunt dc motor connects in parallel, or shunt, with the armature coil.

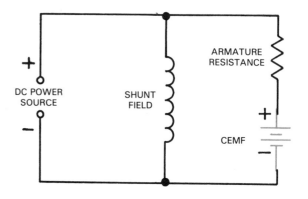

Fig. 12-15. Counterelectromotive force of a shunt dc motor is like a voltage source which is connected series-opposing with the voltage applied across the armature.

a. Compute torque in SI metric units.
$$T = K_T \times I_A = 0.5 \times 24 = 12 \text{ N·m}$$
b. Compute torque in Conventional units.
$$T = K_T \times I_A = 0.37 \times 24 = 8.88 \text{ lbf-ft.}$$

STARTING RESISTOR

Although a shunt motor may draw twice the starting current, its starting torque is less than that of a series motor. This is not the only problem. The large surge of starting current may be more than the power supply can handle.

You can reduce starting current by putting a resistor in series with the armature. However, this also reduces starting torque. Therefore, in selecting a starting resistor, you must be sure there is enough torque to start and accelerate the load.

The starting resistor is automatically cut out of the circuit as soon as armature current has dropped to a normal value. Armature current drops, remember, because the CEMF is opposing applied voltage.

SHUNT MOTOR SPEED

One of the big advantages of a shunt motor is its constant speed. It runs almost as fast fully loaded as it does with no load. Also, it does not "run away" when there is no load.

As the rotor begins to turn, CEMF is induced in the armature windings. In a shunt motor, CEMF is directly proportional to speed. At any given speed, the CEMF permits a specific armature current to flow. If the torque produced by that armature current is greater than that needed by the load, the motor will speed up until this accelerating torque disappears.

Fig. 12-16 shows how the speed of a shunt motor varies with the load placed on it. The table in Fig. 12-17 shows current and CEMF, as well as torque and speed, for four different loads. The motor is rated the same as the previ-

ous series motor: 120 volts, 3.5 amperes, developing 1.6 N·m (1.18 lbf-ft.) at 1800 rpm. Rotational losses require 0.015 N·m (0.011 lbf-ft.).

Although considered constant, you can see that speed does change. Since the field strength does not change, speed must change—to change the CEMF. It is CEMF, remember, that determines armature current and torque.

OVERLOADED CONDITION

If you were to suddenly overload a shunt motor 130 percent of rated load, as shown on the bottom line of the table, Fig. 12-17, you would first see a big drop in speed. You see, the load needs more torque than the motor is delivering. As the speed drops, so does the CEMF. This allows a larger current flow. The larger current, working with the constant field, produces increased torque. This torque will restore some of the lost speed.

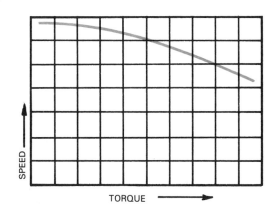

Fig. 12-16. Curve shows how speed of a dc shunt motor decreases as its load increases.

TORQUE DEVELOPED		ARMATURE CURRENT	CEMF	SPEED
N·m	lbf-ft	amperes	volts	rpm
0.015	0.011	0.03	119.85	2075
0.8	0.59	1.6	112	1938
1.6	1.18	3.2	104	1800
2.1	1.55	4.2	99	1713

Fig. 12-17. Shunt motor speed varies with load.

A 30 percent increase in armature current handles a 30 percent increase in load as a shunt motor. The same current increase handled a 60 percent increase in load on a series motor.

NO-LOAD CONDITION

The top line of the table in Fig. 12-17 shows the no-load condition. A shunt motor does not overspeed. The strong magnetic field causes plenty of CEMF to be generated in the armature. That brings current down sooner. Top speed of the shunt motor is 2075 rpm. At that speed, CEMF has reduced armature current to 0.3 amperes. That amount of current produces the torque needed to overcome windage and friction.

LOSS OF FIELD

You need not worry about losing the load of a shunt motor. You must, however, be concerned about LOSING THE FIELD. Loss of field means that power is disconnected from the field coil. The magnetic field caused by current collapses. All that remains is the residual magnetism of the pole pieces. The magnetic field is very weak.

With loss of field, armature current shoots up sharply since there is so little CEMF being generated. The motor begins to speed up. Usually, protective devices—fuses and circuit breakers, for example—turn a motor off before it has a chance to run away.

SPEED CONTROL OF SHUNT MOTOR

As the main magnetic field weakens, speed increases. Thus, you can control speed by controlling field current. Fig. 12-18 shows the circuit. A rheostat is placed in series with the field coil.

It is best to start shunt motors with full field strength. The rheostat, then, is adjusted to zero resistance during start.

After the motor is running, you can add resistance to reduce field current. This, in turn, weakens the field. Since less CEMF is generated, a larger armature current flows. This increases the motor's speed.

Notice that the shunt motor can be made to run faster than normal by field weakening. It does not run slower than normal. Control of speeds below normal is done by a rheostat connected in series with the armature. This arrangement is shown by Fig. 12-19.

COMPOUND MOTORS

A series motor has a high starting torque and the ability to handle overloads. Speed of a shunt motor is relatively constant. A compound motor combines the advantages of both.

The armature of a compound motor is the same as the armature of any dc motor. It is the stator that differs. Each pole piece has two coils, Fig. 12-20. Each of the two coils is different. One is a high resistance shunt coil. The other is a low resistance series coil.

FIELD CONNECTIONS

The series field is connected in series with the armature. The shunt field is connected in paral-

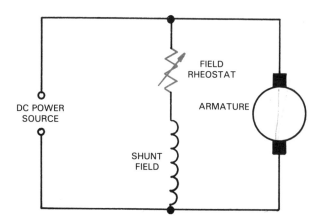

Fig. 12-18. A rheostat, connected in series with the shunt field coil, weakens the magnetic field to increase the motor's speed.

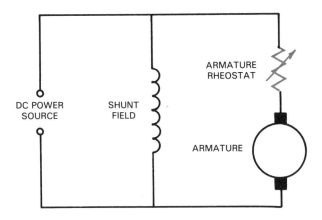

Fig. 12-19. A rheostat, connected in series with the armature coil, reduces armature current. This decreases the motor's speed.

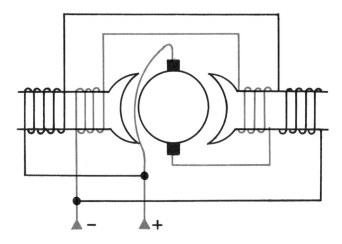

Fig. 12-20. A dc compound motor has both a shunt field coil and a series field coil, wound on the same pole piece.

lel with the armature. Fig. 12-21 shows two ways to do this:

1. In the SHORT-SHUNT method, the shunt field is connected in parallel with the armature alone.
2. In the LONG-SHUNT method, the shunt field is in parallel with the series combination of armature and series field.

There is very little difference in the operation of long-shunt and short-shunt compound motors. The important thing is the magnetic polarity of the series and shunt coils. Current must be passing through them in the same direction. Both coils must produce a north pole in the same pole piece. This is called a CUMULATIVE COMPOUND CONNECTION. If you connected them so that their mmfs (magnetomotive forces) were opposing each other, it would be a DIFFERENTIAL COMPOUND CONNECTION. Differentially compounded motors are seldom used.

OPERATING CHARACTERISTICS

At the instant of start, the magnetic field produced by the series field is more important. Shunt field current is constant, while series field current is very high at start. As the armature spins, however, series field current drops. At this point, the shunt field takes control. The shunt maintains a strong magnetic field at all

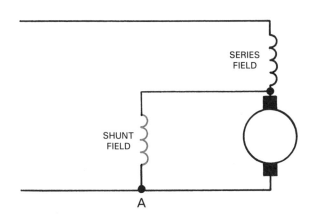

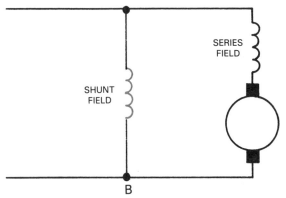

Fig. 12-21. Direct current compound motor field coils may be connected either of two ways. A—Short-shunt. B—Long-shunt.

times. As a result, speed is fairly constant from no-load to full-load range.

Should there be an overload, the series field responds by strengthening the main magnetic field. A stronger field produces the needed torque. Fig. 12-22 shows how the speed of a compound motor varies with the load on it.

ARMATURE REACTION

We know that current through the armature coils produces a magnetic field. It is the magnetic forces between this field and the main field that develop the torque. We also know that magnetic fields tend to combine. Actually, it is their mmf's that combine to produce the field that results.

Fig. 12-23 shows the resultant field in a dc motor. The field's center has shifted. We call this effect ARMATURE REACTION. The amount of shift depends on armature current. In other words, *armature reaction varies with load*.

Armature reaction affects the motor's operation three ways:

1. It reduces torque.
2. It makes the motor less efficient.
3. It causes arcing at the brushes.

As segment after segment of the rotating commutator passes under a brush, the brush short-circuits coil after coil in the armature. At the instant of each short-circuit, the shorted coil should be moving parallel to the main field. There should be no CEMF induced in the short-circuited coil at this point. There will be, however, if armature reaction has shifted the field.

How can we solve the problem? If there were a constant load, the brushes could be fixed at right angles to the main field, wherever it is. This is illustrated by Fig. 12-24. However, if the load and speed are going to be changing, we cannot very well be shifting brushes all the time so they will be at right angles to the field.

One solution is to build a motor with INTERPOLES. Interpoles are sometimes called COMMUTATING POLES. As shown in Fig. 12-25, interpoles are smaller pole pieces, placed halfway between main poles.

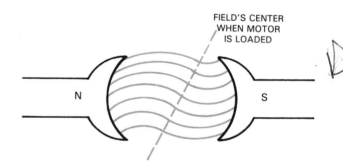

Fig. 12-23. Magnetic field, caused by current through the armature coil, distorts the main magnetic field.

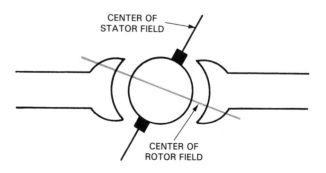

Fig. 12-24. With a constant load on a dc motor, the brushes can be shifted to the actual neutral point.

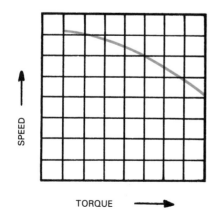

Fig. 12-22. Plotted curve shows how the speed of a dc compound motor decreases as its load increases.

Interpole coils are connected in series with the armature. The strength of the interpole field, then, depends on armature current. Its job is to cancel the effect of the armature field and straighten out the main field.

Another solution sometimes used is to embed a COMPENSATING WINDING in the main pole pieces. The conductors lie parallel with the armature conductors, as shown in Fig. 12-26. The compensating winding is connected in series with the armature winding. This is a very effective way of cancelling out the distorting effects of the armature field.

DIRECTION OF ROTATION

When you turn on a dc motor, which direction will the armature rotate? That depends on two conditions:

1. The direction of the main field.
2. The direction of current through the armature.

You can reverse the direction (magnetic polarity) of the main field by reversing the electrical polarity of the field coil terminals. As long as nothing else is changed, the motor reverses direction. With a compound motor you must reverse polarity of both field coils, Fig. 12-27.

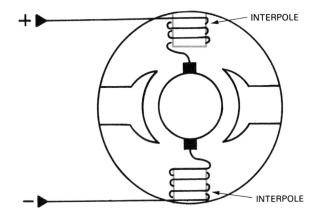

Fig. 12-25. Interpoles are located halfway between the main poles.

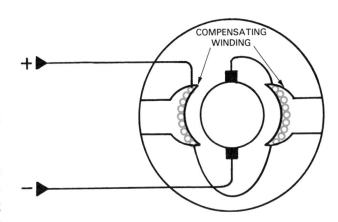

Fig. 12-26. Compensating windings, which are embedded in the pole pieces, are connected in series with the armature.

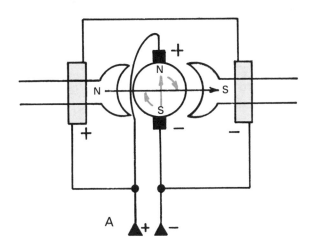

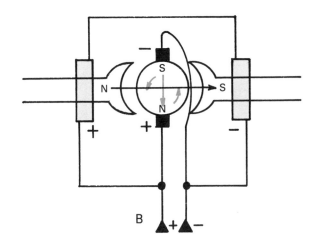

Fig. 12-27. Reversing compound motors. A—Clockwise rotation. B—Reversing terminals will reverse the direction of the armature.

Reversing the polarity of the armature terminals will reverse the direction of current through the armature. As long as nothing else is changed, the motor will reverse direction. Most dc motors are reversed this way.

Use special care if there are interpoles on compensating windings. These must be reversed along with the armature!

You cannot reverse the direction of a dc motor by reversing the polarity of the incoming power lines. That would reverse both the field direction and the armature current direction. As shown in Fig. 12-28, torque would continue to be produced in the same direction.

DIRECT CURRENT GENERATORS

Direct current generators are built very much like dc motors. The main magnetic field may be produced by a permanent magnet, as shown in Fig. 12-29. In larger generators, however, the magnetism is produced by field coils wound on pole pieces as shown in Fig. 12-30. Voltage is generated in the armature coils wound on the rotor.

OPERATION

The generator's rotor is driven by a prime

mover such as a turbine or an engine. As the armature rotates, its conductors cut through the main magnetic field. The changing flux linkages follow a sine wave pattern.

One revolution of a single armature conductor is represented by Fig. 12-31. The generated voltage is at its greatest when the armature conductor is moving at right angles to the main field. Zero voltage is induced when the conductor moves parallel to the field.

Fig. 12-29. Permanent magnet dc generator is similar to this. (Micro Switch Div. of Honeywell)

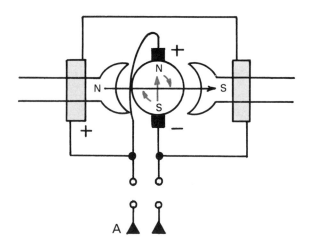 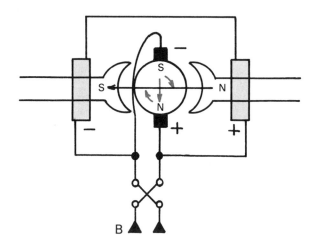

Fig. 12-28. Switching polarity of power source will not reverse motor. A—Clockwise rotation. Note field polarity. B—Motor continues to rotate in the same direction in spite of reversing incoming lines.

There are two important electrical actions to keep in mind:

1. Changing flux linkage causes electrons to be displaced. This creates a *voltage*. There is no current flow until a closed path exists between the generator terminals.
2. Voltage created in the armature conductors is an *alternating* voltage. It must be changed to a *direct* voltage. *A direct voltage is one that always has the same polarity.* A commutator and brushes are used to make the change. As shown by Fig. 12-32, one brush is always positive while the other is always negative.

SMOOTHING OUT THE VOLTAGE

Armature coils are made up of many conductors. These are evenly distributed around the rotor. The terminal voltage, then, is the sum of the voltages induced in every conductor. Fig. 12-33 shows how this changes the voltage output. Not only is the terminal voltage greater, it is smoother. If a pure dc voltage is needed, the ripple or pulsing can be removed with filtering circuits.

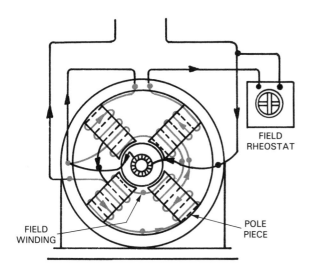

Fig. 12-30. Diagram of stator of "wound stator" dc generator. Colored parts are the wound stator poles. Magnetic field is set up as current flows through coils in the field winding.

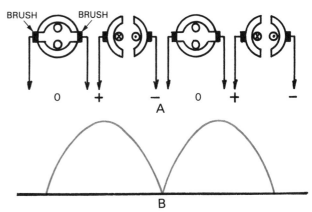

Fig. 12-32. Commutator changes alternating current into direct current.

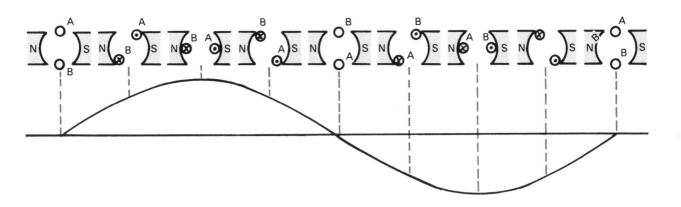

Fig. 12-31. As armature coils rotate, an alternating current voltage is induced in them. Generated voltage is greatest when armature conductor is moving at right angles to main magnetic field.

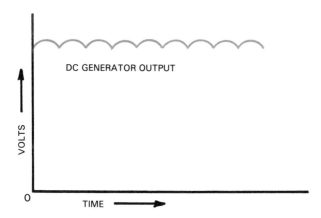

Fig. 12-33. The sum of individual voltages induced into each loop of the armature coil is a relatively constant dc voltage.

SEPARATELY EXCITED FIELD

Direct current may be supplied directly to the field coil. A schematic drawing of a SEPARATELY EXCITED GENERATOR is shown in Fig. 12-34. This arrangement gives precise control over generated voltage. Generated voltage, E_G, is proportional to field strength, ϕ, and speed, S ($E_G = K \times \phi \times S$).

Field strength of a separately excited generator changes only with a change in excitation voltage. This makes it possible to regulate generated voltage precisely. All you need do is control the speed of the prime mover.

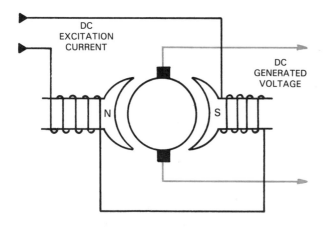

Fig. 12-34. Separately excited dc fields require a voltage source for the field separate from the one that supplies the armature.

Some prime movers have a constant speed which cannot be varied. In such cases, you can control generated voltage by the amount of excitation current reaching the field coil.

SELF EXCITED FIELD

It is not really necessary to supply the field coil from a separate source. You can take some of the direct current you are generating and use it to excite the field. This connection, shown in Fig. 12-35, is called a SELF EXCITED GENERATOR. Notice that the high-resistance field coil forms a closed path between the armature terminals.

If the generator needs a magnetic field to produce the electricity, how can it get started? There is no field current until the generator starts generating.

The answer lies in the residual magnetism of the pole pieces. This may be very little magnetism, but little is needed. As soon as any voltage is induced in the armature, current will begin to flow through the field coil.

VOLTAGE BUILD UP

The flow of field current strengthens the main field. This generates more voltage, resulting in still greater field current. This way, a generator is said to "build up" to rated voltage. It stops building up when the pole piece becomes saturated.

A typical saturation curve is shown in Fig. 12-36. Once the saturation level is reached, further increases in field current cause very little, if any, increases in field strength.

FAILURE TO BUILD UP

If a dc generator fails to build up, one or more of the following things could be wrong:

1. Pole pieces may not have any residual magnetism. This could happen if the field coil

had been connected in reverse briefly. Alternating current applied to a field coil can also wipe out residual magnetism.

You can create residual magnetism by "flashing the field." Flashing the field is done by applying full voltage to the field coil from a separate source for 30 seconds. The connection for this operation is shown in Fig. 12-37.

2. Residual magnetism may have the opposite polarity from that produced by the generated voltage. In this case, as soon as field current starts to flow, voltage goes down. The electromagnetic field cancels out the residual magnetic field.

If it is not practical to reverse the prime mover, or the polarity of the armature terminals, then reverse the polarity of the residual magnetism. Flash the field with the correct polarity.

3. There may be too much resistance in the field circuit. The maximum resistance permitted in the field circuit is called the "critical resistance." Any value greater than the critical resistance will not allow enough field current. Then the field is not strong enough to generate additional voltage. This resistance could come from a field rheostat or a bad connection in the field circuit.

4. The electrical load may be too big. A generator ought to be allowed to build up before any load is applied. You do not want the generator operating in the unsaturated part of the curve. In that range, a small change in field current causes a big change in generated voltage. If the terminal voltage drops because of a load, field current will drop, too. This may actually cause the generated voltage to drop nearly to zero.

VOLTAGE REGULATION

Voltage regulation is the percentage of drop in voltage between no-load and full-load. A curve showing how terminal voltage varies with load is shown in Fig. 12-38.

First, assume there is no load on the generator. The only current through the armature coils is the field current. When a load is connected to

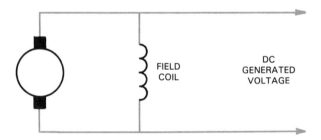

Fig. 12-35. Self excited dc fields use the same voltage that supplies the armature.

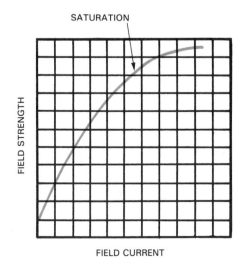

Fig. 12-36. Saturation curve of the stator pole pieces of a dc generator.

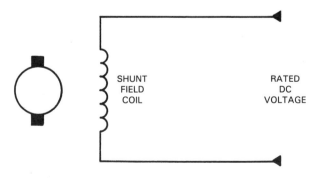

Fig. 12-37. Sometimes, voltage must be applied to the dc field coil to restore the residual magnetism of the core. This is called "flashing the field."

Electrical Power

its terminals, however, a much larger current flows. This makes the terminal voltage drop because of the armature resistance. There is an IR drop across the armature. The larger the load, the greater the drop in terminal voltage.

Terminal voltage, remember, is applied across the field. As this voltage drops, so does field current. This situation does not affect the magnetic field strength very much. The generator is operating in the saturation portion of its magnetization curve. There, changes in field current have little effect on generated voltage.

There is another cause of terminal voltage drop, however. It is the "armature reaction" of the generator. When load current flows, it sets up an armature field. This magnetic field distorts the main field, Fig. 12-39.

Armature reaction produces two bad effects:

1. The armature field weakens the main field, reducing generated voltage.
2. An armature coil undergoing commutation ought to be outside the main field. When this does not happen, voltage induced in that coil opposes voltage induced in other coils. Also, a high short circuit current is present in the coil undergoing commutation.

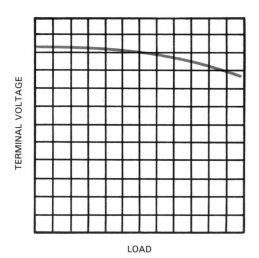

Fig. 12-38. Curve shows how terminal voltage of a dc shunt generator decreases as electrical load increases.

The problem of armature reaction in dc generators is solved the same way as in dc motors. Some generators have interpoles and some have compensating windings. Brush shifting has been used in the past, not to overcome armature reaction, but to control voltage. However, the loss in efficiency and wear on the brushes and commutator make this method impractical.

Example 5

A self excited shunt generator without interpoles rotates at a constant 1800 rpm. Resistance of the armature is 5 ohms. Resistance of the field is 500 ohms. Terminal voltage is 130 volts at no-load and 115 volts at a full-load current of 2.5 amperes. Find the voltage regulation, the voltage drop due to armature resistance and the voltage drop due to armature reaction.

a. Compute voltage regulation.

$$\%VR = \frac{E_{NL} - E_{FL}}{E_{FL}} \times 100$$

$$= \frac{130 - 115}{115} \times 100 = 13.04\%$$

b. Compute the field current at no-load condition.

$$I_f = \frac{E_{NL}}{R_f} = \frac{130}{500} = 0.26 \text{ A}$$

Note: This is also the armature current at no-load.

c. Compute the armature voltage drop,

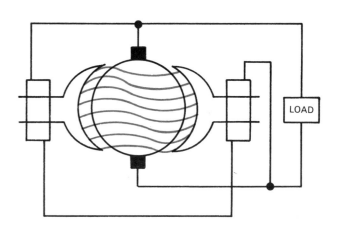

Fig. 12-39. Current flow through armature coil produces a magnetic field that distorts the main magnetic field.

E_A, at no-load.
$$E_{A(NL)} = I_{A(NL)} \times R_A = 0.26 \times 5$$
$$= 1.3 \text{ V}$$

Note: We can neglect armature reaction at no-load. Therefore, the generated voltage is the sum of the terminal voltage and the armature voltage drop.

d. Compute the internally generated voltage, E_g, at no-load.
$$E_g = E_{NL} + E_A = 130 + 1.3$$
$$= 131.3 \text{ V}$$

e. Compute the armature voltage drop at full-load.
$$E_{A(FL)} = I_{A(FL)} \times R_A = 2.5 \times 5$$
$$= 12.5 \text{ V}$$

Note: The terminal voltage is the sum of generated voltage and armature voltage drop.

f. Compute the internally generated voltage at full-load.
$$E_{g(FL)} = E_{FL} + E_{A(FL)} = 115 + 12.5$$
$$= 127.5 \text{ V}$$

Note: If only 127.5 volts is being generated at full-load and 131.3 volts was being generated at no-load, the reduction must be due to armature reaction.

g. Compute the voltage loss due to armature reaction.
$$E_{AR} = E_{g(NL)} - E_{g(FL)} = 131.3 - 127.5$$
$$= 3.8 \text{ V}$$

For curves showing these voltage drops, see Fig. 12-40.

COMPOUND GENERATORS

As load is increased on a shunt generator, the terminal voltage goes down. Many times, this is not acceptable. You may need the same voltage at full-load that you had at no-load.

The solution is quite simple. The generator is built with a second field coil on each pole piece. This low-resistance coil is connected in series with the armature. The result is a compound generator, as shown in Fig. 12-41. Note that the

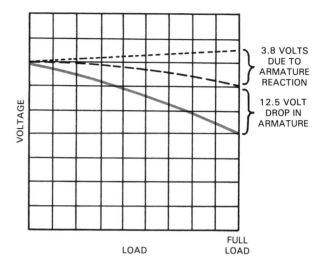

Fig. 12-40. Curves show why terminal voltage of a dc shunt generator is less than generated voltage.

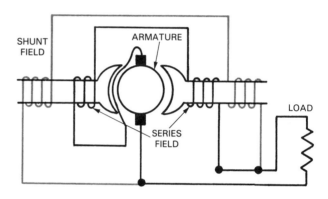

Fig. 12-41. Compound generator has both a series field and a shunt field wound on the same pole pieces.

series field is also in series with the load. Now, when the load increases, the field gets stronger. This generates extra voltage to make up for what was lost in armature resistance and armature reaction.

Under compounding

You can have just a few turns on the series field coil. This would be a slightly compounded generator. It is also known as under compounding. It produces a higher output at full load than a shunt motor. However, its full load voltage is still less than at no-load conditions. This is shown in Fig. 12-42.

Over compounding

The series coil of an over-compounded generator has more turns. Because of the added ampere-turns of mmf, the terminal voltage at full-load is actually larger than at no-load. This is also shown in Fig. 12-42.

Flat compounding

If you want, you can make the terminal voltage at full-load exactly the same at no-load. This is called FLAT COMPOUNDING. Manufacturers build motors with enough turns on the field coil to produce an over-compounded generator. Then the user can connect a diverter rheostat to shunt some of the armature current around the series field coil. The connection is shown in Fig. 12-43.

After the rheostat is connected, the generator is loaded. Terminal voltage will be higher than at no-load. Then the rheostat is adjusted for less resistance. Some of the armature current bypasses the series field. This weakens the magnetic field produced by the series coil. Terminal voltage drops. You can set it for the no-load value. The voltage curve of a flat-compounded generator is also shown in Fig. 12-42.

SUMMARY

Direct current motor operation depends on the interaction of the magnetic flux around the armature conductors with the main field supplied by the stator.

When a motor is running, it is also generating a counterelectromotive force in the armature that opposes the applied voltage.

CEMF times armature current represents the electrical power that is converted to mechanical power by the motor.

The amount of torque developed by a dc motor is directly proportional to the strength of the main magnetic field and the armature current.

The main field may be produced by a permanent magnet or by field coils.

Field coils may be in series, parallel or series-parallel with the armature winding.

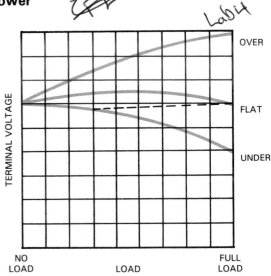

Fig. 12-42. Curves show how terminal voltage changes as electrical load increases for an over-compound, flat-compounded and under-compounded dc generator.

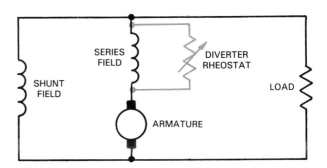

Fig. 12-43. A diverter rheostat can be used to convert an over-compounded dc generator to a flat-compounded or under-compounded generator.

A series motor has a high starting torque but poor speed regulation.

A shunt motor has almost constant speed, but less starting torque than a series motor.

A compound motor combines the high starting torque of a series motor with the good speed regulation of a shunt motor.

Voltage is induced in the armature windings of a generator as it passes through a magnetic field.

The magnetic field may be produced by a permanent magnet or by field coils.

The field coils may be connected to a separate power source or may be connected to the armature coils.

When a load is connected to a generator, current flows through the armature windings.

The armature's magnetic field interacts with the main field to produce a counter torque.

By overcoming the counter torque, the prime mover supplies mechanical power which is converted to electrical power in the generator.

TEST YOUR KNOWLEDGE

1. List two applications of the direct current motor for off-road vehicles.
2. The end of each coil in the armature of a direct current motor is connected to a segment of the _____.
3. Total torque developed by the magnetic forces between the field developed in stator windings and armature windings is proportional to two things:
   a. Field strength and armature current.
   b. Current and CEMF.
   c. Armature resistance and net voltage.
4. What is the relationship between torque and armature current in a series motor?
5. The starting torque of a shunt motor is (less than, more than) that of a series motor.
6. Describe the field connections of a compound motor.
7. What are the advantages of a compound motor over the series and shunt motors?
8. Which of the following reasons would cause a generator to fail to build up voltage:

   a. Pole pieces have no residual magnetism.
   b. Rotor is turning too fast.
   c. Residual magnetism may have polarity opposed to that produced by generated voltage.
   d. Too little resistance in the field circuit.
   e. An electrical overload.
9. What is meant by "flashing the field?" How do you do it?
10. What is the effect of changing the degree of compounding of a generator? How is this done?

PROBLEMS

1. Given: A dc shunt motor having an armature resistance of 5 ohms, with 120 volts applied has a current of 3 amperes. Find the electrical losses in the armature winding, the electrical power that is converted to mechanical power and the CEMF.
2. Given: The field loss in the motor of *Problem 1* is 30 watts. Rotational losses are 15 watts. Find the total power input, total power delivered to the load and efficiency.
3. Given: The motor of *Problems 1 and 2* is running at 1800 rpm. Find the output torque in SI metric units.
4. Given: A separately excited shunt dc generator generates 50 volts when driven at 900 rpm. Find the voltage generated when it is driven at 1800 rpm.
5. Given: The dc generator of *Problem 1* (50 volts at 900 rpm). Find the voltage generated when the field current is reduced to half.
6. Given: A self excited shunt generator is being driven at 1800 rpm. Armature resistance: 4 ohms. Field resistance: 620 ohms. Terminal voltage: 250 volts at no-load, 240 volts at a full-load current of 2 amperes. Find the voltage regulation, the voltage drop due to armature resistance and the voltage drop due to armature reactance at full-load.

13

AC MOTOR CONTROL FUNDAMENTALS

After studying this chapter, you will be able to:

☐ List the purpose of controllers and control systems.
☐ Describe two types of magnetic contactors.
☐ Explain the principle of resistance, reactance, autotransformer, and wye-delta starters.
☐ Define the term "plugging."
☐ Describe the method of controlling speed of a wound-rotor motor.

Sometimes, the only control a motor needs is a switch that turns it on and off. Most appliance motors are like this. They never need to be reversed. They do not have to be braked. They run at a constant speed. Most important, the current inrush during starting is not too great a drain on the power system.

MOTOR CONTROLLERS

Motor controllers serve four main functions:
1. To limit current inrush.
2. To reverse the motor.
3. To control motor speed.
4. To act as a brake.

In addition, some motor controllers protect motors against overheating, overspeeding, or overloading. Motor controllers may be:
1. Manual.
2. Semiautomatic.
3. Automatic.

MANUAL CONTROLLER

A switch, like that shown in Fig. 13-1, is a simple example of a manual controller. Contacts carrying motor current are operated by hand. With a closed switch, full voltage is applied to the motor.

Another type of manual controller is shown in Fig. 13-2. This is known as a drum switch.

Fig. 13-1. A simple on-off toggle switch may be the only control needed for many small motor-driven tools and appliances. (Square D Co.)

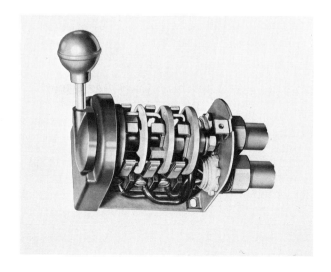

Fig. 13-2. As the lever of a drum controller is turned, internal switches make and break contact. (Square D Co.)

Operating the switch lever causes contacts to "make" or "break." The drum switch can be used for starting or reversing motors.

SEMIAUTOMATIC CONTROLLERS

Most controllers are semiautomatic. The operator does not control application of power directly, but the first move must be made by hand. For example, an operator pushes a start push button. This establishes a current path to a MAGNETIC CONTACTOR, like the one shown in Fig. 13-3.

A magnetic contactor consists of a coil and a pivoted arm, sometimes called an armature. The armature carries one half of a set of contacts. When activated by a current flow, the coil becomes an electromagnet. The armature is pulled toward the coil, closing the contacts. It is these contacts that carry motor current. One armature may have several sets of contacts.

AUTOMATIC CONTROLLERS

Automatic controllers use contactors, too. The difference is that they are operated by PILOT DEVICES. A pilot device will react to conditions in a system. It will turn on a valve or motor, for instance, when conditions are right.

It will shut the valve or motor off when the condition changes.

A home heating system is a good example of this kind of automatic control. The main pilot device is a THERMOSTAT. A typical thermostat, Fig. 13-4, closes a switch when temperature drops. There are many types of pilot devices, operated by pressure, level, or by movement. They all have one thing in common. They open or close a switch to provide a control signal.

CONTROL SYSTEMS

So far, we have considered two parts of a motor control system. Like our senses, pilot devices obtain information. For example, we might touch a hot soldering iron. Our sense of touch immediately sends a signal to the brain. The brain makes a rapid decision; then it tells a muscle to pull the hand away. Electromechanical contactors, the second part of the system, are like the muscles. The third part of the system, like the brain, makes the decisions.

The "brain" of an electromechanical control

Fig. 13-3. Magnetic contactors are used in automatic and semiautomatic control systems. (Square D Co.)

Fig. 13-4. Room thermostats are a popular type of automatic controllers. (Johnson Controls, Inc., Penn Div.)

system is known as RELAY LOGIC. It consists of voltage, current and timer relays.

In recent years, still another type of control called a SOLID STATE CONTROL SYSTEM, has been used. Solid state control consists of electronic switching devices. These are connected into DIGITAL LOGIC systems.

Digital logic, described in Chapter 15, replaces relay logic for many applications.

A further development of digital logic is the PROGRAMMABLE CONTROLLER, Fig. 13-5. Control systems designers still start with relay logic diagrams. Now, however, the programmable controller, which is like a digital computer, makes the conversion to digital logic. The same controller can provide a great many control systems. The programmer operates a keyboard containing relay logic symbols. It is necessary for the programmer to understand electromechanical relay logic diagrams, like those in this chapter and Chapter 14. These drawings are also called "ladder diagrams" because their horizontal lines look a little like the rungs of a ladder.

Now, let us take a look at how some of the common control devices are made. Later on we will see how they are used.

Fig. 13-5. Programmable controller uses computer circuits to perform same control functions as relays. It can be programmed directly from relay ladder diagrams. (Texas Instruments, Inc.)

PUSH BUTTONS

Push buttons are the most popular pilot devices. Usually, they have two sets of contacts:
1. One set, called NORMALLY OPEN (N.O.), make contact only while the push button is being pressed.
2. The other set is called NORMALLY CLOSED (N.C.) contacts. This set is open while the push button is being pressed. The symbols for push button types are shown in Fig. 13-6.

Contacts return to their normal position on MOMENTARY CONTACT type push button. Another type, called the MAINTAINED CONTACT push button, turns on when you push the button. It turns off when pushed a second time.

Fig. 13-7 shown a typical control station having start, reverse, and stop push buttons. The schematic symbol for each is also shown. As you can see from the diagram, the stop push button is normally closed; the forward push button is normally open; and the reverse push button uses both sets of contacts.

Control stations are not always separate. Sometimes, the push buttons are mounted in the cover of the box that holds other controller parts.

CONTACTORS

There are several ways that magnetic attraction forces are used in contactors. View A of

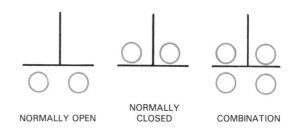

NORMALLY OPEN   NORMALLY CLOSED   COMBINATION

Fig. 13-6. Symbols are used on drawings to indicate types of push buttons.

Fig. 13-8 shows a CLAPPER design. The coil is in the control circuit, while the contacts are in the motor circuit. A SOLENOID, or vertical action contactor is shown in view B. Do you see that the armature is horizontal? When the coil is not energized, its core drops down because of gravity. When the coil becomes energized, however, the core is pulled upward. This action pushes the movable contact points against the fixed contact points.

In schematic drawings, both types of contactors are shown by the same symbols. See Fig. 13-9. A circle represents the coil. Contacts are two parallel lines. The coil is identified by one or more letters and numbers. Its contacts have the same identification.

Along with the main contacts, most contactors contain one or more smaller sets of contacts known as interlocks. Interlocks operate at the same time as the main contacts, but do not carry power current.

SIMPLE CONTROLLER

We are now ready to consider the operation of a simple ACROSS-THE-LINE magnetic starter. See Fig. 13-10. Schematic control

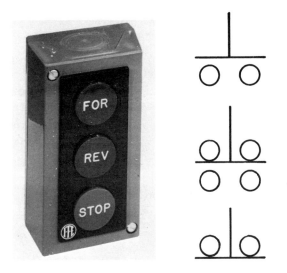

Fig. 13-7. A typical control station contains start, reverse, and stop push buttons. Diagram for control is shown at right. Other combinations are available.

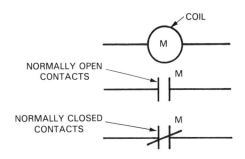

Fig. 13-9. Symbols are used to represent contactor coil and contacts in schematic diagrams.

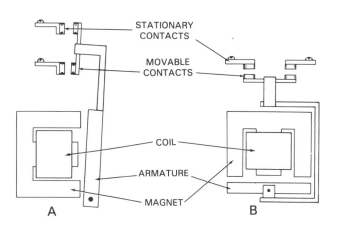

Fig. 13-8. Two contactor designs. A—Clapper type. B—Solenoid or vertical action type. (Square D Co.)

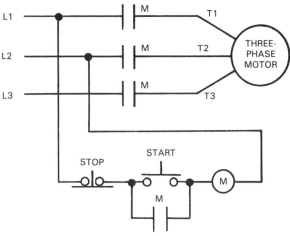

Fig. 13-10. Semiautomatic across-the-line starter diagram. Starter is activated when start switch is pressed.

diagrams must be studied a special way. If you tried simply to trace the circuit from line L1 to line L2, you would not find a closed path. The way to do it is to think in terms of "what happens when." Assume an action. Then figure out how the components will respond. You do this one step at a time:

1. The start push button is pressed and released. During the time its N.O. contacts were closed, there was a current path to the coil of contactor M.
2. Contactor M becomes energized, closing contacts M in the power circuit, applying full voltage to the motor. At the same time, holding interlock M (in parallel with the start push button) closes, so that contactor M remains energized after the push button is released.
3. The N.C. stop push button is pressed and released. Contactor M becomes de-energized. Contacts M open the power circuit, removing power from the motor. At the same time, holding interlock M opens so that the motor will not restart after the push button is released.

Even the most complicated control diagrams can be analyzed this way.

VOLTAGE RELAYS

A relay is used to control one or more circuits according to what is going on in another circuit. Contactors do the same thing. The difference is that contactors carry heavy motor currents and relays do not.

A voltage relay functions at a certain voltage. Usually, they are rated by the current-carrying capacity of their contacts. Contactors, on the other hand, are rated in horsepower, according to the size of the motor they control.

A motor controller may use one or more voltage relays. Their big use is in automated plants, where hundreds of them respond to voltage signals from dozens of pilot devices.

SOLID STATE RELAYS

One way to eliminate the problem of contact wear is not to use contacts. Solid state relays use a device known as a SILICON CONTROLLED RECTIFIER (SCR). The SCR can block current like an open switch. It can also be "triggered" to conduct like a closed switch.

The SCR is used in other devices besides solid state relays. For example, it can carry full motor current like contactors. It also is a part of many speed control systems. Such a system is described in Chapter 14.

CURRENT RELAYS

Fig. 13-11 is a current-sensitive relay. It is usually called a CURRENT RELAY. It looks and acts very much like a voltage relay. The main difference is this: a voltage relay operates (pulls in) at a specific value of voltage through its coil; a current relay operates at a certain value of current flow. It can be an overcurrent relay or an undercurrent relay.

TIMERS

A timer delays closing or opening of a circuit for some specific time. An ON-DELAY timer delays "pulling in" after voltage has been applied. An OFF-DELAY timer delays dropping

Fig. 13-11. Typical current relay looks and functions like a voltage relay.

out after voltage has been removed. Like all other relays, timers may have several sets of contacts. These may be either normally open or normally closed.

Motor-driven timers, Fig. 13-12, are used to time a cycle of operations. A simple example is a washing machine. Solenoid valves are turned on and off. The motor is started and stopped. This is done by cams on the motor shaft. As the cams rotate, switches close and open.

Induction timers delay the opening of dc contacts after power is removed. The cutaway in Fig. 13-13 shows how this works. There are copper rings between the coil and its core. When current stops flowing in the coil, its magnetic field collapses. The collapsing field induces a current in the copper rings. Then the magnetic field from the rings will hold the armature for a short time.

Pneumatic timers, Fig. 13-14, force air through a small hole. This delays the armature's motion after the coil is energized. The air in the

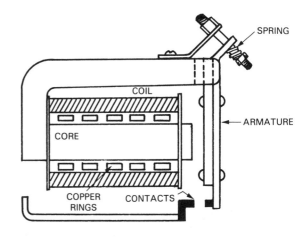

Fig. 13-13. Induction timers depend on current induced into copper rings as the main magnetic field collapses.

Fig. 13-12. Motor driven timers start and stop motors according to a specific time sequence. Top. Reset timer. Bottom. Cycle timer. (Cramer Div., Conrac Corp.)

Fig. 13-14. Pneumatic timer. Air, moving between two chambers in the housing, controls the action of the timing device. Adjusting the knob moves a needle valve against a small opening. Size of opening governs movement of air between chambers to change timing. (Telemechanique)

upper chamber is forced there by a rubber diaphragm. When the coil is energized, the armature pulls the plunger, and with it, the diaphragm. Air is pulled from the upper chamber into the lower chamber through a small hole. This slows the armature's motion.

Electronic timers, Fig. 13-15, may be used to close a set of mechanical contacts or trigger a solid-state switching device. Timing is controlled by a capacitor.

The time lapse created by charging or discharging a capacitor depends on its capacitance and the resistance in series with it. As long as voltage is applied to the coil, the

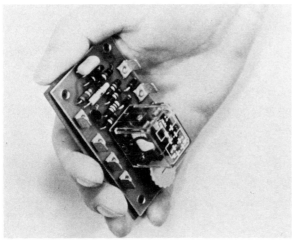

Fig. 13-15. Electronic timers using capacitors are more precise than mechanical types. Top. Solid-state timer. (Cramer Div., Conrac Corp.) Bottom. Partially disassembled electronic timer.

capacitor is charged. When power is removed, the capacitor discharges through the coil. This keeps the coil energized longer. Time lapse can be changed with a rheostat connected in series with the coil. Electronic time delay relays are discussed further in Chapter 15.

OVERLOAD RELAYS

Overload relays are designed to protect a motor from overload during starting and running. An overload may result from:
1. A load heavier than the motor can operate.
2. A load current that is too high during start up.
3. Low line voltage or an open line in a polyphase system resulting in single-phase operation.

An overload relay may be designed to carry an overload temporarily without tripping (cutting off the power). If the overload were to continue on the line, the relay would remove power before the motor sustained damage.

A typical overload relay is shown in Fig. 13-16, view A. The main coil is really a heater. Temperature of the heater depends on the current through it. When the heater gets hot, it causes a bimetal strip to bend. This bending action releases a latch. The normally closed snap switch opens. Fig. 13-16, view B, shows the time it takes to trip an overload relay depending on the load current. The symbol for an overload relay is shown in Fig. 13-17.

The heater carries load current, while its contacts interrupt power to the contactor coil. When the heater has cooled, the overload relay may be reset. Some models must be reset by hand; others do it automatically.

THREE-PHASE INDUCTION MOTOR CONTROLLERS

Squirrel-cage induction motors are widely used in industry. Controllers used on these

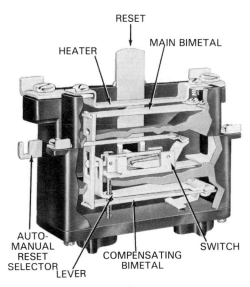

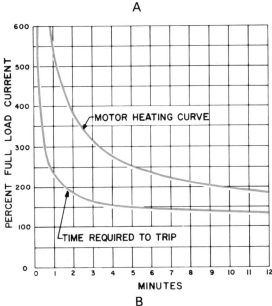

Fig. 13-16. Overload relay protects motor against damage.
A—Cutaway of relay.   (Furnas Electric Co.)   B—Typical
trip curve.

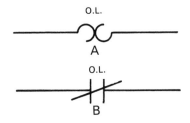

Fig. 13-17. A—Symbol represents heater of overload relay.
B—Contacts in the control circuit.

motors must deal with starting, accelerating, limited speed control and braking.

ACROSS-THE-LINE STARTERS

The simplest way to start an electric motor is to apply full line voltage to the motor windings. Fig. 13-18 is a schematic diagram showing such a circuit for a three-phase motor. A step-down control transformer controls voltage. Motor controls operate on low voltage (120 V) but the motor uses high voltage. Although three heater coils are used, there is only one set of contacts. Across-the-line starters are always preferred where the starting current inrush is not too large for the power supply.

REDUCED INRUSH STARTERS

There are several ways to restrict the starting current. One is to apply a reduced voltage to any three-phase motor. Other ways require specially built motors.

The following reduced inrush starters can be used on three-phase motors:

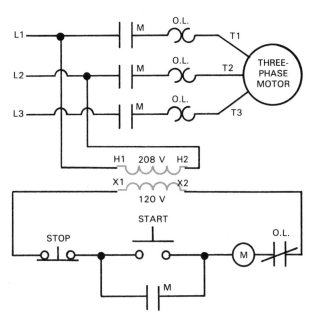

Fig. 13-18. In many ac control systems, a transformer, shown here, reduces voltage for the control components.

1. Resistance starter.
2. Reactance starter.
3. Autotransformer starter (open transition).
4. Autotransformer (closed transition).

RESISTANCE STARTER

Fig. 13-19 shows circuitry for a reduced voltage starter that uses resistors. Starting is in two steps:

1. During the start period, current must pass through the starting resistance to get to the motor. The voltage drop across each resistor reduces the motor's terminal voltage. Meanwhile, the timer is in a timing cycle.
2. After a set period of time, the timer closes its contacts. This energizes the second contactor, which pulls in. Now current bypasses

the resistors going directly to the motor through contacts M2.

Note that this controller has overload and undervoltage protection. A heavy load will trip the overload relay. Low voltage or no voltage will drop out the contactors. They cannot be re-energized except by the start button.

REACTANCE STARTER

Fig. 13-20 has the circuitry for a reduced voltage starter using reactors. Its operation is just like the resistance starter. In fact, the control circuit is identical. Therefore, only the power circuit is shown. The reactors produce an IX voltage drop that reduces motor current during start.

AUTOTRANSFORMER STARTER (OPEN TRANSITION)

Step-down autotransformers are sometimes used to reduce starting voltage. Fig. 13-21 shows the circuit. Notice that two autotransformers are used. They are connected in open-delta. By switching in during start, the autotransformer reduces voltage to the motor.

In the first (start) step, contacts M and C1 are closed. In the second (run) step, contacts C1 open,

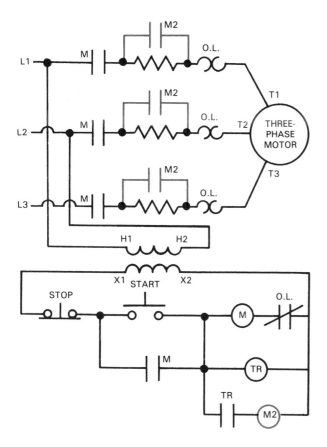

Fig. 13-19. Schematic of resistance type starter. Resistors limit start current inrush.

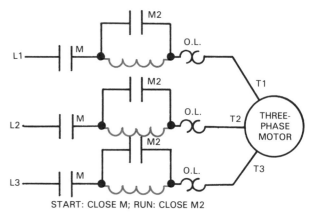

START: CLOSE M; RUN: CLOSE M2

Fig. 13-20. Schematic for reactance type starter. Inductors limit start current inrush in this reactance type starter.

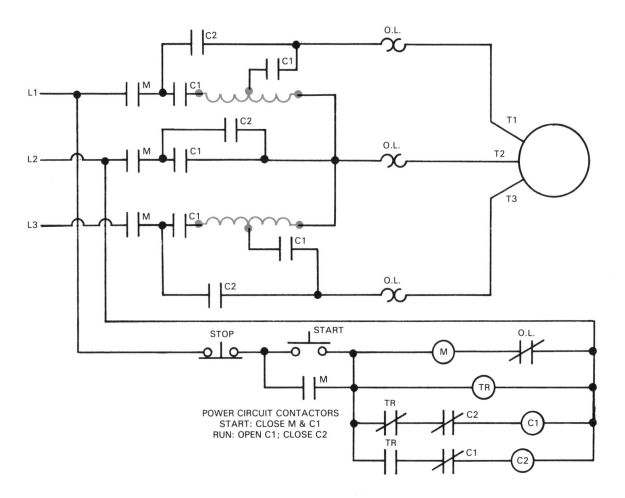

Fig. 13-21. Autotransformers reduce voltage applied to the motor while it is starting. This schematic is for an open transition starter. Circuit opens momentarily between start and run.

disconnecting the autotransformer. Then contacts C2 close, putting full line voltage on the motor. This is called OPEN TRANSITION because the circuit is open during the transition from start to run.

AUTOTRANSFORMER STARTER (CLOSED TRANSITION)

Open transition may produce unwanted line surges. The schematic diagram of Fig. 13-22 shows another way of connecting an autotransformer for starting. This is a closed transition start. The motor has power applied at all times.

In the first step, contacts M and C1 close. This puts full voltage across the autotransformers and half voltage across the stator windings. In the second step, contacts C1 open. Now, half of the autotransformer acts as a reac-

tance in series with the motor for a brief instant. Contacts C2 close, putting full voltage on the motor. This is also known as "Kornodorfer starting," named after its inventor.

OTHER CURRENT LIMITERS

There are two other common ways of limiting current inrush during start. Both of these, as mentioned before, require motors that are specially designed. One is called part winding and the other wye-delta.

PART WINDING STARTING

Motors suited to part winding starting have two sets of stator coils. Fig. 13-23 shows the schematic of such a motor and its control diagram. Terminals, T4, T5, and T6 have been permanently joined so that each set of windings

235

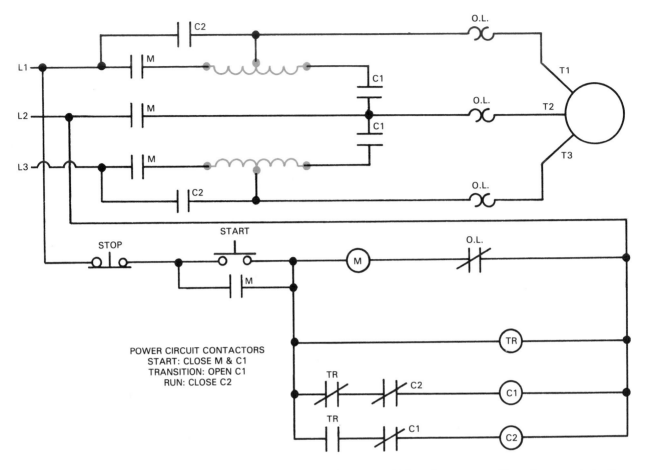

Fig. 13-22. Autotransformers can also be used in a closed transistion system. Voltage is not removed from the motor during the transition from start to run.

is wye-connected. In the first step, full voltage is applied to one set of windings only through contacts M. In the second step, contacts M2 close, connecting the two sets of windings in parallel. This is the normal running condition for this motor.

WYE-DELTA START

To use wye-delta start, the motor must be designed to run delta-connected. Fig. 13-24 shows how wye-delta starting works. Note from the drawing that both ends of the stator windings are brought out.

1. In the first step, contacts M and M2 close, connecting the three windings into a wye. The voltage across each winding is line voltage divided by 1.73. For example,

assume that line voltage is 208 volts. During the start period, the voltage across each coil will be 120 volts ($\frac{208}{1.73}$), approximately 58 percent of line volts.

2. In the second step, contacts M2 open, and contacts M3 close. Now the coils are connected delta. The voltage across each equals line voltage. As more delta-run motors are built, this method of starting will become more popular. It does away with starting resistors, reactors and autotransformers.

JOGGING AND INCHING

There is a slight difference between the terms "jogging" and "inching." To jog a motor you apply short bursts of full power. Inching means

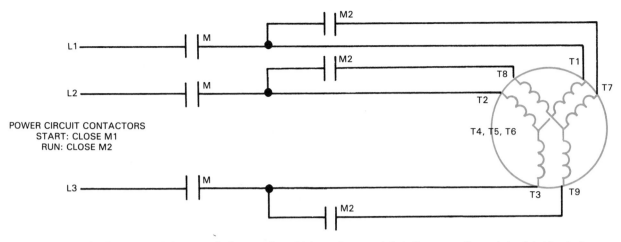

Fig. 13-23. Power circuit for part winding starting which requires specially built motors. Control circuit is identical to that of Fig. 13-19.

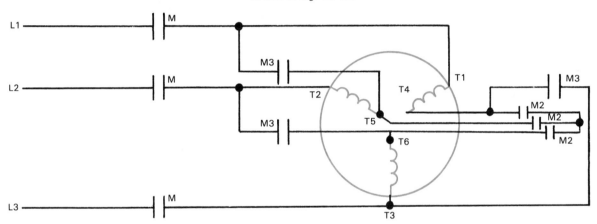

Fig. 13-24. Wye-delta starting requires motors designed to be run with stator coils delta-connected and all terminals brought out. Only the power circuit is shown. The control circuit is identical to the one in Fig. 13-22.

applying reduced power to move a motor slowly to a desired position. Fig. 13-25 shows the circuit for a "start-jog-stop" control system. Note that contactor C1 is energized only by pressing the start pushbutton. Power is applied by the jog pushbutton only while it is being pressed by the operator.

REVERSING CONTROLLERS

A three-phase motor runs in the same direction as its rotating field. The field's direction, in turn, depends on phase rotation. On the power lines, we say that Phase B is the one following Phase A. If coil B on the motor stator is clockwise from coil A, the motor turns clockwise. If we want the coil that is located

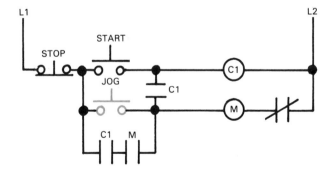

Fig. 13-25. A jogging control system permits application of short bursts of full power.

counterclockwise from coil A to be coil B, all we have to do is swap any two leads. Therefore, if any two of the motor leads are interchanged, the motor reverses direction.

237

A control system designed to do this is shown in Fig. 13-26. One contactor, marked F, makes the motor run in the forward direction. A second contactor, R, makes the motor run in reverse. Each has a normally closed contact in series with its coil. This is to prevent its being energized while the other contactor is closed. This arrangement is called an ELECTRICAL INTERLOCK.

A second electrical interlock is provided by the push buttons themselves. Current for each contactor must travel through the normally closed push button contacts of the opposite push button. For example, assume the motor is running forward and the reversed push button is pressed. The first thing that happens is that the current path to the forward contactor is broken. This removes power from the motor before the reverse contactor is energized.

Forward and reverse contactors have a MECHANICAL INTERLOCK, too. As shown

in Fig. 13-27, both are mounted on the same plate. A pivoted arm is pushed up as one contactor pulls in. Even if the other contactor somehow became energized, it could not pull in.

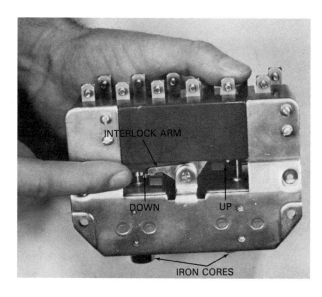

Fig. 13-27. Mechanical interlock prevents both coils from being energized at same time.

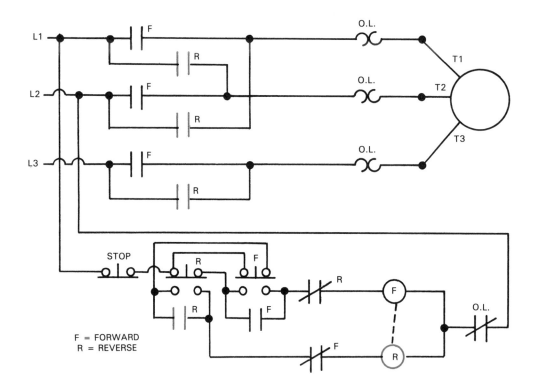

Fig. 13-26. This system reverses a motor by "plugging." Reverse torque is applied while the motor is running.

The reversing control of Fig. 13-26 applies a reverse torque while the motor is still turning forward. This type of reversing is called PLUG-GING, or PLUG REVERSING. Plug reversing may not be used where the mechanical strain would be too great. In those instances, protective devices prevent the motor from starting in the reverse direction until it has come to a complete stop.

SPEED CONTROL

Three-phase induction motors are basically constant-speed machines. Some limited speed control is possible by reducing the applied voltage. However, torque is proportional to the applied voltage squared (T = KV²). With speed reduction comes a severe loss in torque.

Although you cannot vary the motor's speed, you can reduce the speed at which its load moves. This is made possible by a MAGNETIC CLUTCH, also known as MAGNETIC DRIVE.

The magnetic drive is a very simple device. It works through an electromagnet which rests on the motor shaft. Direct current is supplied to the electromagnet through slip rings. When the motor is turning, the electromagnet produces a rotating magnetic field.

A ring, mounted on the load, surrounds the electromagnet. The rotating field induces cur-

rent into this ring. Magnetic forces produce a torque just like the torque in an induction motor. *The load speed is always less than motor speed.*

Torque developed in the magnetic drive is proportional to the direct current applied. This is the way speed is varied. A magnetic drive is shown in Fig. 13-28.

Speed can also be controlled electronically by using the frequency of applied voltage. Electronic circuits rectify the line voltage (allow it to move in only one direction) so it becomes dc.

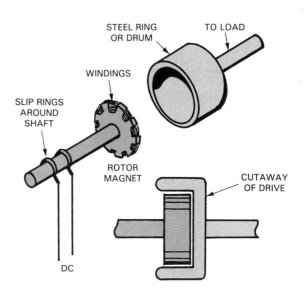

Fig. 13-28, View A. Exploded view of a magnetic clutch. This device is used to vary the speed of an alternating current motor.

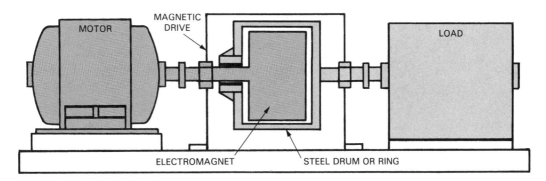

Fig. 13-28, View B. Simplified diagram of magnetic clutch pictured in view A. Motor, magnetic drive and load are indicated. There is no mechanical link between motor and load.

Then they reconstruct a sine wave voltage. The circuit includes controls to vary the frequency. Variable frequency speed controllers are discussed further in Chapter 15.

MULTI-SPEED MOTORS

Although variable speed ac motors are rare, they often operate at two or more fixed speeds. *Speed, remember, depends on both frequency and the number of poles.* The same motor may, for example, run as a two-pole motor, or as a four-pole motor. This method of changing speed is called POLE-CHANGING.

High-speed connection

A two-speed motor is wound as a two-pole machine. Fig. 13-29, diagram A, shows the wir-

ing schematic for the high-speed (3600 rpm) connection. Each winding is centertapped. Points T1, T2, and T3 are joined to produce a parallel-wye connection.

The arrows show instantaneous current flow. The resulting magnetic field is shown in diagram B, Fig. 13-29. There is one north and one south pole for each coil group.

Low-speed connection

The low-speed (1800 rpm) connection is shown at A in Fig. 13-30. Points T1, T2, and T3 have been connected to the line. This arrangement produces a series-delta connection.

The direction of current is such that each coil produces the same pole. Now there are two

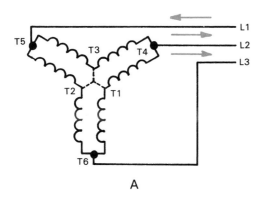
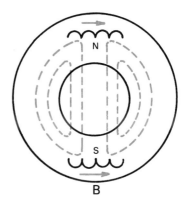

Fig. 13-29. Diagrams of speed hookup of multi-speed ac motors. A—Winding connections. B—Magnetic path of a dual-speed machine connected as a two-pole motor.

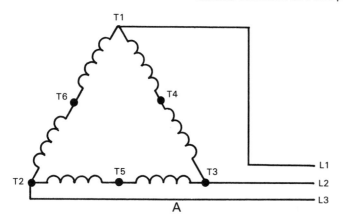
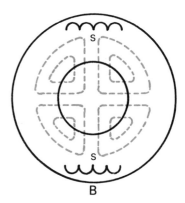

Fig. 13-30. Diagrams of low-speed hookup of multi-speed motor. A—Winding connections. B—Magnetic path of a dual-speed machine connected as a four-pole motor.

fields as shown in B in Fig. 13-30. The magnetic circuit is closed through the stator core. This produces south poles at right angles to the coil-produced north poles. Magnetic poles created this way are called CONSEQUENT POLES. The effect is the same as a four-pole motor.

Some motors are built with two separate sets of windings like these. They can run at four different speeds. It is possible to build multi-speed motors having constant torque, variable torque or constant output power.

BRAKING INDUCTION MOTORS

A motor's load, itself, will help stop the motor. Mechanical brakes also provide stopping power. These may be operated by hand or magnetically.

With electrical braking devices the motor produces the slowing or stopping action. Examples of electrical braking methods are PLUG BRAKING and DC DYNAMIC BRAKING.

Plug braking

To plug a motor is to reverse it while it is still running. In the process of reversing direction, it hits zero speed just for an instant. If power is removed at that instant, the motor comes to a sudden stop.

Since you do not want the motor to actually reverse (plug), you need an anti-plugging switch. This is usually a centrifugal switch mounted on the motor shaft. The switch is open when the motor is at rest. When the motor speeds up, it closes. When the motor is braked, the switch disconnects the main contactor just before zero speed.

The control diagram for a plug brake is shown in Fig. 13-31. The power diagram is the same as the one shown in Fig. 13-26.

Direct current dynamic braking

During normal operation, voltage is induced into the rotor of an induction motor by the revolving stator field. The resulting rotor field produces the torque. A steady stator field, however, would cause rotor current in the opposite direction. This would produce a reverse torque.

The control diagram for a dc dynamic brake is shown in Fig. 13-32. When the brake push button is pressed, contactor B is energized. Its contacts apply dc to one stator winding. As the bars cut through this dc field, current is induced in the opposite direction. A reverse torque on the rotor brings it to a quick stop. The amount of reverse torque determines the braking time. This is controlled by the amount of direct cur-

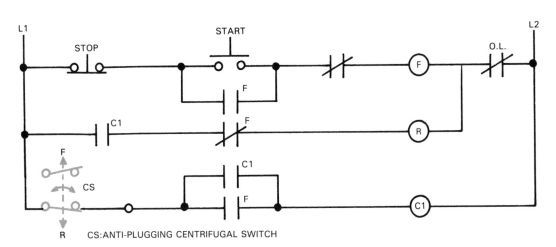

Fig. 13-31. Anti-plugging switch converts plug-reversing system into a plug-breaking system. Power circuit is the same as in Fig. 13-26.

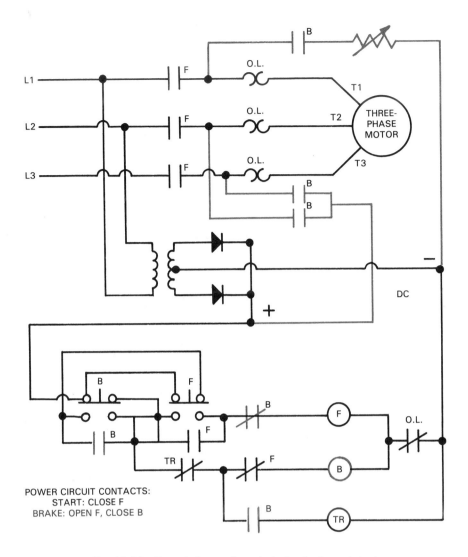

Fig. 13-32. Control diagram is typical of a dc dynamic brake.

rent applied. When the motor stops, there is no more induction. Therefore, the braking torque disappears.

If a load, like an elevator car, must be held at zero speed, a mechanical brake must be used. None of the electrical brakes produce torque at zero speed. Usually, an electrical brake slows the motor and lets a mechanical brake finish the job of braking.

WOUND-ROTOR MOTOR CONTROL

Wound-rotor motors combine high starting torque and low starting current. They accomplish this by having a resistance in series with the rotor winding. As the motor speeds up, the resistance is reduced a step at a time. Normally all external resistance is short-circuited. The controls for this may be manual or automatic.

Manual control is often a drum type device as shown in Fig. 13-33. Usually there are five to seven steps of resistance. Moving the operating handle to the first starting position does two things:

1. It closes the line contactor through an interlock.
2. It places all of the resistance across the slip rings. As the handle is moved, sections of the resistance are short-circuited.

AC Motor Control Fundamentals

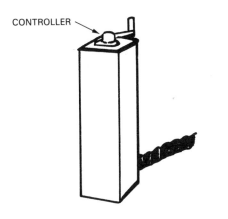

Fig. 13-33. Drum controller for a wound-rotor motor. Usually it has five to seven resistance steps.

Like the manually operated drum controller shown in Fig. 13-33, automatic systems provide control in both the stator (primary) and the rotor (secondary) circuits. The primary control is an across-the-line starter, like that used with squirrel-cage motors. (Refer once more to the diagram in Fig. 13-9.) The contactors that short-circuit resistance sections may be controlled by three types of devices:

1. Timing relays.
2. Current relays.
3. Frequency relays.

Current relays sense load current and act when current falls to a predetermined level. Frequency relays sense the slip frequency of induced rotor voltage. Resistance is cut out as the frequency decreases. Fig. 13-34 shows the circuit for a three-step starter for a wound-rotor motor starter using timing relays.

SYNCHRONOUS MOTOR CONTROL

Synchronous motors start the same way as induction motors. There are six induction motor

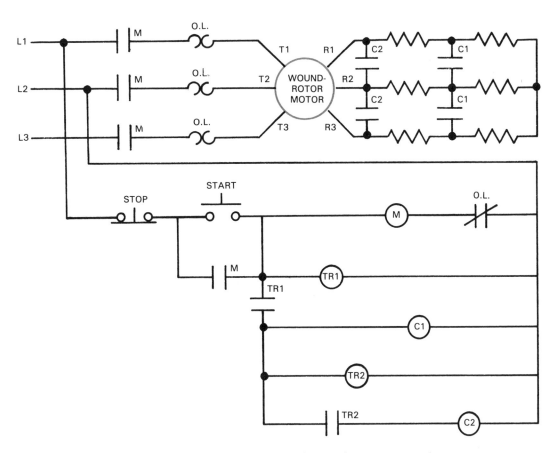

Fig. 13-34. Control diagram of an automatic wound-rotor motor starting system.

starting methods:
1. Across-the-line.
2. Resistor.
3. Reactor.
4. Autotransformer.
5. Part-winding.
6. Wye-delta.

All of these starting methods can be used on synchronous motors. Once they are started, however, the problem of synchronizing requires certain controls.

To synchronize properly, the motor must be rotating as fast as the squirrel-cage bars will let it. This may be 95 to 98 percent of synchronous speed. Usually, a timer delays synchronizing until this speed is reached. The control circuit is shown in Fig. 13-35.

During start periods the rotor's field winding must be short-circuited through a resistor. The resistor's purpose is to dissipate power generated in the field coil. If this circuit is open, the motor will shut down to prevent damage from high induced voltages in the field winding. A normally closed contact of contactor C provides the short-circuiting path.

After timing relay, TR, times out, contactor C opens the short-circuit and applies dc to the field coil through normally open C contacts. This locks the rotor field in step with the revolving stator field. There is pure dc in the rotor if it is traveling at synchronous speed. If the rotor should fail to synchronize, a protective device senses induced current in the field coil. The same situation could develop if the motor later pulled out of synchronism. Either way, power is removed from the motor.

BRUSHLESS SYNCHRONOUS MOTORS

Direct current for field excitation might come from a separate source. Often the motor, itself, drives a dc generator. Either way, current is fed through commutator and brushes.

Brushless excitation has done away with that. A small ac alternator is wound on the rotor

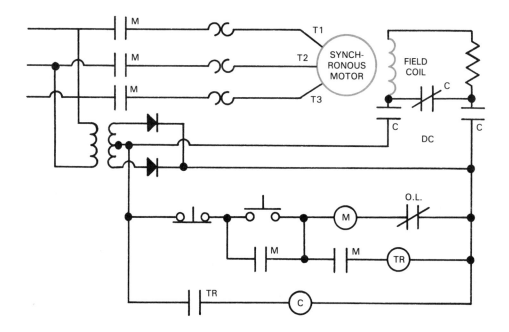

Fig. 13-35. Control system for starting and synchronizing a synchronous motor. Timer delays synchronization until motor is up to speed.

AC Motor Control Fundamentals

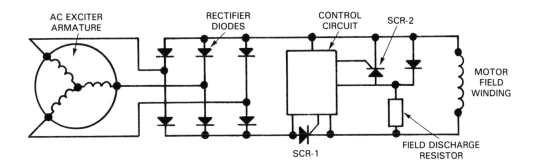

Fig. 13-36. Brushless synchronous motor has field excitation and control components mounted on its rotor shaft.

shaft. Its output is then rectified by a three-phase bridge-type rectifier, also mounted on the rotor shaft. Direct current from the rectifier, then, is used to excite the field. A drawing of the rotor-mounted excitation and control circuit is shown in Fig. 13-36.

The disadvantage of brushless excitation is that you cannot measure or change the value of the excitation current. However, the advantage of not having brushes or commutator to worry about often makes up for it.

SYNCHRONOUS MOTOR BRAKING

Dynamic braking of synchronous motors takes advantage of the counter torque produced by a generator. The circuit for starting and braking a synchronous motor is shown in Fig. 13-37. When the brake push button is pressed, power is removed from the stator. Meanwhile, dc continues to be applied to the field coil through contacts C. Now the synchronous motor is operating as an alternator, driven by the load. The dc field sweeps around, generating an ac voltage in the stator coils.

Immediately, resistors are connected across the terminals by the closing of contacts B. Current flows in the armature coils on the stator. This sets up a stator field which opposes the rotating rotor field. The motor quickly comes to a stop. One set of contacts for contactor B applies voltage across timing relay TD2. When

TD2 times out, its normally closed contacts open. This removes power from B, TD1, and TD2 itself. The circuit is reset, ready for the next start.

SUMMARY

Control systems have three parts: sensing, decision-making, and action.

In electromechanical control systems: 1. Push buttons or pilot devices do the sensing. 2. Relays make the decisions. 3. Contactors produce the action by opening or closing a load circuit.

Across-the-line starters differ from plain switches in that the starters automatically disconnect the motor from the line if power fails.

Timer relays delay the opening or closing of contacts until some time after power is applied to or removed from them.

Among the three types of three-phase motors, induction motors and synchronous motors are considered constant-speed motors. The wound-rotor motor, however, is a variable speed type.

The purposes of reduced voltage starting are to limit the inrush of starting current and to minimize shock to the driven load.

Plugging a motor means to reverse its direction while running. Plug braking means to reverse power but disconnect the motor before the actual reversal takes place.

Wound-rotor motors are started with maximum resistance in the rotor circuit. Resistance is

245

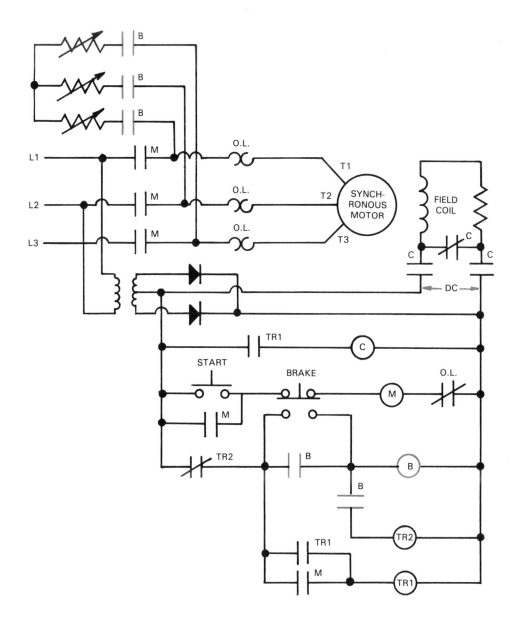

Fig. 13-37. Control diagram of a braking system for a synchronous motor.

cut out to increase speed.

Synchronous motors are started as induction motors. All of the same starting methods for induction motors also apply to synchronous motors.

TEST YOUR KNOWLEDGE

1. Name four main functions of controllers.
2. Explain the difference between a manual controller, a semiautomatic control system and an automatic control system.
3. Describe the operation of an electromechanical relay or contactor.
4. A normally _____ contact makes contact only while the push button is being pressed; normally _____ contacts break contact when the push button is pressed.
5. List four kinds of timer relays and explain their principles of operation.

6. What is the purpose of an overload relay? Describe the operation of one type.
7. Name at least three types of reduced voltage starters for three-phase induction motors.
8. To (jog, inch) a motor, apply short bursts of full power; to (jog, inch) a motor apply reduced power.
9. What must be done to reverse the direction of a three-phase motor?
10. How are different speeds obtained in dual-speed induction motors?
11. Dynamic brakes can hold a motor once it has stopped. (True or False?)
12. What is needed in addition to a standard reduced voltage induction motor starter for the starting of a synchronous motor?

Interior of an autotransformer motor starter. It is sometimes used to reduce starting voltage. This type of starter is discussed on page 234. (General Control Div., Westinghouse Electric Corp.)

14

FUNDAMENTALS OF DC MOTOR CONTROL

After studying this chapter, you will be able to:

☐ Describe the operation of a dc manual starter.
☐ Explain the method of reversing dc motors.
☐ List four methods of reduced-voltage starting of dc motors.
☐ Explain how the speed of dc motors is controlled.
☐ Define regenerative braking, dynamic braking and plug braking of dc motors.

Small dc motors can usually be started by applying full voltage. Units over 1/2 hp (373 watts), however, may need a reduced voltage starter.

The way to reduce voltage is to put a resistance in series with the armature. The resistance is needed only during the start period. After the armature starts turning, its CEMF limits armature current. Therefore, the resistance is cut out as the motor speeds up.

Resistance may be cut out manually or automatically. Manual starters include FACE-PLATE and DRUM CONTROLLERS. Automatic starters may operate from timers, CEMF or armature current.

MANUAL STARTERS

Fig. 14-1 shows a faceplate starter designed for educational use. A THREE-POINT STARTER has three connection terminals. A FOUR-POINT STARTER has four. This one is designed to be used both ways.

The starting resistor is tapped at several places. Making contact with these taps is a wiper arm. The wiper arm moves from tap to tap as the handle is turned.

An iron bar is attached to the wiper arm and moves with it. When all resistance has been cut out, this bar is held in place by the electromagnetic holding coil.

The starter remains in the full ON position as long as the coil is energized. When the coil is de-energized, the bar, wiper arm and handle are pulled quickly to the OFF position by a spring.

THREE-POINT STARTERS

Fig. 14-2 shows a circuit for starting a series motor with a three-point starter. Line L1 is connected to the wiper arm. When the arm reaches the first tap, current begins to flow. Most of it goes through the starting resistor to the armature. The rest is used to energize the holding coil.

To start the motor, you turn on the main

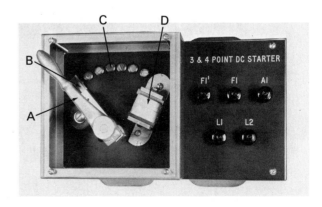

Fig. 14-1. Manual faceplate starter for dc motors. A—Handle. B—Wiper arm. C—Tap. D—Holding coil.
(Hampden Engineering Corp.)

power switch. Then move the handle from the OFF position to full ON. The handle should be moved slowly — about two seconds per tap. Moving it faster produces a current surge. Moving it too slowly may burn out the starting resistor which is rated for short duty.

The correct way to turn the motor off is with the main switch. The starter handle returns to the OFF position by itself. *The handle should never be forced from the holding coil. This produces severe arcing between the wiper arm and taps.*

The circuit of Fig. 14-2 provides NO-VOLTAGE PROTECTION. Should voltage drop to a low level, it turns the motor off.

It does not, however, provide overspeed protection. If a series motor loses its load, its speed can rise to a dangerous level. As it speeds up, armature current goes down. For protection against overspeeding, simply connect the holding coil in series with the armature, as shown in Fig. 14-3. When armature current drops, the holding coil de-energizes and turns the motor off.

The manual three-point starter can also be used for shunt motor starting. A circuit for a shunt motor is shown in Fig. 14-4.

One leg of the parallel circuit consists of the starting resistor in series with the armature. The other leg has a holding coil in series with the shunt field. This connection provides protection against loss of both voltage and field current.

The same connection is used for compound motors, Fig. 14-5. The only difference is in the addition of the series field coil.

FOUR-POINT STARTERS

You can control the speed of shunt and compound motors by changing the field strength. With field at full strength, the motor runs at rated speed. A weaker field makes it run faster.

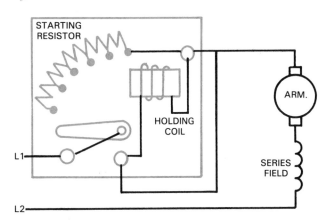

Fig. 14-3. Circuit for a three-point starter which provides overspeed protection on a series dc motor.

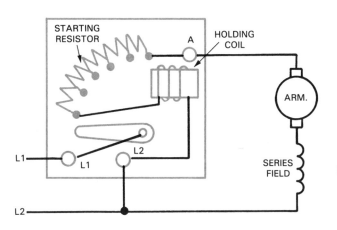

Fig. 14-2. Wiring diagram shows circuit for three-point starter connected to a series dc motor. It also provides no-voltage protection.

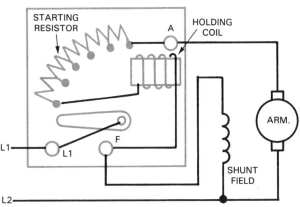

Fig. 14-4. Three-point starter for shunt dc motor. Circuit is wired for no-voltage and field loss protection.

Speed is controlled by a rheostat in series with the field coil. The three-point starters of Fig. 14-4 and Fig. 14-5 could not be used. If you tried to weaken the field, the starter would turn off the motor. However, a four-point starter can be used.

Fig. 14-6 shows a four-point starter circuit. The holding coil now has its own parallel leg along with a current-limiting resistor. The field coil, with its rheostat, forms a second leg. The third parallel leg is the armature and its starting resistor.

This connection permits speed control from field weakening. It does not provide protection against overspeeding if the field circuit becomes open, however. Neither does it prevent an attempt to start the motor with resistance in the field circuit.

COMBINATION STARTER AND SPEED CONTROLLER

Manual starters can double as speed controllers, Fig. 14-7. A heavy-duty resistor is used because it may have to carry current continuously. This controller has two wiper arms. In the OFF position, all of the resistance is in the armature circuit. There is no resistance in the field circuit.

As the handle moves, resistance is gradually cut out of the armature circuit. This does not affect the field circuit. The motor, at this point, runs slower than rated speed. You may set the handle at any tap, according to speed wanted.

At midpoint, there is no resistance in either the armature or field circuit. The motor runs at normal, rated speed. As the arm is advanced, resistance is added to the field circuit. The field weakens and the motor speeds up.

DRUM SWITCHES

Drum switches, Fig. 14-8, can also be used to operate dc motors. Small units are used for

starting, reversing and stopping. Where combined with resistors to control speed, they are called DRUM CONTROLLERS.

REVERSING DC MOTORS

Direction of rotation of a dc motor depends on:
1. Direction of armature current.
2. Polarity of the field.

If either one changes, the motor reverses direction. If both change, the motor continues rotating in the same direction. The normal way to reverse a dc motor is to interchange the armature connections.

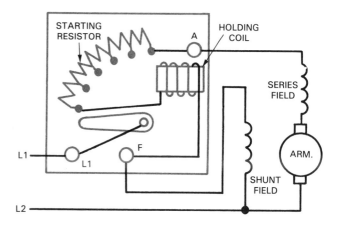

Fig. 14-5. Circuit for three-point starter on compound dc motor. It provides no-voltage and field loss protection.

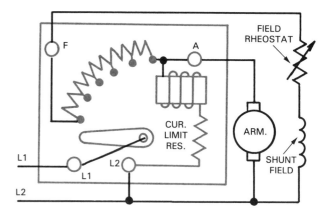

Fig. 14-6. This four-point starter circuit permits speed control by weakening the field.

Fundamentals of DC Motor Control

A toggle switch can be used for reversing. This requires a double-pole three-position switch. In Fig. 14-9, diagram A, the switch is in the forward position. It has been thrown to the reverse position in diagram B. Armature current is reversed. The third position of the switch turns the motor off.

View A of Fig. 14-10 shows a reversing drum switch. Fixed contacts are interconnected as shown in view B. The center lever position is OFF. Normally, motors are allowed to stop completely before reversing. PLUGGING can produce too much strain on the shaft. (Plugging means reversing the motor by means of the switch while it is still running.)

When connecting a compound motor for reversing, it is important not to get the following components reversed while reversing the armature:
1. The series field coil.
2. The interpoles or compensating windings.

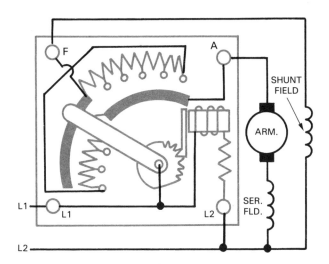

Fig. 14-7. Circuit drawing of a combination manual starter and speed controller. It uses two sets of resistors.

Fig. 14-8. Drum switches are often used to control dc motors. (Square D Co.)

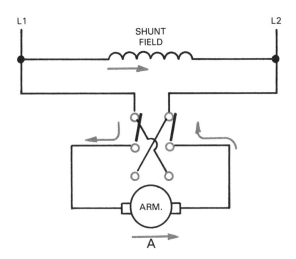

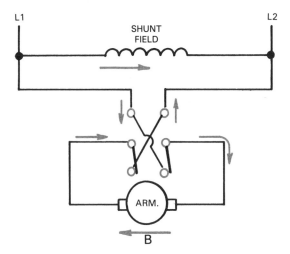

Fig. 14-9. Three-position toggle switch may be used for reversing dc motors. A—Armature current flows left to right. B—Current flow is reversed.

MAGNETIC STARTERS

Magnetic contactors, like the one pictured in Fig. 14-11, can be used to apply power to a dc motor. For large motors, however, current through the contacts is high. As the contacts open, an ARC is drawn which can damage contacts. That is why larger contactors have BLOWOUT COILS. A blowout coil helps extinguish the arc quickly.

An arc is really a stream of electrically charged particles much like current through a wire. As with the current, the arc is surrounded by a magnetic field.

The blowout coil is connected in series with the contactor's coil. It works like the main field coil of a motor. Its magnetic field acts on the arc's field. In that way, it forces the arc off the contacts while increasing the length of the air gap. This is illustrated by Fig. 14-12. An ARC-CHUTE usually covers the contacts to protect surrounding equipment.

Fig. 14-13 is the diagram for a simple controller. It includes start and stop push buttons and overload protection.

OVERLOAD PROTECTIVE DEVICES

Many devices prevent damage from excessive current. Fuses, circuit breakers and bimetal overload relays were discussed in Chapter 13. This chapter will look as three types:

1. Magnetic.
2. Solder-pot.
3. Thermistor.

Magnetic overload relay

This is a current-sensitive relay, consisting of a coil and a normally closed set of contacts. Motor current passes through the coil. The relay's contacts, however, are in series with the contactor coil.

Normal, rated motor current does not energize this relay, but excessive current does. Some

HANDLE END		
REVERSE	OFF	FORWARD
1 o——o 2	1 o o 2	1 o o 2
3 o——o 4	3 o o 4	3 o o 4
5 o——o 6	5 o o 6	5 o——o 6

Fig. 14-10. Reversing drum switch. Top. View of switch with cover removed. Bottom. Diagram of internal switching. (Square D Co.)

Fig. 14-11. Contacts of a contactor carry full load current. When contacts open, an arc is drawn.

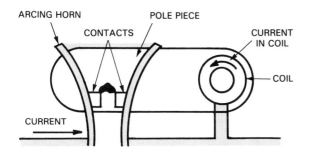

Fig. 14-12. Blowout coil helps extinguish the arc across contacts.

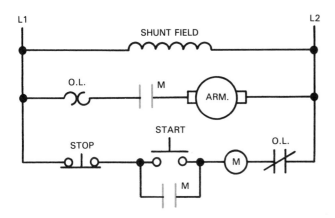

Fig. 14-13. Simple across-the-line starter. Overload protection is included.

overload relays pull in immediately. Others have a built-in timer. When the relay does pull in, its contacts open. This removes power from the contactor.

Solder-pot overload relay

This is a heat-sensitive device. Motor current passes through a small heater. As shown in Fig. 14-14, a ratchet assembly is held in place by a special kind of solder. The solder melts rapidly when heated. It hardens just as rapidly when heat is taken away. A PAWL under spring tension is held by the ratchet. The pawl assembly carries half a set of contacts, which are normally closed. An overload causes the heater to melt the solder. This releases the pawl and opens the contacts. With power off, the solder becomes hard again in about two minutes. The relay can then be reset.

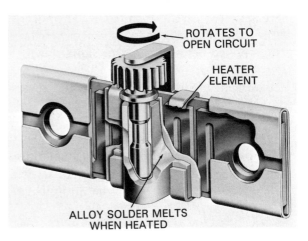

Fig. 14-14. Solder-pot overload relay. Heater melts solder when overload occurs. Relay then opens the circuit. (Square D Co.)

Thermistor overload relay

A thermistor is a resistor whose resistance goes up when it gets hot. Normally, it has a very low resistance. The more current passing through it, the warmer it gets. At some point, its resistance increases sharply.

Thermistors are usually placed in direct contact with the motor winding. Cold thermistors can carry enough current to energize a relay.

The contacts of this relay carry current to the main contactor. If the motor gets too hot, the extra thermistor resistance reduces the current. The relay drops out and so does the main contactor.

AUTOMATIC REVERSING

Two contactors are used to automatically reverse a dc motor. Contacts on the "forward" (F) contactor send current through the armature in one direction. Contacts on the "reverse" (R) contactor send it through the other way. This is shown in Fig. 14-15. The relay marked "AP" is in parallel with the armature. This is an "anti-plugging" relay. For small motors or light loads, plugging may be alright. However, at the instant reverse power is applied, CEMF adds to it. Current is double the normal starting current. Also there is a severe strain on the shaft. The purpose of this control system is to prevent power from being applied before the motor slows down.

A dashed line links the forward and reverse contacts. This means that they are mechanically interlocked to prevent their closing at the same time. Normally-closed contacts marked "F" and "R" provide electrical interlocking. In addition, each has a set of contacts, "AP," in series with it. The contacts open when the anti-plugging relay pulls in.

Operation of anti-plugging relay

Here is the way it works. Pressing the "FOR" push button completes a circuit consisting of:
1. The stop push button.
2. The reverse push button.
3. "AP" contacts.
4. "R" contacts to energize contactor "F."

The "F" contacts in the power circuit then apply power to the armature and relay "AP." Another set of "F" contacts close in parallel with the forward push button and contacts "AP." This maintains the current path after the push button is released and contacts "AP" open.

Contacts "AP," in series with the reverse contactor, are also open. Therefore, if you press the reverse push button, relay "R" does not pull in immediately. Contacts "AP" must return to their normally closed position first. Relay "AP" does not drop out immediately, however. It receives power from the CEMF of the armature.

When armature speed has dropped enough, relay "AP" drops out. At this point, the motor may be reversed.

REDUCED VOLTAGE DC STARTERS

Reduced voltage starters are designed to cut out resistance in the armature circuit in stages. Contactors in the starter cause current to bypass resistors. As the motor accelerates, resistance is cut out in steps. This gives the motor smooth torque without creating large surges of current.

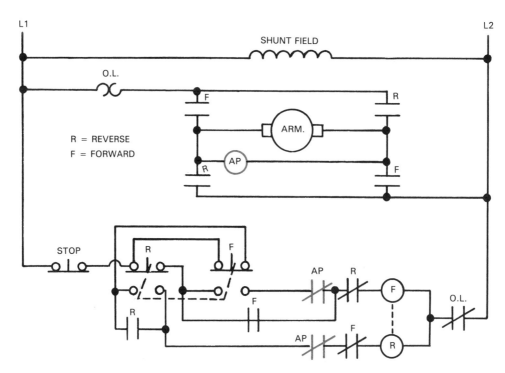

Fig. 14-15. Across-the-line starting system has anti-plugging relay to prevent reverse power being applied while the motor is running full speed.

There are three common types of reduced voltage dc motor starters. Their names indicate the basis on which resistance is cut out:

1. Definite-time acceleration.
2. Current acceleration.
3. CEMF acceleration.

DEFINITE-TIME ACCELERATION

Definite-time acceleration is also known as TIME-LIMIT ACCELERATION. The starter brings the motor up to rated speed in a definite time, regardless of load. It is the simplest and probably the most-often used voltage starter. The timer can be:

1. A part of the accelerating contactor.
2. A separate time delay relay.

Fig. 14-16 shows the schematic of a control system for a two-step starter. Additional steps of starting resistance can be added, if required. A timer and accelerating contactor are needed for each step.

Fig. 14-17 shows a circuit for plugging. When the "forward" push button is pressed, current is routed through the armature in the forward direction. At that point, a resistance is in series with the armature. After a few seconds, contactor "C" pulls in, shorting out this resistance. The "reverse" push button may be pressed while the motor is running in the forward direction. However, contactor "C" drops out first. This puts the starting resistance back in before reverse voltage is applied.

CURRENT-LIMIT ACCELERATION

At the instant you apply power to a dc motor, current is at maximum. This is because there is no CEMF yet. Even with starting resistors, the current is greater than when the rotor is turning.

Fig. 14-18 is the diagram of a current-limit acceleration starter. The starting surge of current energizes a current-sensitive relay. This relay pulls in as soon as power is applied to the armature. That causes its normally closed contacts to open.

Then as the motor picks up speed, its current decreases. At some point, the current will not be able to hold in the relay.

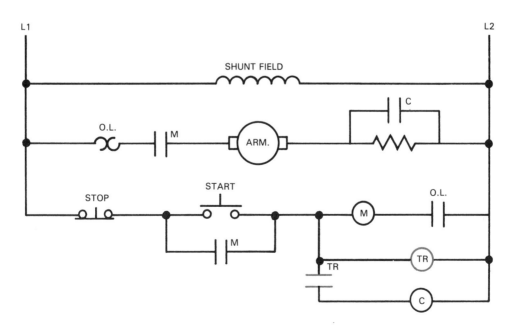

Fig. 14-16. Control circuit of a definite-time acceleration starter. Time delay relay is shown in color.

When the relay drops out, its contacts close. This applies power to the accelerating contactor "C." When the accelerating contactor pulls in, its contacts shunt current around the starting resistance.

Several steps (current levels) must be built into a starter which is designed to limit motor acceleration by changes in current. Each step will have a separate current relay. These relays, then, are set to drop out at different values.

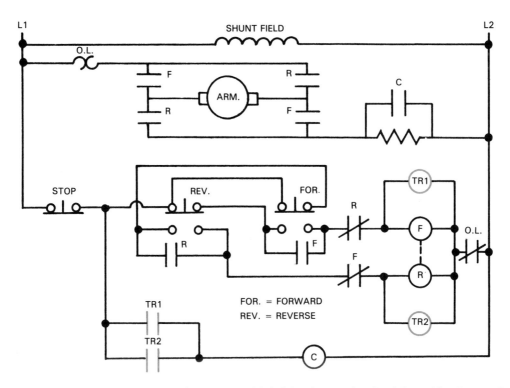

Fig. 14-17. This circuit is found in a reduced voltage starter with definite-time acceleration. It is used for plugging. Reverse push button may be pressed while motor is still running in the "forward" direction.

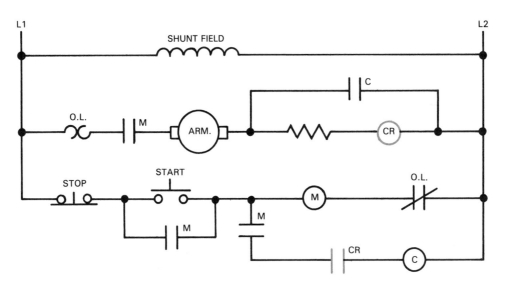

Fig. 14-18. Diagram for control circuit of current-limit acceleration starter. Relay is sensitive to a surge of current. It pulls into the circuit whenever motor is started.

LOCKOUT ACCELERATION

Fig. 14-19 shows another type of relay that works from armature current. Called a LOCKOUT RELAY, it has two coils. One, the HOLDING COIL, closes the contacts. The other, LOCKOUT COIL, opens the contacts.

The holding coil has an iron core that saturates easily. (A saturated core is one holding all the magnetism it can. Flux density, the number of magnetic lines of force in its magnetic field, is therefore, as great as it can be.) The lockout coil has a large air gap which does not saturate. Both coils are connected in series with the armature. See Fig. 14-20.

When armature current is low, the holding coil's magnetic field is stronger than the lockout coil's. The strong field is from the iron core, which has not saturated. As current increases, the holding coil's core becomes saturated. Its field cannot become any stronger. But the lockout coil's field keeps increasing in strength. Finally, the lockout coil's field overrides that of the holding coil.

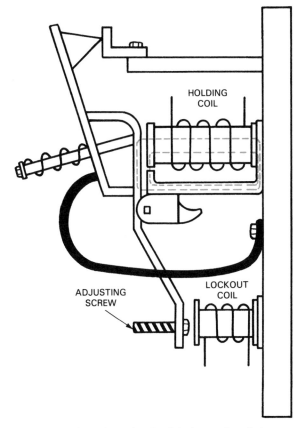

Fig. 14-19. Simplified sketch of lockout relay. It has two coils. One opens contacts; the other closes them.

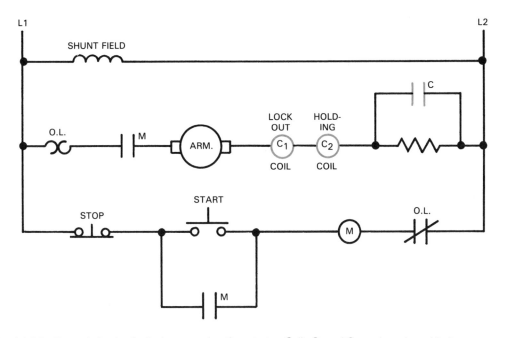

Fig. 14-20. Control circuit of a lockout acceleration starter. Coils C_1 and C_2 are in series with the armature.

High starting currents in the lockout coil keep the contacts open. Then, as CEMF is generated, armature current drops. At some point the holding coil's field becomes stronger than the lockout coil's field. The contacts close, cutting out the starting resistance.

CEMF ACCELERATION

CEMF acceleration starters use a potential relay. The relay's coil is connected in parallel with the armature. Its voltage is, therefore, the same as the armature's. At the instant of start, there is a very small voltage drop across the armature. It is not enough to cause the relay to pull in, however.

As the rotor speeds up, a CEMF builds up in the armature. It is like a resistance whose value is increasing. The voltage drop across the armature increases with speed. At some point, this voltage will be enough to make the relay pull in.

The diagram of this starter circuit is shown in Fig. 14-21. Additional steps of starting resistance may be added. However, each step requires a separate potential relay.

DC MOTOR SPEED CONTROLLERS

Usually, a motor and its speed controls work as a single unit. The object, after all, is to vary the speed of the load. The term DC DRIVE refers to the complete package — motor and speed controls.

High-torque drives often use series motors. This is particularly true if the speed control does not have to be exact. More precise speed control is produced by shunt motors. Standard drives are available which control speed from armature series resistance, armature shunt resistance and field series resistance. There is also the newer SCR drive, which uses solid state components.

ARMATURE SERIES RESISTANCE

A tapped resistor (or rheostat) in series with the armature causes a motor to run at less speed. Drawing A of Fig. 14-22 shows the connection for a series motor. Drawing B shows how speed varies with the load on it for each of the rheostat settings. This type of drive is often used on hoists, cranes and traction machines, such as electric cars.

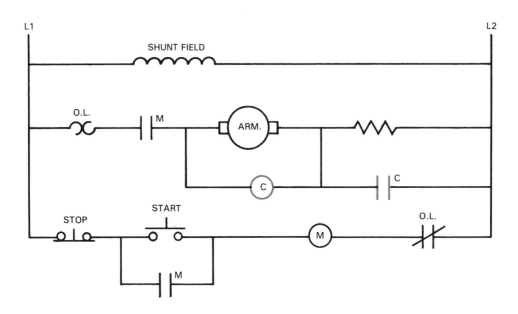

Fig. 14-21. Control circuit of a CEMF acceleration starter. Voltage drop across armature activates the relay, "C."

Armature series resistance can also be used on shunt motors, Fig. 14-23. Drawing A is a diagram of the circuit. The speed torque curves are shown in chart B. Each curve is plotted by setting the resistor on one of the taps. Speed is measured as the load is changed.

If the load is constant, you can determine how each tap affects speed by drawing a vertical torque line. For example, in Fig. 14-23, Chart B, loaded at 100 percent rated torque, the motor will run at approximately 40 percent of rated speed at tap No. 2.

ARMATURE SHUNT RESISTANCE

Another way of varying armature current is by shunting some of the current around it. The connection for a series motor is shown in Fig. 14-24, drawing A. By using a SERIES RESISTOR as well, you can get light-load speeds down to 10 percent of rated speed. The curves are shown in drawing B, Fig. 14-24.

The same type of resistances can be used on shunt motors. See drawing A, Fig. 14-25. At low settings of the shunt resistance, speed is about

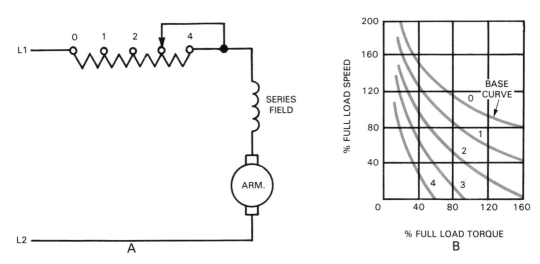

Fig. 14-22. Series armature resistance is used to control speed of series motors. A—Schematic. B—Relationship of speed to torque.

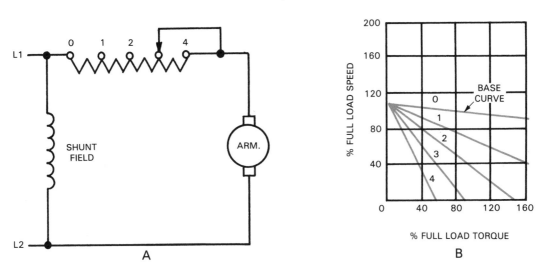

Fig. 14-23. Series armature resistance used to control the speed of shunt motors. A—Circuit diagram. B—Relationship of speed to torque.

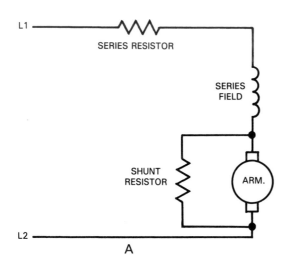

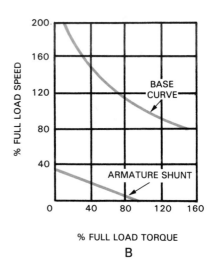

Fig. 14-24. Series and shunt armature resistances are used to control the speed of series motors. A—Typical circuit. B—Relationship of speed to torque.

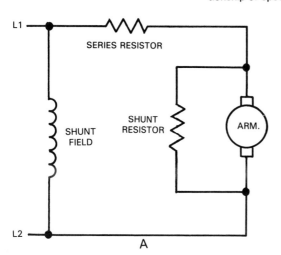

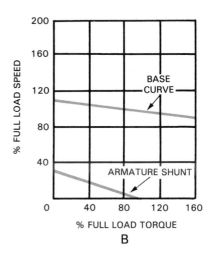

Fig. 14-25. Series and shunt armature resistances are used to control the speed of shunt motors. A—Circuit diagram. B—Speed-torque relationships.

the same as for series motors. This is shown in the graph, Fig. 14-25, drawing B. The armature shunt resistor can also be used as a dynamic braking resistor.

FIELD SERIES RESISTANCE

Shunt motor speed can easily be controlled from a field rheostat. Fig. 14-26, drawing A, shows the connection. Field weakening can cause a typical dc motor to run at three times its rated speed. Special motors can run even faster. The curves are shown in chart B, Fig. 14-26.

SOLID STATE CONTROLS

Originally, gas-filled tubes or vacuum tubes provided electronic control for dc motors. New control devices are made of solid material such as silicon. The term solid state refers to the fact that control action takes place within this solid material.

Most widely used in solid state controls is the SILICON CONTROLLED RECTIFIER (SCR). SCR controllers provide very accurate speed control of dc shunt motors. Basically,

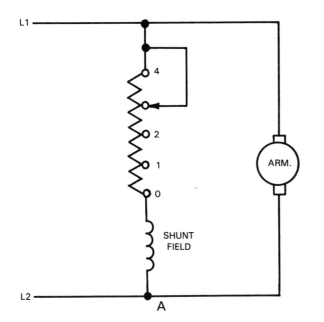

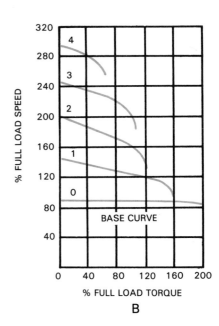

Fig. 14-26. Field rheostat can be used to control the speed of shunt dc motors. A—Wiring diagram. B—Speed-torque relationships.

their job is to control the value of armature voltage.

The SCR, as pictured in Fig. 14-27, (see symbol), has three terminals. When an SCR is "triggered," current flows through it from the anode (+) to the cathode (−). It will not flow in the other direction because the rectifier will not allow a reverse flow.

The current flow starts when a voltage of the proper polarity is applied at the gate. After triggering, however, the gate loses control. Current continues to flow until the anode voltage falls to zero.

Assume that alternating current is applied to an SCR and motor armature in series. Fig. 14-28 shows a half cycle of the voltage drop across the SCR triggered at 45 degrees. From 0 to 45 degrees, it does not conduct. Therefore, the voltage drop follows the sine wave. At 45 degrees, its resistance falls to practically zero. There is no voltage drop across it until applied voltage reaches zero. Fig. 14-28 also shows the voltage across the armature. Voltage is applied

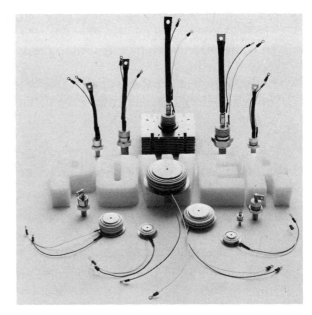

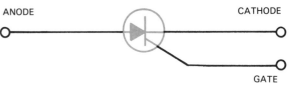

Fig. 14-27. Silicon controlled rectifiers (SCR) control motors with solid state materials. Top. SCRs are made to suit various needs. Bottom. Symbol for SCR.

only between 45 and 180 degrees of each cycle. By changing the angle at which the SCR is triggered, you can change the effective applied voltage.

Fig. 14-29 shows a simple SCR speed control circuit for a dc motor. Alternating current flows through the armature circuit while the field coil has direct current. A rheostat controls the gate-to-cathode triggering voltage. This device applies voltage to the armature at a certain part of each cycle. The rheostat, then, effectively controls the speed of the motor.

SCR speed control holds a motor at constant speed no matter what the load. This is what happens:

1. As a load becomes lighter, the motor starts to speed up.
2. This generates extra CEMF.
3. The CEMF not only reduces armature current, but increases the gate voltage as well.
4. The SCR is now triggered later in each cycle. Thus, the effective voltage applied to the armature is reduced.
5. The motor quickly balances out at the same speed as before the load change. A similar response is triggered when load is increased.

Direct current motor speed controllers using SCRs often control starting as well. They are described in Chapter 15.

BRAKING DC MOTORS

The term "braking" can mean several things.
1. It can mean bring a motor to a complete stop.
2. It can also mean simply slowing it down.

Either way, braking is a retarding action. It works against the force of inertia or, possibly, the force of gravity.

Electrical braking must provide a counter torque. It must create magnetic forces inside the motor in opposition to the turning armature.

Three standard methods of electrical braking are:
1. Regenerative braking.
2. Dynamic braking.
3. Plug braking.

REGENERATIVE BRAKING

You can find the word "generate" in regenerative braking. Generate is what the motor does when, instead of driving, it is driven.

Regenerative braking is the simplest type of electrical braking. It needs no external equipment. It does not even need special connections.

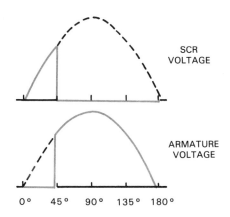

Fig. 14-28. Wave forms of the voltages in an SCR speed control system. Changing angle at which voltage is triggered controls effective applied voltage.

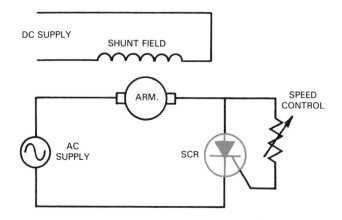

Fig. 14-29. SCR speed control system supplies direct current to field coil.

It happens automatically whenever the load supplies mechanical power to the dc motor coupled to it.

Some examples will show how regenerative braking works.

First, suppose that you are using a dc motor on a hoist. The motor works hard to lift a load. Yet, to lower it takes no work at all. The force of gravity pulls it down. A method is needed to keep it from falling too fast. That is where regenerative braking comes in.

The motor, remember, is connected to the power source at all times. The falling load drives the motor as a generator. Power generated is fed back into the dc power lines. Meanwhile, armature current (from the generated voltage) sets up a magnetic field. This field works against the main field to produce a retarding torque.

Now consider a dc motor driving a large machine like a printing press. It is running fast and has built up inertia. Someone turns the controls to half speed. Suddenly the motor's torque output is far less than the torque produced by the spinning load. What happens? The load begins to drive the motor. The motor acts like a generator. Its CEMF is higher than the applied voltage. This reverses current flow through the armature. A torque is set up in the opposite direction. As a result, the machine's speed drops. Before long it has slowed down to the new setting. This braking action disappears when the new speed is reached and the load is again driven by the motor.

If a number of dc machines take power from the same line, regenerative braking can be an important factor. Any power generated by the motors reduces the amount that the generators have to produce.

DYNAMIC BRAKING

Dynamic braking is much like regenerative braking. The motor runs as a generator driven by its load. There are two differences however:

1. Power is removed from the motor.
2. Power generated in the motor is wasted.

The circuit for a full voltage starter with dynamic braking is shown in Fig. 14-30. When the "brake" push button is pressed, the following sequence of events occurs:

1. The current path is opened to the main contactor.
2. This action removes power from the armature of the motor.

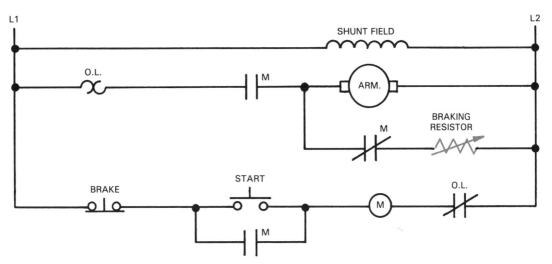

Fig. 14-30. Diagram of full voltage starter with dynamic braking.

3. The "brake" contactor connects a BRAK-ING RESISTOR to the armature terminals.

The braking resistor is adjustable. A high resistance allows only a small armature current. The braking action is, therefore, slight. The smaller the resistance the stronger the braking action. You should avoid shorting the armature, however. Although braking action would be fast, the short circuit current might damage the armature and commutator.

PLUG BRAKING

Unlike the braking methods just described, plug braking reverses armature current outside the motor. Normally, it is not advisable to reverse a motor while it is running. You can only do it on small motors or motors with light loads.

These are two reasons for avoiding it on motors or where loads are heavy:

1. There is severe mechanical strain on the motor shaft. Suddenly it has to withstand double the normal torque. Load torque is twisting one way while motor torque is twisting the other.

2. There is double armature current. When you remove *forward* power, the CEMF is still there. When *reverse* power is applied at the same time, it adds to the CEMF. The current surge is greater than it normally is at start. That is why an additional plugging resistance is often put in series with the armature.

Fig. 14-31 is a diagram of a starter equipped for plug braking. In many ways it is similar to

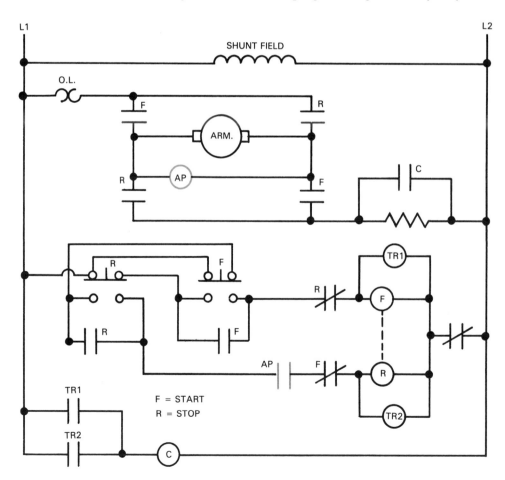

Fig. 14-31. Plug braking circuit for dc motor. Braking resistor is in series with armature.

the reversing starter of Fig. 14-15. However, the differences are important.

The anti-plugging relay has a different job to do in this circuit. Before, it prevented operation of the reversing contactor until after the motor stopped. Now, its job is to disconnect the motor when it reaches zero speed. We do not want to plug (reverse) the motor. We just want to stop it quickly.

When the "brake" push button is pressed the current path is broken to the "forward" contactor. This removes power from the motor. Next a current path is established to the "reverse" contactor. This applies power to the armature with reverse polarity. Note that the anti-plugging relay does not drop out when the "forward" contacts opened. It continues to be energized by the CEMF of the coasting armature.

The reversed current produces a reverse torque on the armature. Speed drops rapidly as it tries to reverse. Meanwhile the current in the anti-plugging relay starts to reverse direction. As this current passes through zero, the relay drops out. This disconnects the motor completely.

Another anti-plugging device is a centrifugal switch mounted on the motor shaft. This speed-

sensitive switch opens at slow speeds, breaking the circuit to the reverse relay.

SPECIAL PROTECTIVE DEVICES

There are a number of devices that are designed either to protect the field coils of a dc motor or to use the field circuit in protecting the armature and commutator.

FIELD LOSS RELAY

If the field coil loses power, the magnetic field becomes very weak. It consists of only the residual magnetism of the pole pieces. CEMF will drop. Armature current and speed will increase. To protect against this, a FIELD LOSS RELAY is placed in series with the field.

Fig. 14-32 shows how the field loss relay is connected. This is a current-sensitive relay. It pulls in as soon as power is applied to the field coils. If field current drops to a low level, the relay drops out. This opens contacts in the main contactor circuit, shutting down the motor.

FIELD ACCELERATING RELAY

A dc motor is started with full field strength. This allows the armature to generate enough CEMF to limit armature current. After it is run-

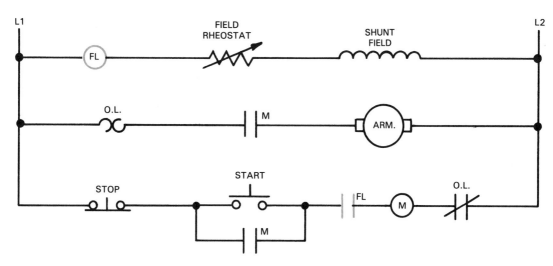

Fig. 14-32. Circuit showing use of field loss relay. This relay is sensitive to current in field coil.

ning, however, the field can be weakened to increase speed.

To double the speed of the motor, you would have to increase the resistance of the field rheostat. When you weaken the field, you also reduce the CEMF. There is a large surge of armature current. The field rheostat has to be cut out to bring the field back to full strength.

A FIELD ACCELERATING RELAY, Fig. 14-33, is a current-sensitive relay in series with the armature. When armature current gets too large, this relay pulls in taking the rheostat out of the circuit. Its contacts are in parallel with the field rheostat.

With the field rheostat out, a sudden shot of CEMF instantly reduces armature current. The relay drops out. This puts the field rheostat back in the circuit. Now, the armature is turning faster. If armature current is still too high, the relay action is repeated.

The field rheostat continues to cut in and out until the motor is up to speed. Because of its rapid on-and-off action, it is often called the FIELD-FLUTTER RELAY.

A field-flutter relay is sometimes used for deceleration, too. To slow down a dc motor,

you decrease the resistance of the field rheostat, increasing field strength. The only problem is that the load begins to drive the motor. This can generate a large voltage in the armature. The resulting current can damage the commutator. The field rheostat has to be cut back in to reduce the regenerative effect.

FIELD DECELERATING RELAY

Fig. 14-34 shows how the FIELD DECELERATING RELAY is used. When the armature current gets too high, this relay pulls in. It puts the full resistance of the rheostat in series with the field. Generating action is reduced and armature current drops. At that point, the field decelerating relay drops out, reducing field resistance. Armature current increases again. If the current is too high, the relay pulls in again. This protective cycle continues until the motor has safely slowed to the desired speed.

SUMMARY

Resistance is placed in series with the armature of a dc motor during starting to reduce current inrush. The resistors carry heavy current but only for a short time.

Speed of dc motors is controlled below base speed also with resistance in the armature circuit. This resistance, however, must be rated

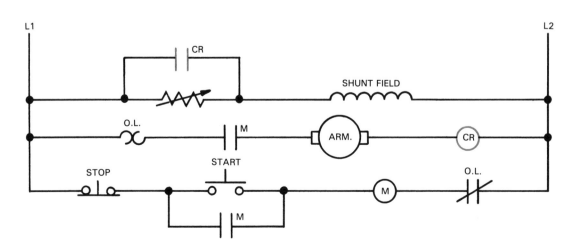

Fig. 14-33. Circuit showing use of field accelerating relay. Its job is to reduce current to the armature if it becomes too high.

Fundamentals of DC Motor Control

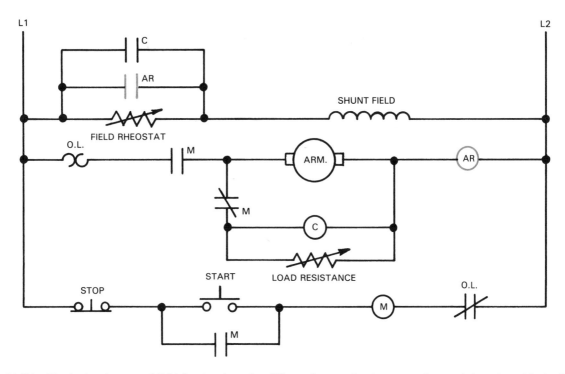

Fig. 14-34. Circuit showing use of field decelerating relay. When relay cuts in, rheostat resistance is in series with the field.

for continuous duty service.

Direct current motors must have full field strength when starting so they can produce CEMF. Otherwise, armature current may be excessive.

Starting resistance may be cut out automatically, based on time, current or CEMF.

Speed of dc motors above base speed may be controlled by a resistance in the field circuit, weakening the field strength.

The direction of dc motors may be reversed by reversing the polarity of the armature connections while maintaining the same polarity of the field connections.

With regenerative braking, energy is returned to the power source. With dynamic or plug braking, energy is wasted in the form of heat.

TEST YOUR KNOWLEDGE

1. Why is starting resistance needed on large dc motors?
2. Two types of manual dc starters are the _____ controller and the _____ controller.
3. Explain how a blowout coil works and why it is needed on large dc motor contactors.
4. Of magnetic, solder-pot and thermistor overload relays, which one(s) work on current and which one(s) work on heat?
5. In what circuit is the coil of an overload relay connected? In what circuit are the contacts?
6. Explain the purpose of a mechanical interlock on forward and reverse contactors.
7. Describe the operation of current acceleration starters. Definite-time acceleration starters. CEMF acceleration starters.
8. A _____ type relay prevents the reversing of a dc motor while it is running.
9. Name two ways of reducing armature current to reduce speed.
10. What is the main advantage of regenerative braking?
11. The purpose of the anti-plugging relay in a plug braking circuit is:
    a. To prevent operation of the reversing contactor until after the motor stops.
    b. To disconnect the motor when it reaches zero speed.

15 SOLID STATE MOTOR CONTROL

After studying this chapter, you will be able to:

☐ Explain the operation of solid state relays and solid state timing relays.
☐ List the basic types of logic gates.
☐ Describe the functions of motor soft starters.
☐ Explain the principle of operation of solid state dc motor speed controllers.
☐ Explain the principle of operation of solid state induction motor speed controllers.

The term, SOLID STATE CONTROLS, refers to many devices and systems. All have one thing in common. They control current inside solid "semiconductor" material.

There are two kinds of solid state components:

1. Those which act as switches.
2. Those which act as amplifiers.

The first type can switch off to on, or on to off very rapidly. Solid state switches include diodes, transistors, silicon controlled rectifiers (SCRs), triode ac switches (triacs), diode ac switches (diacs), and unijunction transistors (UJTs).

The other type, known as AMPLIFIERS, produces an output voltage continuously proportional to its input. The most important control amplifier is the operational amplifier, or "OP-AMP."

Fig. 15-1 shows some of the power semiconductors available. The items shown, along with

Fig. 15-1. Solid state electronic components come in a wide variety of sizes and shapes.
(Westinghouse Electric Corp., Semiconductor Div.)

resistors and capacitors, make up the entire list of control devices. Some are encapsulated (surrounded by protective material) like the relay shown in Fig. 15-2. Others are in the form of a printed circuit board. The motor control in Fig. 15-3 is this type.

This chapter will discuss eight types of solid state controllers:

1. Solid state relays.
2. Electronic timers.
3. Static logic.

Fig. 15-2. The solid state relay performs the same job as an electromechanical relay that has one set of normally open contacts. (International Rectifier)

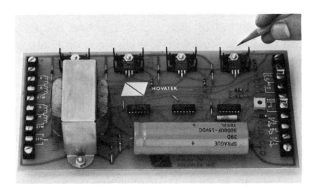

Fig. 15-3. Solid state electronic components of motor control systems may be mounted on printed circuit boards. (Novatek, Inc.)

4. Programmable controllers.
5. Direct current motor starters and speed controllers.
6. Universal motor controls.
7. Induction motor starters.
8. Induction motor speed controllers.

SOLID STATE RELAYS

Solid state relays (SSRs) do the same job as electromechanical relays (EMRs). With an EMR, a control voltage is applied to a coil. The coil becomes an electromagnet. It moves electrical points to make or break contact. This action closes or opens a current path in the load circuit. *The control circuit is isolated from the load circuit. They are two separate circuits with no electrical connection between them.*

The main parts of a solid state relay are:

1. A light-emitting diode (LED).
2. A phototransistor.
3. An SCR or a triac.

A block diagram of a solid state relay is shown in Fig. 15-4. Application of a control voltage turns on the LED. The glow from the LED affects the phototransistor. When light strikes its sensitive surface, its normally high resistance drops to near zero. The transistor turns on like a switch. This, in turn, triggers the

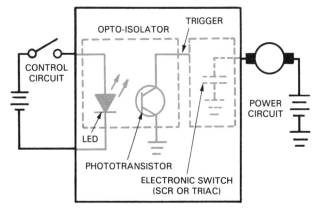

Fig. 15-4. An opto-isolator couples the control circuit to the power circuit with light.

SCR or triac. The action of the SCR or triac closes the path for the load current.

Solid state relays are also available with normally-closed load paths. They open when the control signal is applied. In either case, there is no connection between the control circuit and the load circuit. This use of light is called OPTO (for optical) ISOLATION. The combination of an LED with a phototransistor is often called an OPTO-COUPLER.

ELECTRONIC TIMERS

Some solid state relays are constructed with time delays. There is a time lapse between the application of the control signal and the turn-on (or turn-off).

There are two ways of delaying relay operation:
1. One method is to use a capacitor for timing.
2. The other is to use a digital counting circuit.

CAPACITOR TIMING

A capacitor, a rheostat and a unijunction transistor are in the triggering circuit. This is illustrated by Fig. 15-5, view A. When the trigger is activated, the capacitor begins charging.

The rate at which it charges depends on the amount of resistance in series with it. The use of a rheostat makes this rate adjustable. At some point, the capacitor voltage reaches the threshold value of the unijunction transistor. See Fig. 15-5, view B. *The threshold voltage is measured from the terminal marked E (for emitter) to the terminal marked B1 (for base 1).* When threshold is reached, a current path closes between terminals B1 and B2 (for base 2).

COUNTING

A counting time delay relay (TDR) starts with an oscillator. An oscillator is an electronic circuit that produces a series of voltage pulses.

Refer to Fig. 15-6. The duration (width) and spacing of the pulses are very precise. Setting the delay time is actually setting the number of pulses to be counted. The control signal starts the oscillator. When the right number of pulses have been counted, the SCR or triac is triggered.

The counting type TDR is more accurate than the capacitor type. Capacitor TDRs are good for short delays, from a fraction of a second to an hour. Counting types can produce delays of many hours or even days.

SOLID STATE STATIC LOGIC

In the ordinary sense, logic is a way of thinking. It means coming to a reasonable conclusion. We do it all the time. As we analyzed the control circuits of Chapters 13 and 14, we saw a natural relationship between every cause and effect. We were, in fact, using "relay logic."

There are five basic concepts in relay logic:
1. AND.
2. OR.
3. NOT.
4. TIME DELAY.
5. MEMORY.

AND CONCEPT

For a diagram of AND logic using relay logic, refer to Fig. 15-7, view A. By tracing the current path, you can see that when both push button A and push button B are closed, power is applied to the load. Note that it takes line voltage to energize (power) relays A and B and contactor C.

Solid state logic is also called STATIC logic because there are no moving parts. The job of relays A and B are handled by a single unit called an "AND gate."

Fig. 15-7, view B, shows the symbol for a two-input AND gate. AND gates may have any number of inputs, but two is most common. A logic element that contains 2 two-input ANDs is shown in view D, Fig. 15-7. Not all manufacturers use the NEMA standard logic symbols. Many have developed their own.

Inputs and outputs of logic gates are voltages at a low level. It is called logic-level voltage. This voltage exists between a terminal and

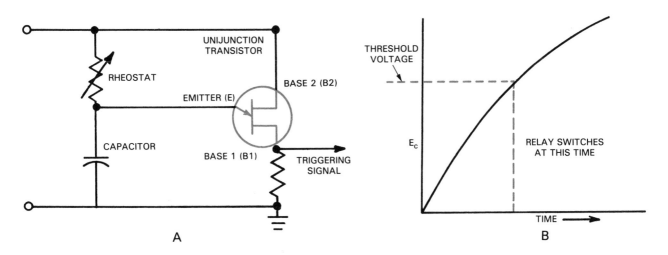

Fig. 15-5. The unijunction transistor is the electronic equivalent of a snap-action switch. It turns on when the capacitor has charged to the transistor's breakover voltage, as applied to the emitter terminal.

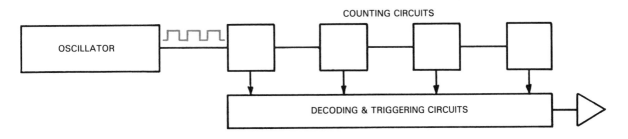

Fig. 15-6. A counting-type time delay relay activates after a set number of pulses from a constant-frequency oscillator have been counted.

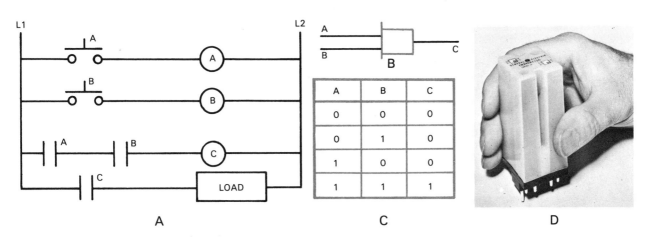

Fig. 15-7. AND gate and its relay logic. A—AND function, using delay logic. B—NEMA symbol for AND logic. C—Truth table for AND logic. D—Static logic AND gate.

ground. However, the ground line is not shown on diagrams.

Different manufacturers use different logic-level voltages. In studying logic control diagrams, therefore, you will not find voltage values. The numerals "1" and "0" are used instead. A logic 1 means that the voltage exists. A logic 0 means that it does not.

AND gate operation

The AND gate operates as follows:
1. The output is a logic 1 only when all inputs are logic 1.
2. The output is a logic 0 if any of the inputs is logic 0.
3. The operation of a logic gate can also be given as a table of all possible inputs and the resulting output. Such a tabulation is called a "truth table." The truth table for AND is shown in Fig. 15-7, view C.

Fig. 15-8 is a logic circuit like the relay logic circuit of Fig. 15-7, view A. Pressing push button A puts a logic 1 at one of the inputs of the AND gate. Pressing push button B puts a logic 1 at the other input. With both inputs at logic 1, the output is logic 1. This output can then be used to trigger a solid state relay.

In logic diagrams, the relay may be represented by a triangle. *This symbol represents an amplifier. Solid state relays are often referred to as amplifiers.* The amplifier closes the current path to contactor C.

OR CONCEPT

The relay logic for the OR function is shown in Fig. 15-9, view A. Trace the current through the circuit. You can see that power will be applied to the load (through contactor C) when either push button A or push button B is pressed.

The symbol for an OR gate is shown in view C, Fig. 15-9. Although a two-input OR is shown, OR gates may have three or more inputs.

An OR gate operates as follows:
1. The output of an OR is logic 1 if any of its inputs is a logic 1.
2. The output of an OR is logic 0 only if all of its inputs are logic 0. The OR truth table is given in Fig. 15-9, view B.

The solid state equivalent of Fig. 15-9, view A, is shown in view D of Fig. 15-9. Pressing either push button puts a logic 1 at one of the inputs of the OR gate. Its output becomes a logic 1. The logic 1 input to the amplifier closes the current path to the contactor coil. Load current flows through contact C.

NOT CONCEPT

The relay logic for the NOT function is shown in Fig. 15-10, view A. There is load current through the relay only when push button A is not pressed. When push button A is pressed, Relay A becomes energized and its normally-closed contacts open. This removes power from contactor C, opening its contacts.

The output of a NOT is always opposite from its input. This is called INVERTING. The NEMA symbol for the NOT gate is shown in view B of Fig. 15-10. Fig. 15-10, view C, is the truth table for a NOT function.

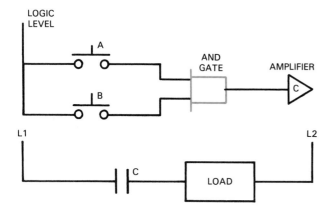

Fig. 15-8. Contactor C will be energized (receive current) only when both push button A and push button B are pressed.

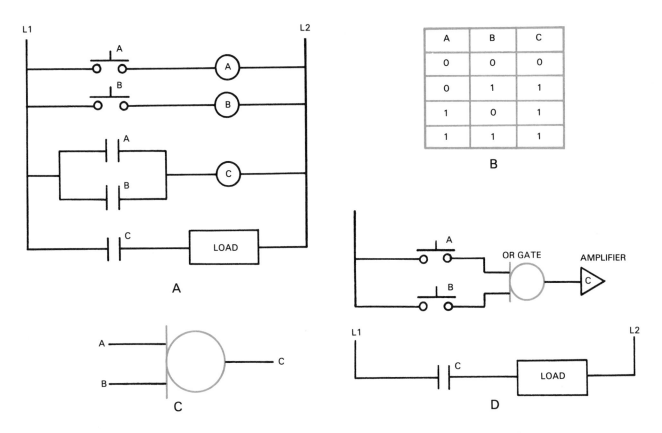

A	B	C
0	0	0
0	1	1
1	0	1
1	1	1

B

Fig. 15-9. The OR gate. A—OR function using relay logic. B—Truth table for OR gate. C—NEMA symbol for OR gate. D—Contactor C will be energized when either push button A or push button B is pressed.

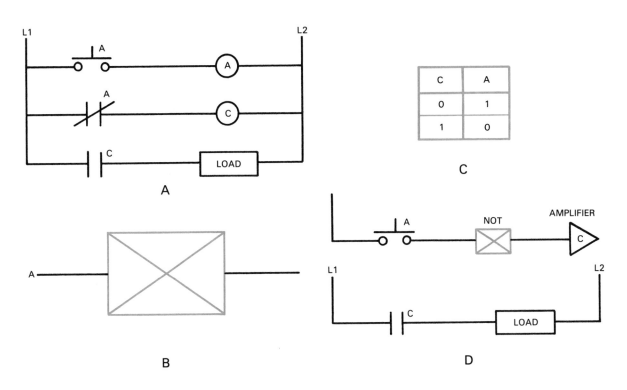

C	A
0	1
1	0

C

Fig. 15-10. The NOT gate. A—NOT function using relay logic. B—NEMA symbol for NOT. C—Truth table for NOT logic. D—Contactor C will be energized only when push button A is not pressed.

View D of Fig. 15-10 shows a solid state logic circuit equivalent to view A. With a logic 0 at its input, the NOT output is logic 1. Therefore:

1. The "contacts" of the amplifier are closed.
2. Contactor C is energized.
3. Power is applied to the load.

Pressing push button A puts a logic 1 on the NOT input. This changes its output to logic 0. A logic 0 at the input of the amplifier opens its "contacts." Contactor C de-energizes and power is removed from the load.

ADDING INVERTER OUTPUT

Some manufacturers add an inverted output to their AND, OR, MEMORY, and TIME DELAY gates. In such cases, the gate symbol has a NOT symbol inside it. Fig. 15-11 shows such AND and OR gates with a truth table.

Note the shorthand way of representing an inverted signal. A signal represented by a letter with a bar over it is always opposite from the signal represented by the letter alone.

TIME DELAY

Logic Time Delay is the same as the capacitance time delay relay described earlier. Fig.

15-12 shows the symbols. Timed opening and timed closing are possible with either type.

MEMORY

A relay memory circuit is shown in Fig. 15-13. This is the standard start-stop line of a motor control ladder diagram. When push button A is pressed, a momentary current path is established through contactor C. An auxiliary C contact is in parallel with the push button. There is a continuous current path when the push button

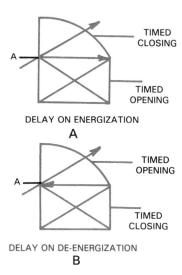

DELAY ON ENERGIZATION
A

DELAY ON DE-ENERGIZATION
B

Fig. 15-12. NEMA symbols for time delay. A—On-delay (for energizing). B—Off-delay (for de-energizing).

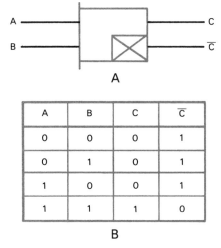

A

A	B	C	$\overline{C}$
0	0	0	1
0	1	0	1
1	0	0	1
1	1	1	0

B

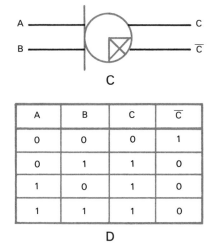

C

A	B	C	$\overline{C}$
0	0	0	1
0	1	1	0
1	0	1	0
1	1	1	0

D

Fig. 15-11. NOT gate. A—AND gate with additional inverted output. B—Truth table showing AND and AND-NOT (NAND) outputs. C—OR gate with additional inverted output. D—Truth table showing OR and OR-NOT (NOR) outputs.

is released. The circuit "remembers" that the push button was pressed.

You can make this circuit "forget" by pressing the OFF push button. The failure of power will de-energize coil C also. When power is restored, push button A must once again be pressed to energize coil C.

OFF-RETURN MEMORY

The symbol for the equivalent logic gate is shown in Fig. 15-14. It is called OFF-RETURN MEMORY because its standard output will go to a logic 0 whenever power is shut off. A logic 1 at the memory's ON input changes its standard output from logic 0 to logic 1. The logic 1 pulse can then be removed from the ON input and the standard output will remain a logic 1. Similarily, a logic 1 at the OFF input changes the standard output from a logic 1 to a logic 0.

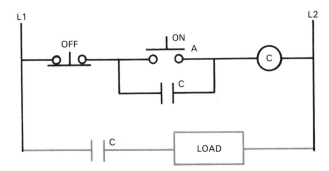

Fig. 15-13. Off-return memory using relay logic. Contacts return to normal condition when power is removed.

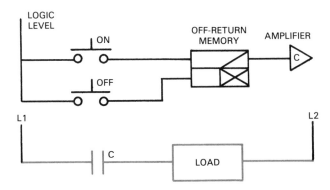

Fig. 15-14. NEMA symbol for off-return memory. This circuit is equivalent to Fig. 15-13.

RETENTIVE MEMORY

A second kind of memory gate, called RETENTIVE MEMORY, is equivalent to a latching relay. A latching relay has two coils. When the ON coil is energized, a mechanical latch (hook) holds the contacts closed (or open). Then, the power's being on or off has no effect on the position of the contacts. The only way to unlatch it is by energizing the OFF coil.

The symbol for retentive memory is shown in Fig. 15-15. If it is in the ON condition when power is removed, it will remain so when power is restored.

CONVERTER

In motor control systems, it is often necessary to use a 120 volt input to a logic system. This is done with a device called a converter. When 120 volts is applied to the converter's input, the converter makes the output a logic 1. The converter symbol is shown in Fig. 15-16.

To see how logic gates are used in control, refer to Fig. 15-17. View A repeats in color the

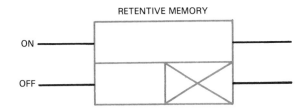

Fig. 15-15. NEMA symbol for retentive memory. It has the function of a latching relay. It remains in the ON position when power is OFF.

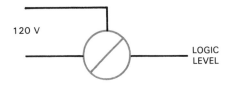

Fig. 15-16. Symbol for logic converter. The converter produces a low-level voltage output when 120 V line voltage is applied to its input.

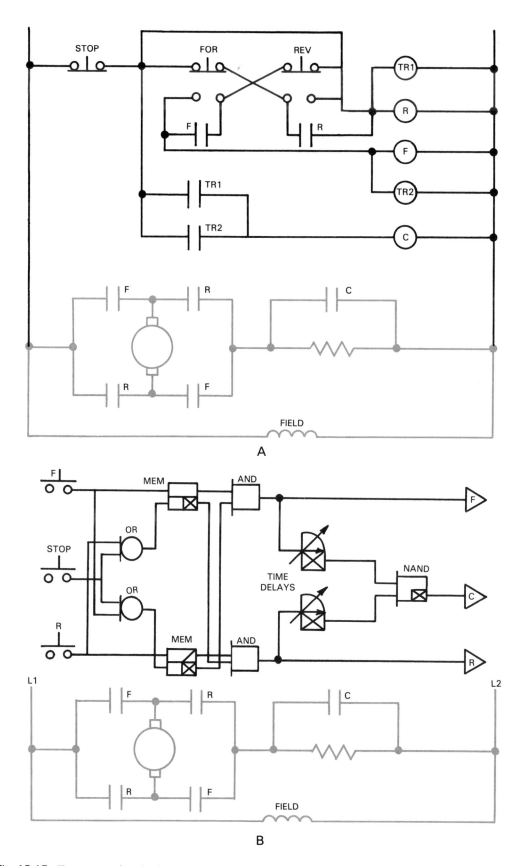

Fig. 15-17. Two-step reduced voltage starter. A—With electromechanical components. B—With logic gates.

electromechanical diagram of Fig. 14-21. This is a two-step reduced voltage motor starter. View B shows the logic circuit that does the same job. The power circuit is the same for both.

INTEGRATED CIRCUIT LOGIC GATES

The logic gates discussed so far have been relatively large. They are about the same size as electromechanical relays. However, there is another family of logic gates. Microengraving techniques can put complex circuits consisting of hundreds (or even thousands) of resistors and transistors on a small chip of semiconductor material. The integrated circuit shown in Fig. 15-18 is only 1 cm long and contains 4 two-input AND gates.

Not only do integrated circuit logic gates look different, their logic symbols are drawn differently. They are generally called MIL (for military) symbols.

Fig. 15-19, view A, shows the MIL symbol for a two-input AND gate. An integrated circuit AND gate can have a standard output or an in-verted output. It cannot have both. The symbol of view A has a standard output and is therefore simply an AND gate. The symbol of Fig. 15-19, view B, on the other hand, has an inverted output and is known as a NAND (for AND-NOT) gate. The small circle (called a bubble) at the output means that the signal is inverted.

The MIL symbol for an OR gate is shown in Fig. 15-20, view A. With an inverted output, like the gate in view B of Fig. 15-20, it is a NOR (for OR-NOT) gate.

The NOT logic function is performed by an integrated circuit known as an INVERTER. The inverter MIL symbol, Fig. 15-21, is similar to the amplifier symbol except for the bubble at the output.

There is one additional gate, the EXCLUSIVE OR (XOR). See Fig. 15-22. The output of an XOR is logic 1 if either input is logic 1, but not both. The output of an XOR is logic 0 if both inputs are the same.

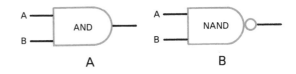

Fig. 15-19. Military (MIL) symbols for AND and NAND gates. A—AND gate. B—NAND gate.

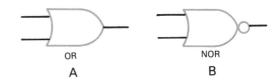

Fig. 15-20. Military (MIL) symbols for OR, NOR gates. A—OR gate. B—NOR gate.

Fig. 15-18. Semiconductor chips like this one (arrow) often contains several logic gates. This one is about the width of a little fingernail. (Hampden Engineering Corp.)

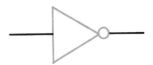

Fig. 15-21. MIL symbol of inverter (NOT gate).

An example shows one use for an XOR gate, Fig. 15-23, view A. With this circuit, a load can be turned on and off by either of two single-pole, double-throw, three-way switches. The logic statement for this circuit is: If switch A and switch B are both "up," or if both switch A and switch B are "down," the load is energized.

The diagram in view B, Fig. 15-23, shows the logic circuit using NEMA symbols. Note that load current does not flow through the logic gates. Only the contactor carries load current. We see in view C, Fig. 15-23, that an XOR gate can replace the AND, NOR and OR gates. In each of the three control methods, the power circuit is the same.

There is no special MIL symbol for time delay. Symbols for delay circuits usually include the components themselves. Likewise, there is no single symbol for memory. However, many types of integrated circuit chips do have memory. They are called LATCHES or FLIP-FLOPS. The R-S (for reset-set) flip-flop is shown in Fig. 15-24. It has both a standard and an inverted output. A logic 1 at the S input puts a logic 1 at the standard output. A logic 1 at the R input changes the standard output to a logic 0. Additional circuitry is needed if the memory is to be the "off-return" type.

PROGRAMMABLE CONTROLLERS

We have seen that electronic switching can replace electromechanical switching in many applications. However, logic gates can do more

than route control signals through a system. They can change function of other gates.

In Fig. 15-25, XOR gates have been added to the A and B inputs of an AND gate. The second input of each XOR gate is a "program" input. A two-digit code changes the relationship between the output and the inputs of the AND gate. If

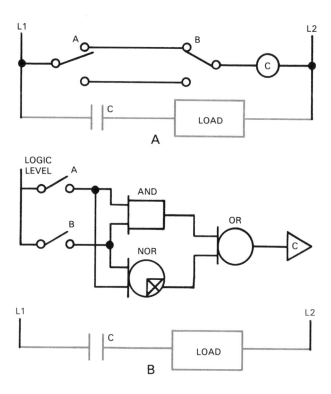

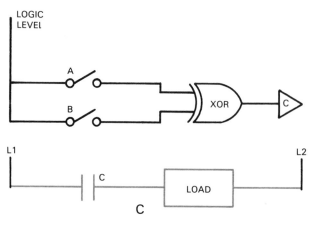

Fig. 15-23. All three control circuits allow load to be turned on and off from two locations. A—Ladder diagram. B—Solid state logic using NEMA symbols. C—Solid state logic using XOR gate.

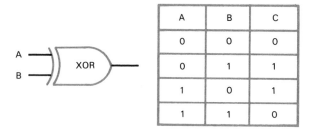

A	B	C
0	0	0
0	1	1
1	0	1
1	1	0

Fig. 15-22. MIL symbol and truth table for Exclusive OR (XOR).

the program code is "00," the circuit follows the truth table of an AND. If, however, the program code is "11," the circuit follows the truth table of a NAND gate.

The heart of a programmable controller is a microprocessor. A microprocessor is a semiconductor chip that contains an enormous number of electronic switches. The actual circuits are far more complicated than the simple example in Fig. 15-25. The result is a solid state controller that can perform the functions of many different controllers. It is up to the programmer to decide which functions are needed.

Manufacturers have made programming easy. The control system is first worked out with electromechanical symbols. The programmer works from a ladder diagram. The program specifies:
1. The location of each device in the circuit.
2. The type of device such as coil, N.O. (normally open) contacts, N.C. (normally closed) contacts or others.

In some cases, a programmer presses a key with a symbol of a coil to produce the electronic equivalent of a coil. Fig. 15-26 shows the logic portion of a typical programmable controller. Fig. 15-27 shows its programming console.

Power level inputs to programmable controllers are often called ORIGINAL INPUTS. Some apply 120 volts to produce action (normally open). Others remove the voltage to produce action (normally closed). In the same

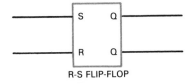

R-S FLIP-FLOP

Fig. 15-24. Symbol for R-S latch. A logic 1 at input S turns it on. A logic 1 at input R turns it off.

PROGRAM CODE

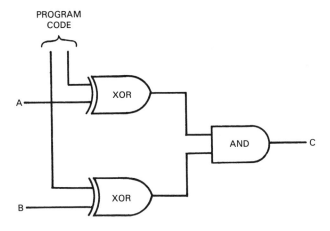

| A | B | PROGRAM CODE | |
| | | 00 | 11 |
		C	C
0	0	0	1
0	1	0	1
1	0	0	1
1	1	1	0

Fig. 15-25. An AND gate can be programmed to also act as a NAND gate by using XORs.

Fig. 15-26. The logic circuits of programmable controllers are typically housed in cabinets like this.   (Square D Co.)

manner, the outputs are equivalent to normally open and normally closed contacts.

DC STARTERS AND SPEED CONTROLLERS

All motor starters discussed so far have been the "on-off" type. The motor is started either in several steps or by connecting it directly across the line.

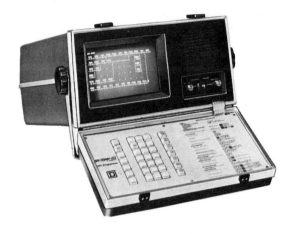

Fig. 15-27. Some programmable controllers use electromechanical symbols and ladder diagrams on a video screen. (Square D Co.)

There is another type of starting called SOFT-START. A motor that is soft-started increases speed smoothly from zero to full speed. There are no sudden jolts to connected equipment. Current inrush is limited. Motor windings do not have a chance to overheat.

Direct current motor soft-starters and speed controllers are usually combined. The device is known as a dc motor drive. Fig. 15-28 is a simplified diagram of a typical dc motor drive. The input is single-phase ac. It may be 120, 208/240 or 380 volts.

A control transformer isolates the control circuit from the power circuit. Part of the 120 volt supply from the control transformer is rectified to provide excitation current to the field coil. There is also a built-in dc power supply for the triggering circuits.

The bridge rectifier supplying the armature with direct current is SCR controlled. The resistance-capacitance network shown in parallel with the SCRs is called a SNUBBER. Any sharp voltage "spike" is absorbed by the snub-

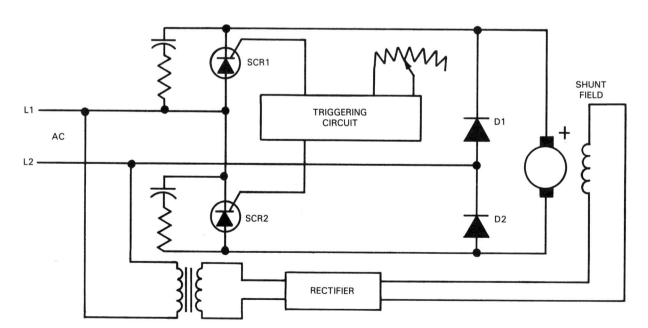

Fig. 15-28. Simplified schematic diagram of solid state dc motor drive. It is designed for smooth starting of motor.

ber. This prevents a false turn-on of the SCR. A spike is a very large change in voltage in a very short time. A spike is often called dv/dt (read DVDT). In higher mathematics, dv/dt is the symbol for "differential of voltage with respect to time." In other words, "rate of change of voltage."

Refer again to Fig. 15-28. During each positive half cycle, current flows through SCR_1, the armature and diode D_2. During each negative half cycle, current flows through diode D_1, the armature (in the same direction) and SCR_2. Each SCR is in control during alternate pulsations. Both SCRs are turned on by the same trigger.

TRIGGERING THE SCR

An SCR is like a diode in that it can conduct current in one direction only. Unlike a diode, it does not conduct at all until it is triggered.

The triggering of an SCR is done by applying a voltage pulse to its "gate" terminal. Once an SCR is triggered, its gate loses control. The SCR continues to conduct until the current falls to a very low value.

An oscillator circuit produces a triggering pulse every half cycle. The control circuit determines the exact point at which the trigger pulse appears. Fig. 15-29 shows the result of triggering at electrical angles of 45, 90 and 135 degrees of each half cycle. The angle at which an SCR is triggered is called the "firing angle."

No matter what the firing angle, the SCR must turn itself off at the end of each half cycle. However, the load it is controlling in Fig. 15-28 is a motor which has inductance. Inductance, as we know, makes current lag voltage. Current (produced by the collapsing magnetic field of the armature) continues to flow after voltage reaches zero. This current is circulated through D_2 and D_1 so that it will not affect the turn off point of the SCRs. Diodes D_1 and D_2 are known as FREEWHEELING diodes.

Two factors determine the exact trigger point. One is the setting of the speed control rheostat. The other is armature current.

Armature current is low when speed is high but high when speed is low. By using armature current as an automatic adjustment of firing angle, motor speed is held relatively constant under changing load. For greater accuracy, a tachometer generator can be used. Its voltage output, which is an exact measure of speed, controls the triggering oscillator.

The solid state device that applies the triggering pulse to the SCR is often a diac. The word diac is an abbreviation for "diode ac switch." The diac is a self-triggering switch that conducts current in either direction. The symbol for a diac is shown in Fig. 15-30. It triggers itself

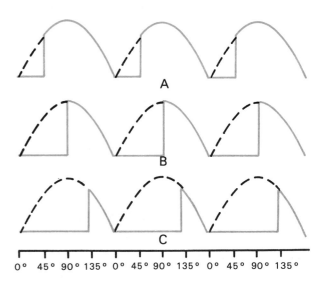

Fig. 15-29. Wave forms of voltage applied to the motor at three different speed settings. The SCR is triggered at different times of the cycle. A—45°. B—90°. C—135°.

Fig. 15-30. Symbol of a diac (diode ac switch).

when the voltage across it reaches the "break-over" value. Once triggered, the diac delivers a sharp triggering pulse to the SCR.

SOFT-START AND OTHER DRIVE FUNCTIONS

When power is first applied, the SCRs turn on late each cycle. A current limit adjustment prevents a large current inrush. The triggering oscillator has a timer that decreases the firing angle smoothly until the motor reaches the set speed. This is called "ramping." Ramp times are usually adjustable, typically from zero to 30 seconds.

Commercial dc drives, like the one shown in Fig. 15-31, have many standard and optional features. Among these are:
1. Protection against low line voltage, field loss, overcurrent and phase loss (on three-phase models).
2. Adjustments for maximum speed limit, minimum speed limit, torque limit and IR compensation. IR compensation boosts the applied voltage to make up for the increased voltage drop due to increased current under load.
3. Jog-run selection that permits jogging at set speed.

4. Reversing.
5. Braking.

Most dc drives provide dynamic braking. Some models combine reversing with regenerative braking. These have reversing SCRs that are connected in inverse parallel with the forward SCRs. See Fig. 15-32. When the speed setting is lowered, the motor becomes a generator for a time, while it is being driven by the load. During this time, the forward SCRs turn off and the reversing SCRs turn on. This returns power to the ac line.

UNIVERSAL MOTOR CONTROL

The speed of an ac universal motor is directly proportional to the voltage applied to it. If voltage is low, speed is low. If voltage is high, speed is high.

The solid state speed control circuit is similar to that of a dc motor. The main difference is that a "triac" is used instead of an SCR. The symbol for a triac (triode ac switch) is shown in Fig. 15-33.

A triac combines features of the diac and the SCR:
1. Like a diac, the triac conducts in both directions.
2. Like an SCR, a triac conducts only when triggered.

Fig. 15-31. Typical dc motor drive.
(Hampton Products Co., Inc.)

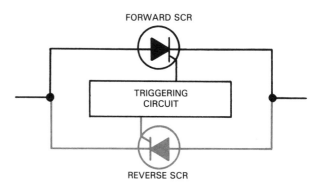

Fig. 15-32. Forward SCRs conduct while running. Reverse SCRs conduct while braking.

Fig. 15-34 shows a simplified control circuit that could be used for an ac universal motor or a shaded-pole motor. Similar ac power controls are used for light dimmers and electric heater regulators.

During each cycle, as the sine wave crosses zero in the positive-going direction, the capacitor begins charging through the rheostat. A certain number of electrical degrees into the cycle, the capacitor will be charged to the breakover voltage of the diac. The diac triggers, applying this voltage to the gate of the triac. The capacitor discharges through the diac and triac, ready to be charged again during the next half cycle. Meanwhile, the triac remains on until the sine wave reaches zero again in a negative-going direction. There, the sequence begins again, but with the opposite polarity.

The resistance setting of the rheostat determines how long it takes the capacitor to reach the diac's breakover voltage. This, in turn, determines how long the triac will remain on

Fig. 15-33. Symbol for triac (triode ac switch).

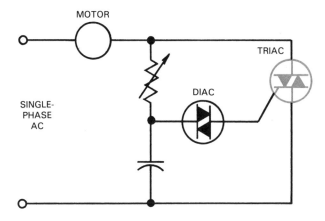

Fig. 15-34. Simplified diagram of ac motor (universal or shaded-pole) speed control.

during each cycle. The average applied voltage, therefore, depends on the rheostat setting.

INDUCTION MOTOR STARTERS

Large three-phase induction motors are never started across the line. The large current inrush would put too great a strain on the power supply, the load and the motor itself. Many types of electromechanical induction motor starters are available. Among these are:

1. Resistance start.
2. Reactance start.
3. Autotransformer start.
4. Part-winding.
5. Wye-delta.

All of these involve one or more steps from a reduced voltage to a full voltage. Solid state ac motor starters, however, produce soft starting. The stator voltage increases smoothly from zero to full voltage.

Alternating current motor starters use SCRs as do the dc motor drives. They are not used for rectification, however. Each line has two SCRs connected in inverse parallel. This connection is sometimes called BACK-TO-BACK. Refer to Fig. 15-35.

One SCR controls the positive alternation while the other controls the negative. The control operation is similar to that of a triac. However, SCRs are available with much more power-handling capability than triacs. The result is full control of ac, as shown by the wave forms of Fig. 15-36.

There are three ways that the triggering circuits can control the smooth application of power. One is based on time only. The second is based on time and current. The third is based on time and torque.

A block diagram of a solid state induction motor starter is shown in Fig. 15-37. Note that,

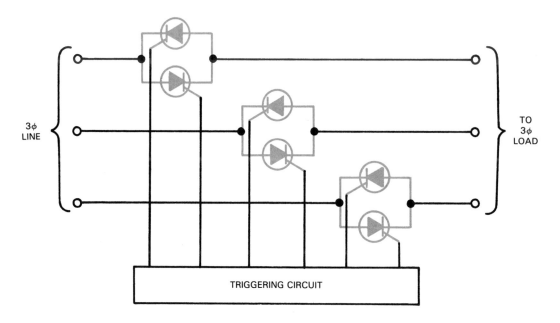

Fig. 15-35. Alternating current motors often use SCRs connected in inverse parallel.

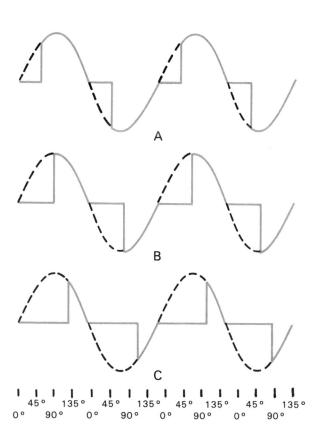

Fig. 15-36. Wave forms of voltage applied to the motor as the starter ramps the voltage up. Both halves of each cycle are triggered at the same number of degrees from the zero crossing. A—45°. B—90°. C—135°.

for control, it uses a combination of time, current and torque.

TIME

After the start push button is pressed (at t_0), voltage increases at a constant rate. See Fig. 15-38, view A. The rotor starts turning (at t_1) only when the voltage reaches the value that produces breakaway torque. Speed increases sharply, reaching rated speed at t_2.

CURRENT

The triggering circuits include current sensors. The maximum value of current is preset. Current limit is set well above the full-load current value.

After the start push button is pressed (at t_0), voltage ramps up until the preset current is reached (at t_2). The motor speeds up as shown in Fig. 15-38, view B.

When the motor approaches full speed, current to it decreases. It falls below the set current level shown at t_2. At t_3 the motor reaches full speed at normal current.

Solid State Motor Control

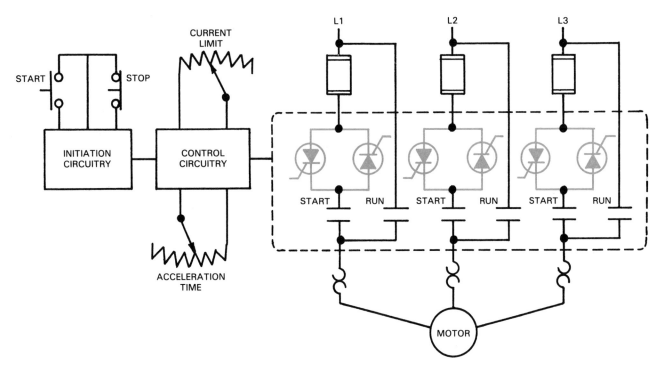

Fig. 15-37. Solid state soft starter for ac induction motor.

TORQUE

Control based on torque receives a tachometer signal representing motor speed. It can supply needed torque (up to full locked-rotor torque) the instant the push button is pressed. Then, when rotation is sensed, it reduces motor current immediately. This fast response assures breakaway, but eliminates shock to the driven load. Fig. 15-38, view C shows how voltage and speed vary with time. Fig. 15-39 compares the speed-torque curve of a NEMA design B motor using soft-start with the standard speed-torque curve.

VARIABLE FREQUENCY SPEED CONTROL

Normally, ac induction motors are run at constant speed. This speed is based on the motor's natural synchronous speed. *Synchronous speed is the speed of the stator's revolving magnetic field.*

The equation for synchronous speed follows:

Synchronous speed (rpm)
$$= \frac{\text{frequency (Hz)} \times 60 \text{ (sec/min)}}{\text{N (pairs of poles)}}$$

A four-pole motor running on 60 Hz ac, therefore, has a synchronous speed of 1800 rpm. With a full-load slip of 50 rpm, actual motor speed is 1750 rpm.

The speed of induction motors cannot be controlled by controlling applied voltage. Wound-rotor motors are controlled with resistance in the rotor circuit. The speed of induction motors can only be controlled by controlling frequency. Solid state variable frequency drives can provide NEMA design B squirrel-cage motors with the speed characteristics of a wound-rotor motor.

A problem that arises with varying the frequency is the inductive reactance, X_L, of the stator windings. The lower the frequency, the lower the reactance. Lower reactance means higher motor current, which could cause over-

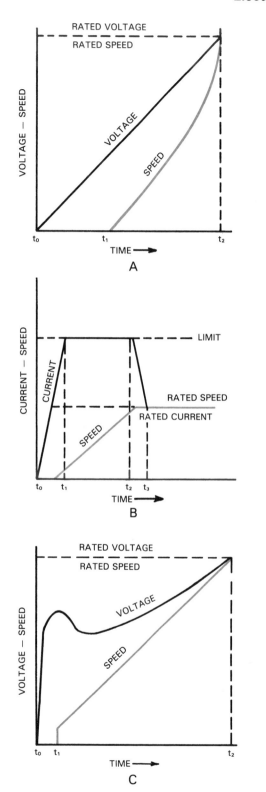

heating. Therefore, variable frequency drives must also vary applied voltage. Voltage must be reduced as frequency goes down. As long as the ratio between voltage and frequency is constant, current will be constant. Constant current produces a constant torque.

Industrial three-phase motor drives are powered by single-phase ac, typically 230 volts. This is shown by the block diagram of Fig. 15-40. Incoming power is rectified to a direct current. The rectifier is a standard SCR-controlled full-wave bridge of the type shown in Fig. 15-28. The rectifier output is filtered by an LC filter. The result, an adjustable dc voltage, is fed to the "dc bus."

Direct current from the bus is fed to three switching modules of the inverter, Fig. 15-41. Each module contains two POWER TRANSISTORS. The transistors are turned on and off by a circuit called a RING COUNTER. This circuit is controlled by an oscillator. The ring counter distributes firing pulses to the transistors in the

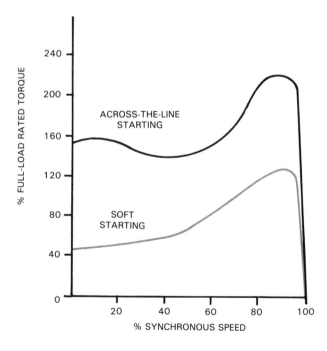

Fig. 15-39. Speed-torque curves of a NEMA B induction motor with and without soft starting.

Fig. 15-38. Curves for ac motor soft starters. A—Simple voltage vs time type starter. B—Current limit type starters. C—Torque-based starter.

Solid State Motor Control

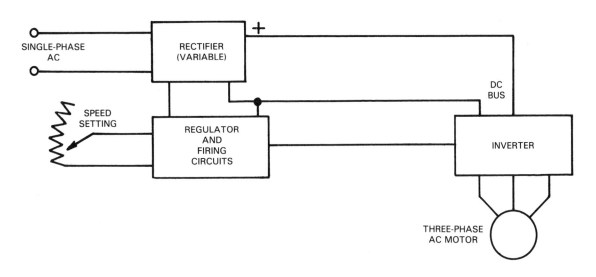

Fig. 15-40. Block diagram of variable frequency ac motor speed controller.

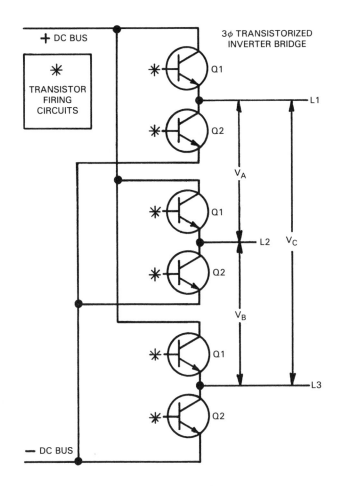

Fig. 15-41. This inverter module changes direct current into an alternating current by turning the transistors on and off at the proper time.

correct order to produce three wave forms 120 electrical degrees apart.

Consider one module. When both transistors are off, output is zero. When transistor Q_1 turns on, output voltage steps up to the value of the upper line. The upper line is positive with respect to zero. When transistor Q_2 turns on, output voltage steps down to the value of the lower line. The lower line is negative with respect to zero. The output, then, is an alternating voltage that is not a sine wave. It is more like a rectangular wave. This voltage, along with the resulting current is shown in Fig. 15-42. This variable

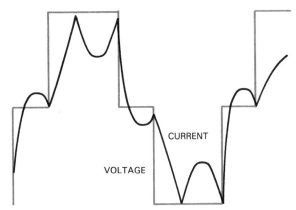

Fig. 15-42. Voltage and current wave forms from one inverter module.

frequency drive design is often referred to as a "six-step" inverter.

The speed adjustment controls two factors at the same time. First, it sets the value of the output voltage. Second, it sets the frequency of the control oscillator. As a result, voltage and frequency go up and down together. When the motor is started, stopped or when the speed setting is changed, the "accel/decel" circuit takes over. The output power is ramped up or ramped down at a preset rate until the desired speed is reached.

Many variable frequency drives contain the following protective circuits:

1. Overcurrent. There is usually a maximum current adjustment. If a mechanical or electrical overload causes current to exceed the value set, a trip will open the circuit immediately.
2. Low voltage. If voltage falls below a predetermined value, a trip occurs after a 10-cycle delay.
3. Overvoltage. Overvoltage can occur if the load begins to drive the motor. The system trips instantly.

Braking, reversing and power factor correction are functions that may be added. The power factor corrector automatically reduces applied voltage when there is no load. Many factory machines (drill presses, for example) run continuously. They are loaded only when used, which is a small percentage of their running time. An unloaded induction motor has a very poor power factor. Reducing voltage reduces reactive current. This improves power factor, resulting in energy savings of up to 50 percent.

MICROPROCESSOR-BASED DRIVES

The odd-shaped wave forms of Fig. 15-42 will operate most motors satisfactorily. However, a true sine wave current produces less wasted energy. Fig. 15-43 shows the wave form of a microprocessor-based drive. Extremely fast electronic switches produce up to 1200 pulses per cycle. The amplitude (size) and duration of these pulses are under control of a microprocessor. The microprocessor computes the number of pulses per cycle needed to generate a sine wave output.

SUMMARY

The main characteristic of solid state control devices is that switching takes place within solid material.

Typically, a solid state relay opens or closes a single load current path when a signal between 3 and 24 volts is applied to its input. The load circuit is electrically isolated from the control circuit.

Some electronic time delay relays use a resistor-capacitor network. Others produce pulses at a fixed rate and count the number of pulses.

Solid state logic controllers use low-voltage, low-current signals to operate "gates" for the decision-making part of the control system. The basic logic functions are AND, OR, NOT (inverter), TIME DELAY and MEMORY.

Logic converters change a power level input to a logic level. Amplifiers convert a logic level output to a power level.

Programmable controllers use logic circuits to produce functions that are equivalent to electromechanical control.

The principle component of both ac and dc motor starters and drives is the silicon con-

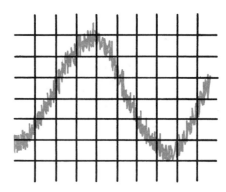

Fig. 15-43. Current wave form produced by a microprocessor-based variable frequency drive.

trolled rectifier (SCR). SCRs can be turned on at any point during an ac cycle or dc pulsation. They turn off when the current through them is near zero.

A soft starter is a type of reduced voltage starter, except that there is a smooth, instead of a step, increase in voltage.

DC motor starters and drives increase voltage by increasing the amount of time during each pulsation that the SCR is turned on.

Speed of ac induction motors is controlled by controlling frequency. Sixty hertz ac is rectified to dc, then inverted into three-phase ac. The ac, however, has a square wave form instead of a sine wave.

TEST YOUR KNOWLEDGE

1. List at least four categories of solid state control devices and systems.
2. Give an example of AND logic; OR logic; NOT logic; Memory.
3. A _____ provides the OFF-RETURN memory function in an electromechanical control system.
4. Draw the NEMA and MIL symbols for AND, OR, NOT, NAND and NOR.
5. How is the average value of dc voltage varied in motor starters and speed controllers?
6. Explain how SCR turn off at zero voltage is accomplished in circuits having an inductive load.
7. The term "ramping" means:
   a. A late firing of the SCR on a dc speed controller.
   b. Supplying excitation current to field coil of a dc starter.
   c. Timing action or timing device on the triggering oscillator for a soft-start SCR which smoothly decreases the firing angle of the oscillator until the electric motor reaches set speed.
   d. None of the above.
   e. All of the above except "b."
8. What is the advantage of using two SCRs connected in inverse parallel over using a triac?
9. _____ and _____ determine the synchronous speed of an induction motor.
10. Why must voltage be reduced along with frequency in a variable frequency speed controller?
11. What is an inverter? How is it used in variable frequency ac induction motor drive?
12. Sketch the motor voltage wave forms from a six-step inverter type ac motor drive and a microprocessor-based drive.

METRIC UNITS OF MEASUREMENT

The Conventional system of measurement was developed over many centuries in England. In fact, it is hardly a system at all. It is a mixture of definitions. We do not even know how some of the units were determined.

About 150 years ago, scientists wanted a more precise way of making measurements. They wanted units based on natural laws, which never change. Over the years, several other measurement systems have been developed. To a large degree, these will all be replaced by the "Systemé International d'Unités," SI metric for short.

Fortunately, the most common electrical units: the volt, the ampere, the ohm, the watt, the henry, and the farad have not changed. They mean the same today as they did 100 years ago.

In SI metric units, multiples and submultiples carry prefixes according to Table 1:

TABLE 1: PREFIXES IN METRIC

MULTIPLIER	EXPONENT FORM	PREFIX	SYMBOL
1 000 000 000 000	10^{12}	tera	T
1 000 000 000	10^9	giga	G
1 000 000	10^6	mega	M
1 000	10^3	kilo	k
10	10^1	deka	da
0.1	10^{-1}	deci	d
0.01	10^{-2}	centi	c
0.001	10^{-3}	milli	m
0.000 001	10^{-6}	micro	u
0.000 000 001	10^{-9}	nano	n
0.000 000 000 001	10^{-12}	pico	p.

For example a kilowatt is 1000 watts while a microfarad is one millionth (0.000 001) of a farad.

There are two areas of electrical power where older systems are still in common use. The first is mechanical torque and power. The second is magnetic units.

In the Conventional system, the unit of mass is the POUND (lb.). The unit of distance is the FOOT (ft.). In technical terms, work is accomplished when mass is moved through some distance. One FOOT-POUND of work is done when a mass of 1 lb. is lifted a distance of 1 ft. If this takes place in a time of 1 second, 1 ft.-lb. per second of power is used. Now the unit "1 ft.-lb. per second" is a mouthful. Besides that, it does not represent much work. Through experience, it was determined that a work horse could lift 550 lbs. a distance of 1 ft. in 1 second. That is where we get the term HORSEPOWER for mechanical power.

Motors in the United States are rated in horsepower. In SI metric units, mechanical power is expressed in WATTS. There are 746 watts per horsepower. This conversion (from horsepower to watts and vice versa) had to be done often, even when working in Conventional units. It is not new. What is new is the way in which torque is expressed. Torque is a force applied in a way that tends to cause rotational motion.

In the Conventional system, force was expressed in POUNDS. Distance from the pivot (or shaft center) was expressed in FEET. Therefore, the basic unit of torque was the POUND-FORCE-FOOT.

In SI metric units, the unit of force is the NEWTON. The newton is that force which gives a mass of one kilogram an acceleration of one metre per second. That is hard to visualize. It works out to about 4.45 newtons per pound force. The unit of distance is the METRE (which is why the system is called "metrics"). One newton applied one metre from the center produces one NEWTON METRE of torque. One newton metre per second, then, is one watt of power.

A pound-force-foot is a little larger than a newton metre. Therefore, you multiply pound-force-feet by 1.356 to find torque in newton metres.

MAGNETISM

For those working with magnetism, the problem is more complicated. Magnetism, itself, was discovered long before anyone knew that it can be caused by electricity. Early experimenters thought that they could isolate a "pole," like they isolated an electrical charge. Now we know that magnetic lines are in closed loops and that poles, as such, do not exist.

Meanwhile, a whole system of magnetic units were developed. The starting point was the "unit pole." The system is the CGS (for centimetre-gram-second) EMU (for electromagnetic units).

The unit of magnetic field intensity (H) was the OERSTED. A field having an intensity of 1 oersted exerted force of 1 dyne on a unit pole. A flux line was called a MAXWELL. A flux density (B) of one maxwell per square centimetre was called a GAUSS. The magnetomotive force (mmf) was the GILBERT, which is one dyne of force per unit pole.

The starting point for SI metric magnetic units, also called MKS (for metre-kilogram-second) units is the WEBER. A weber of flux is that amount of flux that induces 1 volt into 1 turn of a conductor when it changes uniformly in one second. This is equivalent to 10^8 maxwells.

Instead of defining mmf in terms of actual force, the SI metric unit is AMPERE-TURNS. One ampere through one turn produces one ampere-turn of mmf. One gilbert equals 0.796 ampere-turns.

The oersted was equivalent to one gilbert per square centimetre. The SI metric unit for field intensity is the AMPERE TURN PER METRE.

One oersted equals 79.577 ampere turns per metre.

Flux density in SI metric is given in TESLAS. One tesla is one weber per square metre. One tesla equals 10,000 gauss. Table 2 summarizes these relationships.

TABLE 2: CONVERSION FACTORS

	CGS UNITS	SI UNITS	SYMBOL	CONVERSION
mmf	gilbert	ampere-turn	F	1 gilbert = 0.796 ampere turns
Field Intensity	oersted	ampere turn/metre	H	1 oersted = 79.577 ampere turns/metre
Flux	maxwell	weber	ϕ	1 weber = 10^8 maxwells
Flux Density	gauss	tesla (weber/metre²)	B	1 tesla = 10^4 gausses

SI metric units may never be adopted completely. While some units make computations easy, they are hard to use. An example is the radian. The radian is an angle—a portion of a circle. We are used to thinking of angles in degrees. There are 360 degrees in a circle. There are 360 electrical degrees in a cycle of alternating current, as well.

The SI metric definition of a radian is: When the arc of a circle is equal to its radius, the subtended angle is equal to one radian. One radian is approximately 57.3 degrees. There are approximately 6.28 (2π) radians in a circle.

In SI metric units, rotational speed is measured in radians per second. Since each revolution is 360 degrees, there are 2π (approximately 6.28) radians in each revolution. One hundred revolutions per minute (100 rpm), for example, equals 11.62 radians per second. This makes the power computation very simple. Mechanical power in watts equals torque in newton metres times speed in radians per second. We are so used to measuring speed in rpm, however, that radians per second may never be used.

Electrical Power

DRAWING SYMBOLS

Drawings are a way of communicating. Like a spoken language, drawings are made up of individual symbols put together in a way that makes sense. Very often, the symbols on a diagram of an electrical circuit say more to the person who understands them than several pages of words of explanation.

Following are some of the most commonly used diagram symbols. In some cases, more than one symbol is shown for the same device. There may even be variations, not shown here, that are also used. The differences are slight, however, and will be understandable if you are familiar with these.

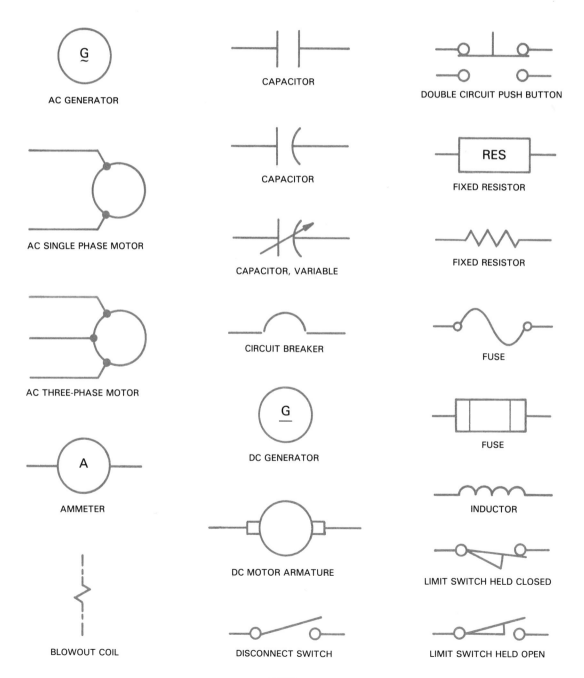

AC GENERATOR

AC SINGLE PHASE MOTOR

AC THREE-PHASE MOTOR

AMMETER

BLOWOUT COIL

CAPACITOR

CAPACITOR

CAPACITOR, VARIABLE

CIRCUIT BREAKER

DC GENERATOR

DC MOTOR ARMATURE

DISCONNECT SWITCH

DOUBLE CIRCUIT PUSH BUTTON

FIXED RESISTOR

FIXED RESISTOR

FUSE

FUSE

INDUCTOR

LIMIT SWITCH HELD CLOSED

LIMIT SWITCH HELD OPEN

Reference Section

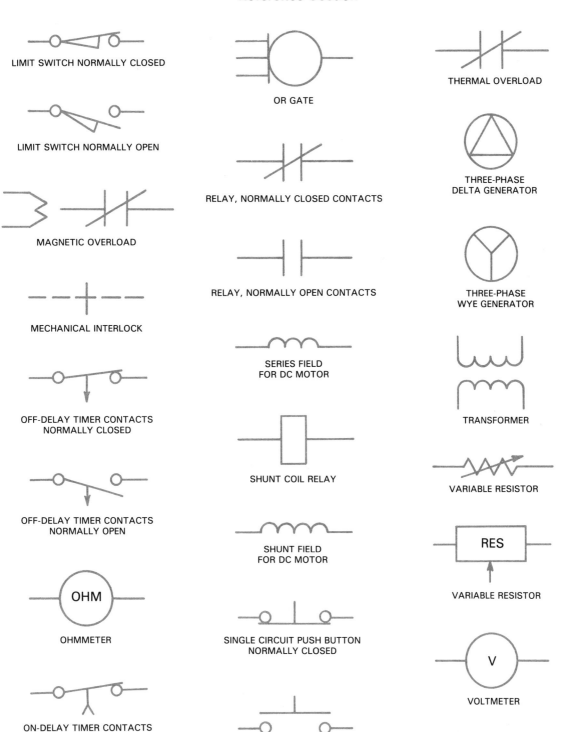

LIMIT SWITCH NORMALLY CLOSED

LIMIT SWITCH NORMALLY OPEN

MAGNETIC OVERLOAD

MECHANICAL INTERLOCK

OFF-DELAY TIMER CONTACTS
NORMALLY CLOSED

OFF-DELAY TIMER CONTACTS
NORMALLY OPEN

OHMMETER

ON-DELAY TIMER CONTACTS
NORMALLY CLOSED

ON-DELAY TIMER CONTACTS
NORMALLY OPEN

OR GATE

RELAY, NORMALLY CLOSED CONTACTS

RELAY, NORMALLY OPEN CONTACTS

SERIES FIELD
FOR DC MOTOR

SHUNT COIL RELAY

SHUNT FIELD
FOR DC MOTOR

SINGLE CIRCUIT PUSH BUTTON
NORMALLY CLOSED

SINGLE CIRCUIT PUSH BUTTON
NORMALLY OPEN

THERMAL OVERLOAD

THERMAL OVERLOAD

THREE-PHASE
DELTA GENERATOR

THREE-PHASE
WYE GENERATOR

TRANSFORMER

VARIABLE RESISTOR

RES

VARIABLE RESISTOR

VOLTMETER

WOUND ROTOR AC MOTOR

293

Electrical Power

NATURAL TRIGONOMETRIC FUNCTIONS

Angle	sin	cos	tan	Angle	sin	cos	tan	Angle	sin	cos	tan
0°	.0000	1.0000	.0000	30°	.5000	.86603	.57735	60°	.86603	.5000	1.73205
1	.01745	.99985	.01745	31	.51504	.85717	.60086	61	.87462	.48481	1.80404
2	.03490	.99939	.03492	32	.52992	.84805	.62487	62	.88295	.46947	1.88072
3	.05234	.99863	.05241	33	.54464	.83867	.64941	63	.89101	.45399	1.96261
4	.06976	.99756	.06993	34	.55919	.82904	.67451	64	.89879	.43837	2.05030
5	.08715	.99619	.08749	35	.57358	.81915	.70021	65	.90631	.42262	2.14450
6	.10453	.99452	.10510	36	.58778	.80902	.72654	66	.91354	.40674	2.24603
7	.12187	.99255	.12278	37	.60181	.79863	.75355	67	.92050	.39073	2.35585
8	.13917	.99027	.14054	38	.61566	.78801	.78128	68	.92718	.37461	2.47508
9	.15643	.98769	.15838	39	.62932	.77715	.80978	69	.93358	.35837	2.60508
10	.17365	.98481	.17633	40	.64279	.76604	.83910	70	.93969	.34202	2.74747
11	.19081	.98163	.19438	41	.65606	.75471	.86929	71	.94552	.32557	2.90421
12	.20791	.97815	.21256	42	.66913	.74314	.90040	72	.95106	.30902	3.07768
13	.22495	.97437	.23087	43	.68200	.73135	.93251	73	.95630	.29237	3.27085
14	.24192	.97029	.24933	44	.69466	.71934	.96569	74	.96126	.27564	3.48741
15	.25882	.96592	.26795	45	.70711	.70711	1.00000	75	.96592	.25882	3.73205
16	.27564	.96126	.28674	46	.71934	.69466	1.03553	76	.97029	.24192	4.01078
17	.29237	.95630	.30573	47	.73135	.68200	1.07236	77	.97437	.22495	4.33147
18	.30902	.95106	.32492	48	.74314	.66913	1.11061	78	.97815	.20791	4.70463
19	.32557	.94552	.34433	49	.75471	.65606	1.15036	79	.98163	.19081	5.14455
20	.34202	.93969	.36397	50	.76604	.64279	1.19175	80	.98481	.17365	5.67128
21	.35837	.93358	.38386	51	.77715	.62932	1.23489	81	.98769	.15643	6.31375
22	.37461	.92718	.40403	52	.78801	.61566	1.27994	82	.99027	.13917	7.11537
23	.39073	.92050	.42447	53	.79863	.60181	1.32704	83	.99255	.12187	8.14435
24	.40674	.91354	.44523	54	.80902	.58778	1.37638	84	.99452	.10453	9.51436
25	.42262	.90631	.46631	55	.81915	.57358	1.42814	85	.99619	.08715	11.4300
26	.43837	.89879	.48773	56	.82904	.55919	1.48256	86	.99756	.06976	14.3006
27	.45399	.89101	.50952	57	.83867	.54464	1.53986	87	.99863	.05234	19.0811
28	.46947	.88295	.53171	58	.84805	.52992	1.60033	88	.99939	.03490	28.6362
29	.48481	.87462	.55431	59	.85717	.51504	1.66427	89	.99985	.01745	57.29
								90	1.0000	.0000	∞

ABBREVIATIONS USED IN ELECTRICITY AND ELECTRONICS

C — Capacitance
CEMF — Counterelectromotive force
d — Distance
E_G — Generated voltage
E_L — Voltage across inductance
F — Force
f — Frequency
I_A — Armature current
I_L — Current through inductance
K — Dielectric constant
K_E — Voltage constant combining built-in constant and magnetic flux of a P-M generator
K_G — A constant built into generators
K_M — A constant built into generators which affects counter torque
K_T — Torque constant

L — Inductance
P — Active power
P_{CL} — Copper losses
P_M — Mechanical power
P_Q — Reactive power
P_S — Apparent power
R_A — Armature resistance
RMS — Root mean squared
S — Rotor speed
T — Torque
X_C — Capacitive reactance
X_L — Inductive reactance
Z — Impedance
ϕ — Flux caused by magnetic action of magnets in motors and generators

294

DICTIONARY OF TERMS

A

ACTIVE POWER: True electrical power or real power going out to the user.

AC RIPPLE: A pulsating direct current produced by a half or full-wave rectifier on alternating current. Defined as the RMS value of the ac component superimposed on the average voltage.

ACROSS-THE-LINE STARTER: A device or switch which starts an electric motor by connecting it directly to the supply line. Also called full voltage controller.

AIR GAP: The air space between two magnetically related or electrically related parts. For example, the space between poles of a magnet or poles of an electric motor.

ALTERNATING CURRENT: Electrical current that alternates in flowing first with a positive polarity and then with a negative polarity.

ALTERNATOR: Rotating machine whose output is alternating current.

AMPERE: Unit of measure for current flow. One ampere equals a flow of one coulomb of charge per second.

AMPLIFIERS: Devices which produce an output voltage or current that is continuously proportional to input voltage or current. Its purpose is to obtain either control or enhancement of signals.

APPARENT POWER: The phasor sum of active power and reactive power. Symbol is P_s. It can be found by multiplying voltage times current. Unit of measure is the volt-ampere (VA).

ARC-CHUTE: Cover around contacts to protect surrounding parts from arcing.

ARMATURE REACTION: The reaction of the magnetic field produced by the current on the magnetic lines of force produced by the field coil of an electric motor or generator.

AUTOMATIC CONTROLLER: Motor or other control mechanism which uses automatic pilot devices such as pressure switches, level switches or thermostats as activating devices.

AUTOTRANSFORMER: A transformer with but a single coil. The entire coil acts as a primary winding while part of it acts as a secondary winding. Or: The entire winding acts as a secondary winding and part of it as a primary winding. It is simple and of relatively low cost but does not provide electrical isolation.

AVERAGE VALUE: In electrical current the value obtained by dividing the sum of a number of current value quantities by the number of quantities. Average voltage in a sine wave is equal to 0.637 times the peak voltage. The average value of a wave form is usually specified over its period.

AWG: American Wire Gauge; A standard for wire sizes relating to their diameter.

B

BACK VOLTAGE: An induced voltage in a coil which opposes voltage applied to it. It is proportional to the number of turns on the coil times the time rate of change of the magnetic flux.

BATTERY: Device which stores electricity in the form of chemicals. The chemical material can be in a semisolid or liquid form. The voltage is a function of the difference in the battery electrode materials.

BLOWOUT COIL: Electromagnetic coil used in contactors and starters to extinguish or deflect an arc (spark) when a circuit is interrupted.

BREAKDOWN POINT: Point at which the insulation medium in a capacitor becomes a conductor of electricity, i.e. its dielectric strength is exceeded.

BRUSH: Sliding contact, usually made of carbon, located between a commutator and the outside circuit in a generator or motor. Its purpose is to conduct electric current from a rotating armature to an electrical circuit or vice versa.

C

CAPACITANCE: Ability to store electricity in an electrostatic field. This property is found in specially designed storage devices called capacitors. It is also found to exist in parallel current-carrying wires which are close together. Capacitance is the quotient of the total charge divided by the voltage.

CAPACITOR: Electrical device for storing electrical energy and blocking current flow. It has two conducting surfaces insulated from each other.

CAPACITOR-START MOTOR: An alternating current split-phase induction motor that has a capacitor connected in series with an auxiliary winding to provide a way for it to start. The auxiliary circuit disconnects when the motor is up to speed.

CHARGING CURRENT: Current flow from source being stored in a capacitor.

CHOKE: A coil used in a direct current circuit to smooth out a pulsating wave form or reduce ripple.

CIRCUIT: An electrical or electronic path between two points. The path consists of a conductor or conductors such as an insulated wire or wires.

COMMUTATOR: Bars of metal grouped around the shaft of an electric motor or generator. Their purpose is to provide an electrical connection between armature coils and brushes. They maintain current flow in one direction by reversing the magnetic polarity of the rotor every revolution.

COMPENSATING WINDINGS: Windings embedded in the main pole pieces of a compound dc motor. Their main purpose is to stop arcing in the brushes and to reduce armature reaction.

COMPOUND GENERATOR: A generator built with two field coils on each pole piece. One is connected in parallel with the armature circuit; the other is in series.

COMPOUND MOTOR: Direct current motor with two types of field coils:
1. Shunt field coil which is connected in parallel with the armature. 2. Series field coil which is wired in series with the armature.

CONDUCTOR: A material in which electrons (electrical currents) flow easily.

COPPER LOSS: Electrical power lost through resistance of coils to current movement through the wire of the coils.

CORE: Magnetic path through the center of a coil or transformer. It may consist of air, iron, or other magnetic materials. Sometimes used to mean the core of laminated material iron for example that makes up the center of a coil or transformer.

CORE LOSSES: Loss of power in a transformer or coil core due to eddy currents and hysteresis.

CORE TYPE TRANSFORMER: Electrical transformer which has the core inside of the coils.

COSINE FUNCTION (of the angle Θ): In trigonometry, the ratio of the side adjacent (next to) the angle to the hypotenuse.

CONTROLLER: A device for starting a motor in either direction of rotation or adjusting speed of rotation or torque.

COULOMB: The amount of electrical charge which passes any point in an electrical circuit during one second where a current of one ampere flows. It is the fundamental unit of charge.

COUNTERELECTROMOTIVE FORCE: Voltage induced in the armature coil of an electric motor.

COUNTER TORQUE: Repulsion force between two magnetic fields. It is proportional to the strength of the magnetic fields.

CURRENT: Flow of electrons through a conductor. In magnitude it is the amount of charges per unit of time.

CURRENT RELAY: A device which looks and acts much like a voltage relay. It operates at a certain value of current. It can be an overcurrent or an undercurrent relay.

D

DAMPER WINDINGS: In electric rotating machines, a permanently short-circuited winding, usually uninsulated, and arranged to oppose rotation or pulsation of the magnetic field with respect to the pole shoes. Windings consist of copper bars located just below the surface of the spider poles.

DC DYNAMIC BRAKING: Braking of a motor accomplished by stopping the revolving stator field. In effect, the motor is used like a generator. An energy-dissipating resistor is applied to the armature.

DELTA CIRCUIT: Three-phase circuit in which the windings are connected in the form of a closed ring. Instantaneous voltages around the ring are equal to zero. Transformer coils and electric motors are often connected this way.

DELTA-DELTA CONNECTION: A transformer coil connection in which the primary and secondary coils may be delta connected.

DELTA-WYE CONNECTED: A transformer connection in which the primary is delta connected while the secondary is wye connected.

DIAC: Two-lead alternating current semiconductor. It will conduct current when its breakdown voltage is reached. It has properties of both diodes and transistors.

DIELECTRIC: Insulating material which separates and insulates the conducting plates in a capacitor. It can be a gas, a liquid, plastic, glass, paper, etc. or a combination of these.

DIELECTRIC BREAKDOWN: Failure of an insulating material to separate electrical charges. Breakdown occurs when the insulating material changes and conducts the electrical charge between plates.

DIGITAL: Relating to a class of devices or circuits in which the output varies in discrete (distinct) steps such as pulses or "on" and "off" operation.

DIRECT CURRENT (dc): Current that always flows in only one direction.

DISTRIBUTED-FIELD ALTERNATOR: A revolving-field type alternator with field coils buried in the face of the rotor.

DOMAIN THEORY: A theory which says that each atom of a material is a small magnet because of electrons orbiting their nuclei. Many atoms lining up north to south make up a domain creating a magnetized material.

DOT NOTATION: A system drafters use to show relative polarity of coils in a transformer. Terminals of coils with dots have the same polarity at the same time.

DRUM SWITCH: Type of motor controller. It has electrical connecting parts in the form of fingers held by springs against contacts or surfaces. These contacts are on the perimeter of a rotating cylinder.

E

EDDY CURRENTS: Electrical currents circulating in the core of a transformer as the result of induction. These currents produce unwanted heat in the iron core and copper windings of the transformer.

EFFICIENCY: Ratio of the output power to the input power of a device.

ELECTRICITY: Electrical charges in motion. The movement is called current.

ELECTROLYTIC CAPACITOR: A capacitor which uses a liquid or paste (electrolyte) as one of its electrical storage plates.

ELECTROMAGNET: Temporary magnet created by passing an electric current through a coil wound around a soft iron or other magnetic core.

ELECTROMAGNETISM: Magnetic field which occurs around a wire or other conductor when a current is passing through it.

ELECTROMECHANICAL: Any device which uses electrical energy to create mechanical motion or force.

ELECTROMOTIVE FORCE: Voltage or force which causes free electrons to move in a conductor. Unit of measure is the volt (V).

ELECTRON: A negative electric charge.

ELECTRON FLOW: Movement of electrons from a negative point to a positive point in a conductor such as a metal wire. Electrons may also move through liquids, gases, or vacuum.

ELECTRON THEORY: Belief that all matter is made up of atoms. Atoms have a nucleus of positive charges called protons and neutral charges called neutrons. Negative charges called electrons orbit around the nucleus.

ELECTROSTATIC CHARGE: The electrical charge stored by a capacitor.

ELECTROSTATIC FIELD: Stored electrical charges on the surface of an insulator.

EXCITATION: Creating a magnetic field; used to create electromagnetics by passing an electric current through a coil.

EXCITATION CURRENT: The current in the shunt field of a motor that results when voltage is applied across the field.

F

FACEPLATE CONTROLLER: Motor starter having a wiper arm and several taps which are contacted by the wiper arm which is manually controlled.

FARAD: Unit of capacitance. A capacitor is said to have a capacity of 1 farad when a current of one ampere is produced by a rate of change of 1 volt per second. Or: One coulomb per volt.

FARADAY, MICHAEL: English physicist and chemist. He invented first electric generator, formulated the laws of magnetic induction and plane of rotation of polarized light in a magnetic field. Unit of electrical capacity (farad) was named for him.

FIELD ACCELERATING RELAY: A relay which weakens field strength in a motor to increase speed after it has started with full strength field.

FIELD DECELERATING RELAY: Relay used to slow down an electric motor. It is pulled into operation by armature current. It creates resistance in the motor circuit. This resistance slows the motor.

FIELD INTENSITY: The amount of magnetizing force available to produce flux (lines of force) in the core of a magnet. Its SI metric units are amperes per metre.

FIELD LOSS RELAY: Also called a motor field failure relay. A relay which acts to disconnect the motor armature from the line in case field excitation is lost.

FLASHING THE FIELD: Method of producing residual magnetism in the pole pieces of a dc generator. It is done by applying full voltage to the field coil from a separate source for 30 seconds.

FLAT COMPOUNDED GENERATOR: Compound wound generator with series field winding adjusted so that output voltage is nearly constant for currents between no-load and full-load.

FLUX DENSITY: The number of lines of flux in a cross-sectional area of a magnetic circuit.

FLUX LINKAGE: The magnetic lines of force linking a coil of wire. Whenever the linkage changes, voltage is induced in the coil.

FREE ELECTRONS: Electrons that move from one atom of a substance to another without becoming strongly attached to any particular atom.

FREQUENCY: In electricity, the number of times alternating current changes direction during a second. Frequency is measured in Hertz (cycles per second).

G

GATE: One of the leads on a thyristor. Usually, this lead is the one that controls output when it is properly biased.

GALVANOMETER: Meter that can measure small currents and voltages.

GENERATOR: Rotating machine which changes mechanical energy into direct current electrical energy.

GENERATOR ACTION: Inducing of voltage into a wire that is cutting a magnetic field.

H

HARD NEUTRAL POSITION: Describing a position of repulsion motor brushes in which the brushes are 90 degrees from the center of the stator field. Rotor currents are high but there is no torque.

HENRY: Electrical unit of inductance. Symbol

is "H." It is named for Joseph Henry, an American scientist. A henry is an induced voltage of 1 volt when the current is changing at a rate of one ampere per second.

HERTZ: Measurement of frequency. It means "cycles per second," of alternating current.

HIGH SIDE: In a transformer, the designation indicating which is the high voltage coil.

HORSEPOWER: Conventional unit of measure for power. It indicates the result of force times distance times time. One horsepower (hp) equals 746 watts, or 33,000 ft. lb. per minute, or 550 ft. lb. per second.

HYPOTENUSE: Longest side of a right angle triangle.

HYSTERESIS: Property of a magnetic substance that causes the magnetization to lag behind the magnetizing force.

I

INCHING: Applying reduced power to move a motor or its load slowly to a desired position.

INDUCED CURRENT: Current which flows in a conductor because of a changing magnetic field.

INDUCTANCE: Electromotive force resulting from a change in magnetic flux surrounding a circuit or conductor. Property of a circuit or two neighboring circuits which determines how much emf will be induced in one of the circuits by a change in current in either of them.

INDUCTION: Generation of electricity by magnetism.

INDUCTIVE KICK: Voltage produced by the collapsing field in a coil when the current through it is suddenly cut off. This voltage is much higher than the applied voltage.

INDUCTIVE REACTANCE: Opposition (in ohms) to an alternating current as a result of induction (voltage resulting from cutting lines of magnetic force). The component of impedance in a circuit not due to resistance or capacitance.

IN PHASE: Two waves of alternating current which are changing polarity at the same time and reach their positive (negative) peak voltage or current at the same time.

IN-PHASE CURRENT: The part of current in a coil which is due to resistance. It cannot be measured alone since it is combined with current caused by inductive reactance.

IMPEDANCE: Total opposition to current flow in a circuit. It is measured in ohms. Its symbol is Z. Also: The phasor sum of reactance and resistance in series.

INSTANTANEOUS VALUE: The magnitude of a current or voltage at any point when it is a changing current or voltage as in ac.

INTEGRATED CIRCUIT: A circuit made from transistors, resistors and the like. All are placed in a package referred to as a "chip," since all circuits are on one piece of semiconductor material.

INTERLOCKS: Small contacts in a push button switch which operate at the same time as the main contacts. They are used to govern succeeding operations of the same or allied devices. They may be either electrical or mechanical.

INTERPOLE: Small auxiliary (helper) pole placed between the main poles of a dc generator or motor to reduce arcing (sparking) at the commutator.

INVERTER: Circuit which receives a positive signal and sends out a negative one or vice versa. Also, a device which changes alternating current to direct current or vice versa.

J

JOGGING: Rapid and repeated opening and closing of a motor starting circuit under full power to produce successive slight rotation of the motor.

K

KINETIC ENERGY: Ability of all objects in motion to do work (for example, the impact of a pile driver).

KIRCHOFF'S LAWS (concerning electricity): 1. Current flowing to a given point in a circuit is equal to the sum of the current flowing away from that point. 2. The algebraic sum of the voltage drops in any closed path in a circuit is equal to the algebraic sum of the electromotive forces in that path.

KVAR: The reactive power (kilovolt-amperes) in a circuit.

L

LAMINATION: One thickness of the sheet material sandwiched together to construct a stator or rotor of a rotating machine. The laminations are stamped out of sheet metal. Each is cut to shape and notches and holes are cut before the pieces are laminated.

LENZ'S LAW: Inductance always tends to oppose whatever causes it. The direction of the induced voltage is such as to produce a current opposing the flux change.

LEYDEN JAR: A glass jar used to store electrical charges in early experiments with electricity. The jar was lined with metal foil inside and out. One end of a conductor was in contact with the foil inside. Electric charge was stored on the inside foil. Current flowed when there was an electrical path connecting the inside foil and the outside foil.

LIGHT-EMITTING DIODE (LED): A pn (positive-negative) junction that sends out light when biased in a forward direction. (Forward bias means that the anode is positive with respect to the cathode.)

LOCKED-ROTOR TEST: A test of an electric motor in which the shaft is prevented from turning while power is applied. It can be used to determine such things as fixed and variable losses in a motor.

LOCKED-ROTOR TORQUE: The least torque a motor will develop at rest for all positions of the rotor when the rated voltage is applied at rated frequency.

LOCKOUT RELAY: A relay which uses armature current. It has two coils. One, the holding coil, closes the contacts. The other, called the lockout coil, opens the contacts. It is used to control starting windings in electric motors.

LOGIC: A method of using the symbols AND, OR, NOT, NAND and NOR to represent complex circuit functions.

LOGIC CIRCUIT: A circuit, usually electronic, which provides an input-output relationship that is in agreement with a Boolean-algebra logic function.

LOGIC SYMBOL: A device that performs a logic function; a gate or flip-flop or sometimes a combination of these devices treated as a single element.

LONG SHUNT FIELD CONNECTION: A method of wiring up the compound motor. The shunt field is wired in parallel with the series combination or armature and series field.

LOW SIDE: In a transformer, the designation indicating which is the low voltage coil.

M

MAGNETIC CONTACTOR: A part of a control device for a motor. When a push button is pressed, the contactor, which consists of a coil and a pivot arm, closes the contact. This completes a circuit and causes the motor to run.

MAGNETIC DRIVE: Electromagnetic device connected between a three-phase motor and its load for purpose of regulating speed at which load is rotated. It causes slippage between motor and load. Also called a magnetic clutch.

MAGNETIC FIELD: The invisible lines of force found between the north and south poles of a magnet.

MAGNETIC LINES OF FORCE: In a magnetic field, an imaginary line which has the direction of the magnetic flux at every point. Also: The magnetic line along which a compass needle will align itself, i.e. the direction of the force on the north magnetic pole of a magnet.

MAGNETOMOTIVE FORCE: The force which produces the lines of magnetic force in a magnetic circuit. It is the "push" behind the magnetic field.

MAGNETIC STARTER: A starter (switch) that uses electromagnetic components to make it operate. For example, a push button activates an electromagnetic device which then moves to close the contacts or open them.

MAINTAINED CONTACT PUSH BUTTON: Switch closes circuit when the push button is pressed the first time. It opens the circuit when the push button is pressed a second time.

MECHANICAL ENERGY: In moving objects, the force of motion they possess.

MEMORY: The equipment and media (such as magnetic tape) used to hold machine-language information in electrical or magnetic form.

MICROFARAD: One millionth of a farad. The farad is the unit of capacitance.

MICROPROCESSOR: A computer central arithmetic and logic unit and control unit on a single integrated circuit chip. One single IC chip may hold tens of thousands of transistors, resistors, diodes and connecting circuitry.

MOTION EMF: Electromotive force induced when a wire cuts through a magnetic field.

MOTOR ACTION: Mechanical forces that exist between magnets. When two magnets approach each other, one will be either pulled toward or pushed away from the other.

MOTOR CONTROLLER: Device in a motor circuit for starting, stopping, reversing, changing speed, or for protection of the motor against overheating.

MOTOR EFFICIENCY: Ratio of output power to input power.

MOTOR FIELD FAILURE RELAY: See FIELD LOSS RELAY.

MULTI-SPEED MOTOR: A motor which can operate at two or more fixed speeds. This is done by changing either frequency or number of poles.

MULTIVIBRATOR: An oscillator in which the in phase feedback comes from two electron tubes or transistors.

N

NEMA: National Electrical Manufacturers Association.

NEUTRON: One of the three principle particles making up an atom of matter. It has no electrical charge.

NEWTON: Metric unit of measure for force. Symbol is N. Its equation is kg per m per sec² (the force that causes a kilogram of mass to accelerate at one meter per second per second). Equal to approximately one-fourth pound.

NO-LOAD TEST: Operating a motor at full speed with no load to determine rotational power losses.

NORMALLY CLOSED CONTACTS: A set of motor control contacts that are open only while the push button is depressed.

NORMALLY OPEN CONTACTS: A set of contacts which are closed only while the push button is depressed.

O

OHM: Electrical, electronic unit of resistance. It is the resistance an element carrying one ampere has when there is a one volt drop across this element.

OHM'S LAW: $I = \dfrac{E}{R}$ or Current is equal to voltage divided by resistance.

OPERATIONAL AMPLIFIER (OP-AMP): An amplifier that will perform mathematical operations. Given the proper feedback it can add, subtract, average, integrate, differentiate, or amplify (multiply).

OSCILLATOR: Electrical device which generates alternating current at a frequency determined by the values of certain constants in the circuit. It may be considered an amplifier with positive feedback.

OUT OF PHASE: Two or more phases of alternating electrical current that are changing direction at different times (contrast with "in phase").

OVER COMPOUNDING: In a compound wound generator, using enough series turns in the field coil to raise the voltage as load increases. Purpose is to compensate for increase in voltage drop caused by the increased load.

OVER EXCITED: Condition of a synchronous motor when the dc field supplies more magnetization than is necessary.

OVERLOAD RELAYS: A device for protecting a motor or other electrical device from an overload current or voltage.

P

PARALLEL CIRCUIT: Circuit in which all positive terminals are connected at a common point while all negative terminals are connected to a second point. Voltage is the same across each element of the circuit.

PEAK VALUE: Highest value present in a varying or alternating voltage. Value can be either positive or negative.

PERMANENT-CAPACITOR MOTOR: A single phase electric motor which uses a phase winding in conjunction with the main winding. The phase winding is controlled by a capacitor which stays in the circuit at all times and is rated for continuous running. The capacitor improves starting and running power factors.

PERMANENT MAGNETISM: A magnet that will keep its magnetic properties indefinitely.

PERMEABILITY: The ease with which the domains in a magnetic core can be made to line up to create magnetism. It is the ratio of the magnetic flux density to the magnetic field intensity.

PHASE ANGLE: The difference in angle between two sine wave vectors. In equations it is represented by the symbol, theta (Θ).

PHASE: Relationship of two wave forms having the same frequency.

PHASE SHIFT: Creation of a lag in or advance in voltage or current in relation to another voltage or current in the same circuit. Used with some motors to make single phase current act like two phase current.

PHASE VOLTAGE: Voltage across a coil.

PHASOR DIAGRAM: Plotting of phase angles, voltage values, resistance, and current values on paper. There are four types: voltage, current, power and impedance. Used to show relationships between different phases of alternating current. Made up of lines or vectors each pointing in the direction corresponding to the phase angles of the voltages.

PHOTOTRANSISTOR: A junction transistor with its base exposed to light through a lens or window in the housing. Collector current increases as the light increases.

PIEZOELECTRIC: Property of certain crystals which permits them to: 1. Produce a voltage when under mechanical stress. 2. Experience mechanical stress when subjected to electrical

pressure (voltage).

PLATE: One of the electrodes in a capacitor.

PLUGGING: Braking a motor by reversing the line voltage or phase sequence.

POLARITY: A condition of a magnet (or a circuit) to have north and south poles (positive and negative charge). Also: Property of electrical device having positive and negative terminals.

POLAR NOTATION: A method of keeping track of phase angle in equations for solving alternating current circuit current values.

POLE-CHANGING: Method of changing speed of an electric motor by changing the number of poles the motor is running on.

POLE SHOE: Part of the stator of an electrical rotating machine.

POLYPHASE ELECTRICAL POWER: Power source in which there is more than one phase. Also, the several voltages change direction and amplitude (strength) at different times.

POUNDS FORCE: English unit of conventional measure for force.

POWER FACTOR: Ratio of the active power of an alternating or pulsating current (when measured by a wattmeter) to the apparent power indicated by an ammeter and voltmeter.

POWER FACTOR CORRECTION: Addition of a capacitor to circuit which includes an electric rotating machine (motor) to reduce useless reactive current. Current is made to shift so it is more nearly in phase with voltage.

PRIMARY COIL: One of two coils in a transformer. It is connected to the source of power.

PRIME MOVER: First or primary power source. For example a turbine or a propeller may be the prime mover that provides the power for a generator.

PROGRAM: Sequence of instructions that tells a computer how to receive, store, process, and deliver information.

PROGRAMMABLE CONTROLLER: Solid-state controller which receives and follows a sequence of instructions. This type of controller is useful in control of processes, materials handling, and certain machine functions.

PRONY BRAKE: Device for measuring torque.

PROTON: One of the particles making up an atom. It has a positive electrical charge.

PULL-OUT TORQUE: In a synchronous motor, the maximum torque developed by a motor for one minute, before it pulls out of step due to overload.

PULSATING CURRENT: Direct current that will rise and fall in value but generally will not fall below the zero reference point on a sine wave.

PUSH BUTTON CONTROL: Control of equipment through push buttons which activate relays.

PUSH BUTTON SWITCH: A switch using a button for manually activating a contact to open or close a circuit.

PYTHAGOREAN THEOREM: Mathematical rule for finding the length of an unknown side of a triangle. The theorem states: The square of the hypotenuse of a right angle triangle equals the sum of the squares of the other two sides.

Q

QUADRATURE CURRENT: That part of the current in a coil which is due to inductive reactance.

R

REACTIVE LOAD: Load caused by capacitance or inductance.

REACTIVE POWER: Reactive voltage times the current, or voltage times the reactive current, in an alternating current circuit. Its unit of measure is the VAR. It may be thought of as "shuttle power" or the power moving back and forth between the source and a coil in a circuit.

RECTIFIER: An electrical device which is to electricity what a one-way valve is to liquids in a pressure supply system. The rectifier converts alternating current to direct current by allowing the current to move in only one direction.

REGENERATIVE BRAKING: System of electric motor braking which uses energy of inertia and of gravity. It causes motor to act as a generator and retards rotational speed or stops the motor altogether. At the same time, power

generated can be fed back into the circuit.

RELAY: Electromechanical device with contacts that are opened and closed by changes in the condition of an electric circuit. The result of the opening and closing is to affect the operation of other devices in the same or other circuits.

RELAY CONTACTS: Contacts closed or opened by movement of a relay armature.

RELAY LOGIC: System of controls and operations that make up an electromechanical controller.

REVOLVING ARMATURE ALTERNATOR: An alternating current generator which has a stationary magnetic field and a set of alternating current windings.

REVOLVING-FIELD ALTERNATOR: An ac generator in which the armature is stationary while the field windings rotate.

REDUCED INRUSH STARTER: A group of motor starters which apply a reduced voltage to a three-phase motor. They may use resistors, reactors or autotransformers to reduce the voltage.

RELUCTANCE: The ratio between the magnetomotive force and the resulting flux. In a sense, reluctance is the opposition to magnetic flux.

RESIDUAL MAGNETISM: That magnetism remaining in the core of a coil or an electromagnet after the current to it has ceased to flow.

RESISTANCE: A property of conductors which makes them resist the movement of an electrical current and to create heat. Dissipative versus reactive impedance.

RESISTORS: An electrical-electronic device attached to a circuit to produce a certain amount of resistance to current.

RETENTIVITY: Ability of a material to retain magnetism after magnetizing force is removed.

RHEOSTAT: A variable resistor. It has one fixed terminal and a movable contact.

RIPPLE: Series of peaks in a current value after ac has been rectified (changed to dc) but not completely filtered.

ROOT-MEAN-SQUARE: The value of

alternating current that would produce the same heating effect as an equal amount of direct current. In ac, the value is 0.707 times the peak value. This is the same as effective value.

ROTATIONAL LOSSES: Losses in a generator or alternator caused by windage and/or friction.

ROTOR: Rotating part of an electric rotating machine. In a motor it is connected to and turns the drive shaft. In an alternator or generator it is turned to produce electricity by cutting magnetic lines of force.

ROTOR IMPEDANCE: In an electric motor, phasor sum of resistance and inductive reactance. Resistance is constant but inductive reactance changes with slip.

RMS: Abbreviation for root-mean-square.

S

SALIENT-POLE ALTERNATOR: A revolving-field alternator in which the rotating member has projections to hold the field coil windings.

SATURATED: A condition or point where an electrical or magnetic component can receive no more electrical current or magnetism.

SATURATION: Point at which a magnet will take on no more flux density.

SATURATION CURVE: A magnetization curve on a chart for a ferromagnetic material (also called hysteresis or B-H curve).

SECONDARY COIL: In a transformer, the coil that is connected to the load in the electrical circuit.

SELF EXCITATION: In a generator, supplying excitation voltages by a device on the generator itself rather than from an outside source.

SELF-INDUCTION: A counterelectromotive force which is produced in a conductor when the magnetic field produced by the conductor collapses or expands after a change in current.

SEPARATE EXCITATION: Producing generator field current from an independent source.

SERIES AIDING: A condition in which power sources are connected so that the positive terminal of one is contacting the negative terminal of another. Total voltage is the sum of the separate voltages.

SERIES CIRCUIT: A circuit in which all resistances and other components are connected end to end so that the same current flows throughout the circuit.

SERIES FIELD: In a rotating machine, the part of total magnetic flux due to the series winding. Also: A type of field coil, in a direct current motor, which is wired in series with the armature.

SERIES MOTOR: Motor in which the field and armature circuits are connected in series. Frequently called a "universal" motor.

SERIES OPPOSING: Power sources which are connected together so that terminals are connected positive to positive.

SERIES-PARALLEL NETWORK: Combination of components where there are many paths for current in a circuit. Some are in series followed by others connected in parallel.

SHADED-POLE MOTOR: Single phase squirrel cage induction motor with stator poles slotted to create two sections in each pole. The smaller of the two sections is wrapped with one or more loops of copper to create an auxiliary coil. Magnetic field of shaded section combines with main section to produce starting torque.

SHADING COIL: In a small shaded-pole electric motor a copper ring or coil set into a section of the pole piece. Its purpose is to produce the lagging part of a rotating magnetic field for starting torque.

SHELL TYPE TRANSFORMER: Electrical transformer so constructed that the core surrounds the coil as well as going through the center of the coil.

SHORT SHUNT FIELD CONNECTION: Method of making field connection in a compound motor. Shunt field is connected in parallel with the armature only.

SHUNT FIELD: A type of field coil for a dc motor. It is wound with many turns of fine wire. It connects in parallel with the armature.

SHUNT or SHUNT-WOUND MOTOR: Direct current motor whose field circuit and armature circuit are connected in parallel. Speed is controlled by controlling the applied voltage.

SHUTTLE POWER: Power stored in the magnetic field of an inductance or the electric field of a capacitance and returned to the source twice during each cycle.

SIDE ADJACENT: In trigonometry, the side next to the angle formed by that side and the hypotenuse.

SILICON CONTROLLED RECTIFIER (SCR): A semiconductor device which, in its normal state, blocks a voltage applied in either direction. It can be made to conduct current in a forward direction on a signal applied to the gate. It will continue to conduct after the control signal has stopped. Its purpose is to control the value of armature voltage. It, therefore, provides very accurate speed control of dc shunt motors.

SINE FUNCTION (of the angle Θ): In trigonometry, the ratio of the side opposite Θ to the hypotenuse.

SINE WAVE: Alternating current action plotted on paper as a time-value wave. (Amplitude versus time expressed in angles.)

SINGLE PHASE: Having only one alternating current or voltage in a circuit.

SLIP: The difference between the synchronous speed of a motor and the speed at which it operates.

SLIP RING MOTORS: Another term for wound-rotor motors. The rotor in such a motor has the same number of magnetic poles as the stator. Rotor windings are wye-connected and terminate (end) at slip rings.

SLIP RINGS: Circular bands on a rotor which transmit current from rotor coils to brushes.

SLIP SPEED: The difference between rotor speed and synchronous speed in an induction motor.

SOFT NEUTRAL POSITION: When brushes of a repulsion electric motor are lined up with the stator field. There is no rotor current and no torque.

SOLDER POT OVERLOAD RELAY: Protective device which uses a solder to open a circuit. The solder melts and releases a ratchet assembly which rotates and breaks contact.

SOLID STATE: Referring to circuits and components using semiconductors.

SOLID STATE CONTROLS: Devices that control current to motors through semiconductors. Solid state switches include diodes, transistors, silicon controlled rectifiers (SCRs), triode ac switches, diode ac switches and unijunction transistors.

SOLID STATE RELAY: A relay that uses semiconductor devices.

SPIDER: The core of a salient-pole rotor for an alternator.

SPLIT-PHASE (RESISTANCE-START) MOTOR: Single-phase induction motor which has an auxiliary winding connected in parallel with the main winding. However, its magnetic position is not the same as the main winding. Thus, it can produce the required rotating magnetic field needed for starting.

SQUIRREL-CAGE ROTOR: Rotor made up of metal bars which are short circuited at each end.

STALL TORQUE: Torque which the rotor of an energized motor produces when the rotor is kept from rotating.

STARTER: Control device designed for accelerating a motor to full speed in one direction of rotation.

STATIC ELECTRICITY: Electricity at rest. Electric charge stored up in an electrical device such as a capacitor. Also called static charge.

STATOR: Stationary (not moving) part of an electric rotating machine. In a motor it usually contains the primary winding.

STATOR FIELD: Magnetic field set up in the nonmoving part of an electric motor when the motor is energized and electric current is flowing through the motor.

STATOR POLES: Shoes on an electric motor stator which hold the windings and make up the magnetic poles of the stator.

STROBOSCOPIC EFFECT: Used to measure speed of a rotating shaft. As a strobe flash flickers on the shaft, the shaft will appear to slow down and then to stop if the flash rate equals the rotational rate. Dial reading at point where shaft appears to stop is its rotational speed.

SYNCHRONISM: Applied to a synchronous alternator or electric motor, the condition under which the rotating machine runs at a speed directly related to the frequency of the power applied or generated.

SYNCHRONOUS: In step or in phase as applied to currents, voltages or two different rotating machines.

SYNCHRONOUS IMPEDANCE: Combined effect of armature resistance, inductive reactance and armature reaction.

SYNCHRONOUS MOTOR: Induction motor which runs at synchronous speed. Its stator windings have the same arrangement as in non-synchronous induction motors. However, rotor does not lag behind the rotating magnetic stator field.

SYNCHRONOUS SPEED: Rate of travel of the stator field of a three-phase electric motor.

SYNCHROSCOPE: An instrument for determining the phase difference or degree of synchronism of two alternators.

T

TACHOMETER: Device which measures rotational speed of rotating machines.

TANGENT FUNCTION: In trigonometry, the ratio of the side opposite to the side adjacent is called the tangent function of the angle Θ.

TAP CHANGERS: Mechanical device for changing the voltage output of a transformer.

TAPS: Fixed electrical connections at different positions on a transformer's coil. Each tap takes off a different voltage from the transformer.

TESLA: Unit of measure for flux density. One tesla is equal to one weber per square metre of flux.

THERMISTOR: A resistor whose resistance goes up when it gets hot (positive coefficient type). Or a resistor whose resistance goes down when it gets hot (negative coefficient of resistance).

THERMOCOUPLE: Device which consists of two unlike metals which are joined at a point where heat is applied. Heat creates electron im-

balance in the two metals and a current will flow.

THERMOSTAT: Temperature-sensitive device. When used in a motor controller it is activated by temperature change to close the motor's electrical circuit.

THREE-PHASE ALTERNATORS: Rotating machines which generate three separate phases of alternating current. Such power is more efficient to transmit than single phase.

THREE-PHASE ELECTRIC MOTOR: A motor which operates from a three-phase power source. In three-phase power three voltages are produced which are 120 electrical degrees apart in time.

TIME CONSTANT: The duration of time required for current in an inductor or capacitor to increase to 63.2 percent of maximum value or to drop to 36.7 percent of maximum value.

TIME-LIMIT-ACCELERATION: Starting and bringing a motor up to speed in a definite time span regardless of its load.

TIMER: Device which delays closing or opening of a circuit for a specific time. They can be operated by a clock mechanism, voltage, air pressure, current, or a motor.

TORQUE: Torque is a force that produces a rotating or twisting action.

TRANSFORMER: Electrical device designed to change the voltage in an alternating current electrical circuit. Step-up transformers increase the voltage. Step-down transformers decrease the voltage.

TRANSFORMER EFFICIENCY: Ratio of input to output power.

TRIAC (triode ac switch): A three terminal thyristor or semiconductor device that is triggered into conducting by applying a small current to its gate.

TURNS RATIO: Ratio of the number of turns in the primary winding of a transformer to the number of turns in the secondary winding.

TURBOCONSTRUCTION ALTERNATOR: See Distributed-Field Alternator.

TWO-CAPACITOR MOTOR: An induction motor which uses one capacitor for starting and one for running. Starting capacitor is in parallel with the running capacitor as motor is starting. At 75 percent of speed starting capacitor is cut out of the circuit.

TWO-PHASE VOLTAGE: Two alternating voltages in a circuit. They are 90 degrees out of phase.

U

UNDERCOMPOUNDED: A generator in which the output voltage drops as the load is increased.

UNDER COMPOUNDING: In a compound dc generator, a series field coil with only a few turns. It produces a higher output at full load than a shunt motor. Output voltage drops as load increases.

UNDER EXCITED: Term used to describe the magnetizing power of a synchronous motor when only part of the magnetizing is done by the alternating current.

UNITY POWER FACTOR: A power factor of 1. This is the best power factor that can be achieved. It is attained only when current and voltage are in phase; when all the power supplied is used.

V

VECTOR: In phasor diagrams, lines having a specific length and specific direction.

VOLTAGE: Electrical pressure; the force that causes current in an electrical conductor.

W

WATT: Unit of electrical power. One watt is the power of a current of one ampere when voltage drop is one volt.

WATTMETER: Instrument for measuring electrical power.

WEBER: Unit of measure for flux in a magnetic field. A weber equals 100,000,000 lines of force.

WINDAGE: Rotational losses in a generator or alternator which are due to the friction the rotor creates with the surrounding air.

WYE or STAR CONNECTION: An electrical connection in which all of the terminals having the same instantaneous polarity are joined at the neutral junction. It takes the shape of the letter Y or of a star.

WYE-WYE CONNECTION: Transformer coil arrangement in which both the primary and the secondary coils are wye-connected.

Y

YOKE: Framework or housing in a dc motor to which the magnetic field pole pieces are attached.

INDEX

Index

U

V

W